Understanding Position Sensors

As the definitive resource on position-sensing technology, *Understanding Position Sensors* encompasses all aspects necessary for a full understanding of the field, with topics of background, operational theory, design, and application.

While grasping the theory of technologies used in the measurement of linear and angular/rotary position sensors, the reader will also learn about terminology, interfacing, testing, and other valuable concepts that are useful in the understanding of sensors in general.

The first three chapters provide readers with the necessary background information on sensors. These chapters review the working definitions and conventions used in sensing technology; specification of position sensors and the effect on performance; and sensor output types, plus an extensive section covering communication protocols. The remaining chapters describe each separate sensor technology in detail. These include resistive sensors, cable extension transducers, capacitive sensors, inductive sensors, LVDT and RVDT sensors, distributed impedance sensors, Hall effect sensors, magnetoresistive sensors, magnetostrictive sensors, linear and rotary encoders, optical triangulation position sensors, and ultrasonic position sensors.

The book:

- Presents sensor specification, theory of operation, sensor design, and application criteria;
- Reviews the background history of position sensors as well as the underlying engineering techniques;
- Includes end-of-chapter exercises.

Understanding Position Sensors is written for electrical, mechanical, and material engineers, as well as for engineering students who are interested in understanding sensor technologies, and can be used as a textbook for an engineering course on sensor technology.

Understanding Position Sensors

David S. Nyce

CRC Press
Taylor & Francis Group
Boca Raton London

CRC Press is an imprint of the
Taylor & Francis Group, an **informa** business

Designed cover image: David S. Nyce

Seventh edition published 2023
by CRC Press
6000 Broken Sound Parkway NW, Suite 300, Boca Raton, FL 33487–2742

and by CRC Press
4 Park Square, Milton Park, Abingdon, Oxon, OX14 4RN

CRC Press is an imprint of Taylor & Francis Group, LLC

First edition published by Willan 2008

Sixth edition published by Routledge 2009

ISBN: 978-1-032-43699-9 (hbk)
ISBN: 978-1-032-43824-5 (pbk)
ISBN: 978-1-003-36899-1 (ebk)

DOI: 10.1201/9781003368991

Typeset in Times
by Apex CoVantage, LLC

Dedication

This work is dedicated to my very talented, understanding, and amazing wife, Gwen, and to our three wonderful children, Timothy, Christopher, and Megan. To friends, colleagues, and readers who might find this book helpful in their endeavor to understand sensing technology. And thanks to my parents, Jonathan and Emma Nyce, who allowed me to build and operate a lab in the basement of our home where I grew up. Much was learned by experiments in the lab involving electronics, chemistry, physics, mechanics, aerodynamics, rockets, model airplanes, and other areas of interest to me (although some experiments proved to be a little on the dangerous side).

Contents

About the Author

David S. Nyce, a scientist, engineer, and inventor, has developed sensors of many types for over 40 years. He has worked as an electronics, mechanical, and chemical engineer as well as technical manager, operations manager, technical consultant, and business owner. His duties have ranged from development engineer and project engineer, to chief engineer, director of technology, general manager, vice president, and owner. He founded the Revolution Sensor Company in 2003 and has developed industrial, commercial, medical, military, aerospace, undersea, and automotive products, including an automated production line for onboard automotive sensors.

His experience comprises the design of transducers, sensors, and instrumentation for many types of measurement, including temperature sensing using a thermocouple, RTD, thermistor, or semiconductor; pressure sensing with LVDT, resistive, strain gage, and diffused semiconductor sensing elements; resistive and strain gage flow sensors; linear and angular position, velocity, and acceleration transducers and sensors with LVDT, Hall effect, inductive, capacitive, optical, magnetostrictive, and distributed impedance technologies; resistive, inductive, and force-balance accelerometers; liquid level sensors based on ultrasonic waves, magnetostriction, capacitive, LVDT, and distributed impedance technologies; densimeters; gas analysis using flame ionization, chemiluminescence, infrared, paramagnetic, zirconia, and electrochemical techniques; and intrinsic safety, explosion-proof, purging, and inerting safety systems.

Nyce holds a Bachelor of Science degree in Electrical Engineering, and Master of Business Administration. He is the inventor on 29 US patents.

He is an FAA-certificated airplane and helicopter pilot and an NRA-certified firearms instructor and has other hobbies, including competitive shooting, amateur rocketry, photography, riding Harleys, teaching science classes at local schools, teaching children to build and fly model airplanes and amateur rockets, playing beach doubles volleyball, practicing archery, knife and axe-throwing, brewing beer, winemaking, hang gliding, hunting, fishing, boating, kayaking, and flying model aircraft.

Preface

The ubiquitous use of sensing technology in almost all realms of our daily lives becomes more apparent every year. Sensors are abundant in automobiles, appliances, security systems, our homes, manufacturing facilities, medical equipment, and so on.

Position sensors are an important sector of sensing technology, with applications in each of these areas. This book explains the theory and application of the technologies used in sensors for the measurement of linear and angular/rotary position, providing information important to sensor design, and how they function. A chapter on sensor outputs and communication protocols is included and is applicable to all types of industrial sensors. Also included is information on electromagnetic interference, electrostatic discharge, intrinsic safety, temperature effects, and reliability.

Compared with the predecessor text *Position Sensors*, this *Understanding Position Sensors* includes updated information, and extensive new information has been incorporated. A new chapter has been included on ultrasonic position sensors. Many new figures and more detailed explanations have been added, plus new sections on EtherNet/IP and MODBUS.

Chapter 1 presents a working definition of terminology used in the field of sensing technology. Chapter 2 explains how the performance of position sensors, and sensors in general, is specified. Chapter 3 presents the various protocols for analog and digital signal output types. The remaining chapters present the theory supporting the prominent technologies in use in linear and angular/rotary position sensor and transducer products. Application guidance and examples are also included.

There is also a laboratory notebook *Understanding Sensors* available from Revolution Sensor Company. The notebook presents experiments in which sensors of various types are constructed and tested by the student. In addition, a set of parts to support the experiments is available.

Trademarks

The following are the owners of the trademarks as noted in the book:

17–7 PH	AK Steel, West Chester, OH
Beldsol	Belden Corporation, Chicago, IL
CANbus	Robert Bosch GmbH, Stuttgart, Germany
DeviceNet	Rockwell Automation, Milwaukee, WI
EtherNet/IP	ODVA, Inc, Ann Arbor, MI
Excel	Microsoft Corporation, Redmond, WA
HART	HART Communications Foundation, Austin, TX
Kynar	Arkems, Inc., Colombes, France
Loctite	Henkel AG & Company KGaA, Düsseldorf, Germany
Manganin	Isabellenhütte Heusler GmbH & Co. KG, Dillenburg, Germany
Ni-Span C	Huntington Alloys, Incorporated, Huntington, WV
Norsorex	Startech Advanced Materials, GmbH, Wien, Austria
NyceWave	Revolution Sensor Company, Apex, NC
Permalloy	B&D Industrial Mining Services, Inc., Macon, GA
Polaroid	PLR IP Holdings, LLC, Minnetonka, MN
Pomux	Sick-Stegmann, Minneapolis, MN
PowerPoint	Microsoft Corporation, Redmond, WA
PROFIBUS	PROFIBUS International, Karlsruhe, Germany
Ryton	Phillips Petroleum Company, Houston, TX
Solidon	Superior Essex Corporation, Atlanta, GA
SSI	Sick-Stegmann Corporation, Minneapolis, MN
Teflon	The Chemours Company, Wilmington, DE
Temposonics	MTS Systems Corporation, Eden Prairie, MN
Terfenol D	Extrema Products, Inc., Ames, IZ
Thermaleze	Phelps Dodge Copper Products, Norwich, CT
Torlon	Amoco Performance Products, Inc., Alpharetta, GA

All photos were taken by the author. All drawings were drawn by the author, using Microsoft PowerPoint™.

About the Support Material

This book is accompanied by Support Material:
https://www.routledge.com/Understanding-Position-Sensors/Nyce/p/book/9781032436999

The website includes:

- Answer guide

1 Sensor Definitions and Conventions

1.1 IS IT A SENSOR OR A TRANSDUCER?

There has been an ongoing evolution of the accepted use of the words transducer and sensor, and the differentiation between them. Especially since "smart sensor" (and not smart transducer) has become an often-used term. So, this text will address the terms as they are more commonly used, which relies partly on the original definition of "transducer"—*a transducer is a device that changes energy from one form into another*—and relies partly on the widespread acceptance of "smart sensor", and rarely, if ever, a smart transducer. This may conflict with some other treatments of the subject but will clarify the further use of these terms within this book.

As mentioned, a *transducer* is generally defined as a device that changes energy from one form into another or, more specifically, a device that converts input energy into output energy. Typically, the output energy may be in a different form from the input energy but is related to the input. This includes, for example, converting mechanical energy into electrical energy as well as converting from one form of mechanical energy into another form of mechanical energy, and so on. Accordingly, a convoluted thin metal diaphragm converts a differential pressure change into a linear motion change with a force, and a bimetal strip converts a temperature change into a motion with a force. Besides electrical and mechanical energy, forms of energy also include heat, light, radiation, sound, vibration, and others. Sometimes there is no external application of energy in addition to the input energy that is being transduced or changed. The transducer output is often, but not necessarily, in the form of a voltage or current directly converted from the input energy. A transducer that does not require external application of energy (other than the energy that is being transduced) in order to produce a desired output is called an *active transducer* [1, pp 2–4]. Some active transducer examples include the following: a loudspeaker converts a varying electrical energy input into a varying pressure wave output, a piezoelectric microphone converts a varying pressure wave input into an electrical output, a thermocouple pair converts a temperature difference into a voltage and current, a stepper motor converts an electrical input into a change of rotary position with a force, and an antenna converts an electromagnetic field into a voltage and current.

A transducer that requires an external supply of energy is called a *passive transducer*. A passive transducer produces an output signal that is usually a variation in an electrical parameter, such as resistance, capacitance, or inductance. For example, a photocell responds to a variation in light level by producing a relative change in the electrical resistance across two terminals (this is different from a solar cell that produces an electrical output from a light input). An external power supply can be used to convert this resistance change into a change in voltage or current. Other examples

DOI: 10.1201/9781003368991-1

of passive transducers include a coil with a moveable iron core so that moving the core further into the coil causes an inductance increase, a thermistor has a changing resistance with temperature, and others.

A *sensor* is generally defined as an input device that provides a usable output signal or information in response to a specific physical quantity input. The physical quantity input to be measured is called the measurand (such as a measured pressure, temperature, or position) and affects the sensor in a way causing an output that is indicative of the input quantity. The output of most modern sensors is an electrical signal but, alternatively, could be a motion, pressure, flow or another usable type of output. Some examples of sensors include the following: a pressure sensor typically converts a fluid (gas or liquid) pressure into an electrical output signal indicative of the amount of pressure, a magnetostrictive position sensor converts a position into an electrical output signal indicative of the measured position, and many other types of sensors are in common use.

A sensor may incorporate several transducers [2, p 4, fig. 1.2]. In the general case, a sensor is the complete assembly required to detect and communicate a particular event, while a transducer may be the element within that assembly that accomplishes a detection and/or quantification of the event.

For example, a diaphragm may be the transducer that changes a differential pressure into a linear motion or force, but a pressure sensor would include that plus additional transduction and circuit elements as needed in order to provide a desired electrical output, such as an output of 0–5 VDC.

In the example of a bimetal strip temperature transducer, adding a needle and a calibrated scale can form a complete sensor. Or adding a Linear Variable Differential Transformer (LVDT) and signal conditioning could make it a sensor having an electrical output.

Obviously, according to these definitions, a transducer can sometimes be a sensor and vice versa. For example, a microphone or a thermocouple can each fit the description of both a transducer and a sensor. This can be confusing, and many specialized terms are used in particular areas of measurement (e.g. an audio engineer would seldom refer to a microphone as a sensor, preferring to call it a transducer).

Although the general term "transducer" refers to both input and output devices, we are concerned only with input devices in this book. Accordingly, we will use the term transducer to signify an input transducer (unless specified as an output transducer, such as a speaker or a stepper motor).

So, for the purpose of understanding sensors and transducers in this book, these terms will be more specifically defined as they are typically used in developing sensors for industrial, commercial, factory automation, medical, automotive, military, and aerospace industries, as follows:

> *An input transducer produces a usable output that is representative of the input measurand.* Its output would then typically be conditioned (that is, amplified, detected, filtered, and scaled) before it is suitable for use by the receiving equipment (such as an indicator, controller, computer, or PLC—programmable logic controller). The terms "input transducer" and "transducer" can be used interchangeably, as will be done in this work. So for example, a pressure sensing diaphragm could be the input transducer

that becomes part of a complete pressure sensor. An input transducer is sometimes called the sensing element, primary detector, or primary transducer.

A sensor is an input device that provides a desired electrical output in response to the input measurand. A sensor provides a signal that is conditioned and ready for use by the receiving equipment. A sensor is sometimes able to send its signal over long distances by wire, fiber-optic cable, or sometimes can transmit the signal information wirelessly.

One final note on transducers and transducers versus sensors: a review of the literature by the author revealed some sources that agree with the definitions presented here regarding active and passive transducers. But a significant number of the works proposed the opposite meanings, with the same definitions but the words active and passive switched. The same thing happens when checking the definitions of a sensor versus a transducer: some say a sensor is the complete device that includes a transducer plus conditioning electronics, while others say the opposite (switching the words transducer and sensor). So, a reader would be well advised to avoid getting caught up with the discrepancies. But the definitions presented here will be utilized in this book.

A *smart sensor* is a term commonly used since the mid-1980s referring to sensors that incorporate one or more microcontrollers in order to provide increased quality of information as well as additional information. This may include such increased quality as provided by linearization, temperature compensation, digital communication, remote calibration, and sometimes the capability to remotely read the model number, serial number, range, and other information.

Intelligent sensor is the term commonly used when a smart sensor includes additional functionality, such as self-calibration, self-testing, self-identification, adaptive learning, and taking a particular action when a pre-determined condition is present.

The usages of the smart sensor and intelligent sensor monikers also reinforce our working definition of the word "sensor", since these are not called smart transducers or intelligent transducers.

Sometimes, common usage will have to override our theoretical definition in order to result in clear communication among engineers in a specific industry. The author has found, for instance, that some actual manufacturers of pressure sensing devices that include internal voltage regulator, amplifiers, filters, and other signal conditioning electronics, call their product a transducer. That is, the term transducer is sometimes used to name what our definition defines as a sensor. In any case, we will rely upon the definition presented here, because it now seems to apply to most modern uses.

An example of a transducer as part of a sensor may help one to understand our present definition of a sensor being a device that provides a desired electrical output (such as 0–5 VDC) in response to the input measurand (such as 0–5 PSIG) and a transducer being a device that changes (or transduces) energy from one form into another. Figure 1.1 is an example of a pressure sensor (designed by the author) on left, and at right is the same sensor with the top and bottom covers removed.

The pressure capsule (i.e. a set of circular convoluted thin metal diaphragms welded together at their edge) acts as a transducer to change the potential energy of

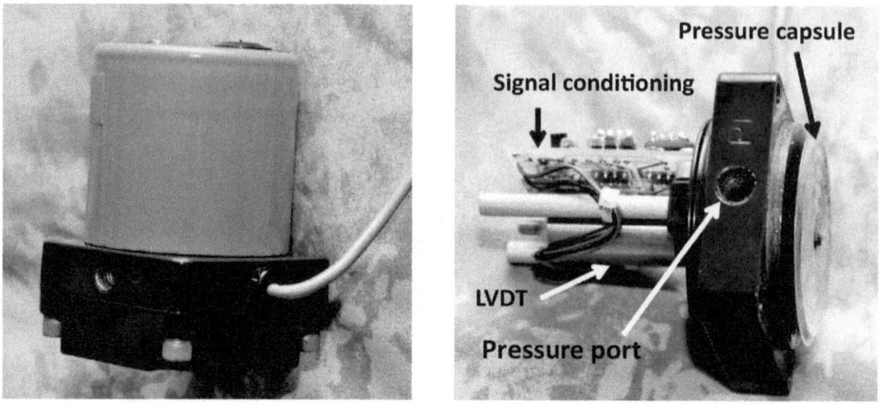

FIGURE 1.1 Example of a pressure sensor, on the left. Partly disassembled at right, to show a pressure capsule that transduces pressure variation into movement of the core of an LVDT.

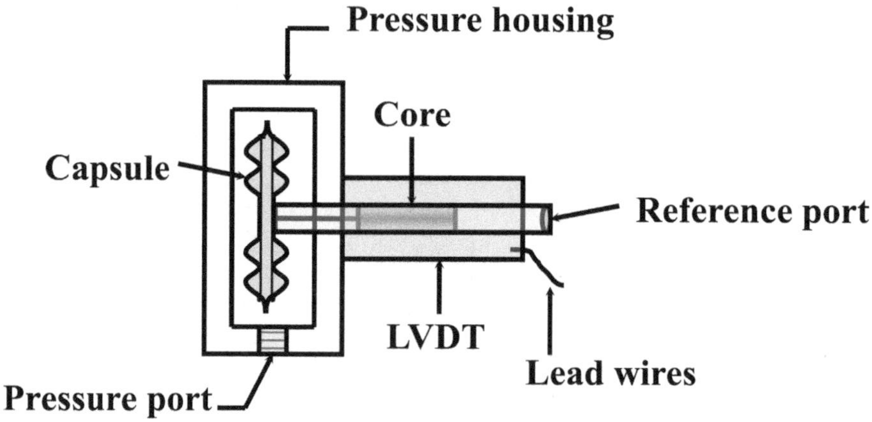

FIGURE 1.2 Pressure capsule and LVDT, some of the components of a pressure sensor.

a pressure difference into the mechanical energy of a linear motion at a force. The motion is internal to the housing, and also moves a ferromagnetic core within an LVDT (LVDTs are explained in Chapter 8). The pressure capsule expands with an increasing internal pressure, and compresses with increasing external pressure. In this example, the pressure to be measured is introduced through a pressure port into the inside of the capsule. The outside of the capsule is exposed to atmospheric air, thereby enabling the sensor to respond to gauge pressure (see Figure 1.2).

The pressure capsule is mechanically coupled with the core of an LVDT. The LVDT also acts as a transducer so that movement of its core affects the inductive coupling among its three internal coils. An associated electronics module powers the LVDT, demodulates the LVDT coil output voltages, and provides an electrical output signal indicative of the measured pressure.

One electrical configuration of sensors has its own descriptive term: a two-wire current loop transmitter, also called a loop transmitter, or often, just shortened to transmitter. Popular transmitter types include temperature, pressure, flow, and position, among others. Such a transmitter has two wires over which both power and signal are transferred. This is explained further in Section 3.1.

1.2 POSITION VERSUS DISPLACEMENT

Since linear and angular position sensors are presented in this work, the difference between position and displacement should be understood. A *position* sensor measures the distance between a reference point and the location of a target. The word target is used in this case to mean that element of which the position or displacement is to be determined. The reference point can be one end, the face of a flange, or a mark on the body of the position sensor (such as a fixed reference datum in an absolute sensor), or it can be a movable point, as in a secondary target, or programmable reference datum.

As an example, consider Figure 1.3, showing the measuring range and external components of a magnetostrictive linear position sensor designed by the author. This is an absolute sensor, measuring the location of a position magnet with respect to the face of the mounting flange (Chapter 12 presents more detail on magnetostrictive position sensors). Some position sensors may have an unusable area near the end of their measuring range, called a dead zone, as noted in Figure 1.3.

A magnetostrictive sensor may have another unusable area near the other end of its measuring range, called the null zone. This is not to be confused with the null of an LVDT, in which the null normally falls within the measuring range (LVDT null is explained in Chapter 8 on LVDTs).

A *displacement* sensor measures the distance between the present position of a target and the previously recorded position of the target. An example of this would be an incremental magnetic linear encoder as shown in Figure 1.4.

Displacement sensors typically send their data as a series of pulses, or sets of pulses on two lines, in which the pulses are time-related so that both the amount

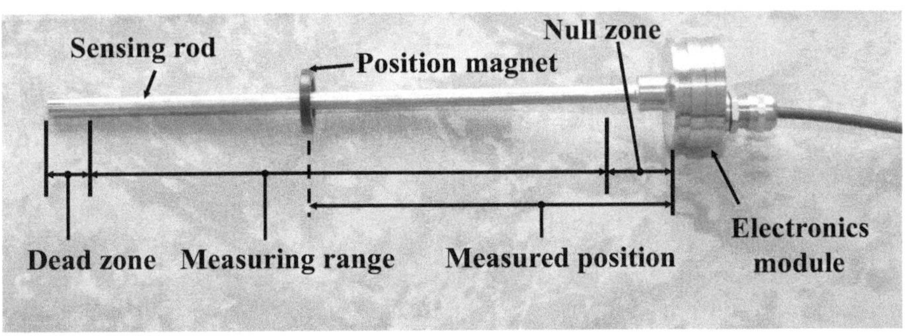

FIGURE 1.3 Magnetostrictive linear position sensor with position magnet.

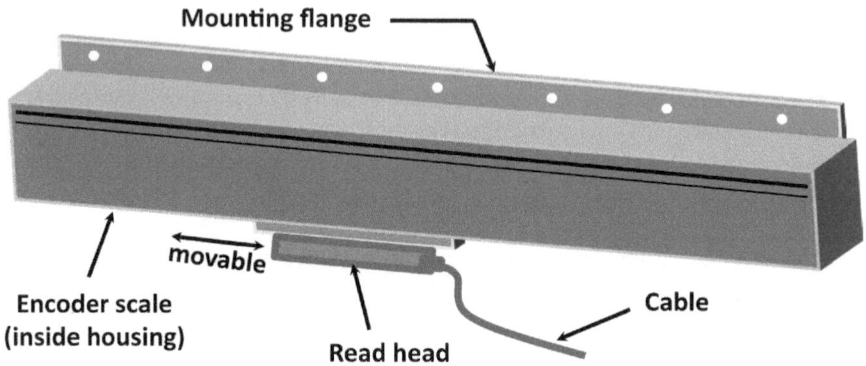

FIGURE 1.4 Incremental magnetic linear encoder.

of displacement and its direction can be encoded. This is known as square wave in quadrature (or just quadrature), and is explained further in Chapter 13 on encoders.

Position sensors can be used as displacement sensors by adding circuitry to remember the previous position and subtract the new position, yielding the difference as the displacement. Alternatively, the data from a position sensor may be recorded into memory by a microcontroller, and differences calculated as needed to indicate displacement. Unfortunately, it is common for many manufacturers of position sensors to call their products displacement sensors or transducers.

To summarize, position refers to a measurement with respect to a reference datum, while displacement is a relative measurement indicating the amount (and sometimes the direction) of movement from a previously noted location.

1.3 ABSOLUTE OR INCREMENTAL READING

An *absolute*-reading position sensor indicates the measurand with respect to a constant, or reference, datum. The reference datum is usually one end, the face of a flange, or a mark on the body of the position sensor. For example, an absolute linear position sensor may indicate the number of millimeters from one end of the sensor, or a datum mark, to the location of the target (the item to be measured by the sensor). If power is interrupted, or the position changes repeatedly, the indication when normal operation is restored will still be the number of millimeters from one end of the sensor, or a datum mark, to the location of the target. If the operation of the sensor is disturbed by an external influence, such as by a power outage or by an especially strong burst of electromagnetic interference (EMI), the correct reading will be restored once normal operating conditions return.

To the contrary, an *incremental*-reading sensor indicates only the changes in the measurand as they occur. An electronic circuit in the receiving equipment is used to keep track of the sum of these changes (the count) since the last time that a reading was recorded and/or the count was zeroed. If the count is corrupted due to a power interruption, or the sensing element is moved while power has been interrupted, the count when normal operating conditions are restored will not represent the present magnitude of the measurand. For example, if an incremental encoder is first zeroed, then moved upscale 25 counts, followed by moving downscale 5 counts, the resulting position would be represented by a count of 20. If there are 1000 counts per millimeter (mm), the displacement is 0.02 mm. If power is lost and regained, the position or displacement would probably be reported as 0.00 mm. Also, if the count is corrupted by an especially strong burst of EMI, changing the count, the incorrect count will remain when the EMI subsides. When an incorrect count is suspected, it becomes necessary to re-zero the sensor. Problems can arise in a control system when it remains undiscovered that the count is incorrect.

1.4 CONTACT OR CONTACTLESS SENSING AND ACTUATION

One classification of a position sensor pertains to whether it utilizes a *contact* or *noncontact* (also called *contactless*) type of sensing element. Another aspect of contactless sensing is whether or not the sensor also uses contactless actuation.

In a contact-type of linear position sensor, one or more parts of the device that make the conversion between the physical parameter being measured (such as a movable arm, or a rotating shaft) and the sensor output, incorporates a sliding electrical and/or mechanical contact. A primary example is the linear potentiometer (see Figure 1.5). An actuator rod is connected internally to a wiper arm. The wiper arm incorporates one or more flexible metallic contacts, which press against a resistive element that is inside the housing and extends over most of the housing length. The potentiometer is powered by applying a voltage across the resistive element from one end to the other. Changing position along the motion axis causes the wiper(s) to rub against the resistive element in respective positions along the resistive element, thus functioning as a voltage divider circuit and producing an output voltage as an indication of the measurand. A more complete description of the linear potentiometer is provided in Chapter 4.

The movable mounting feet shown in Figure 1.5 can be moved to their respective positions, as desired, along the mounting foot rail for mounting. Then they lock into place on the rail when tightened.

It is because of the rubbing contact between the wiper and the resistive element that a linear potentiometer is called a contact sensor. The primary advantages of

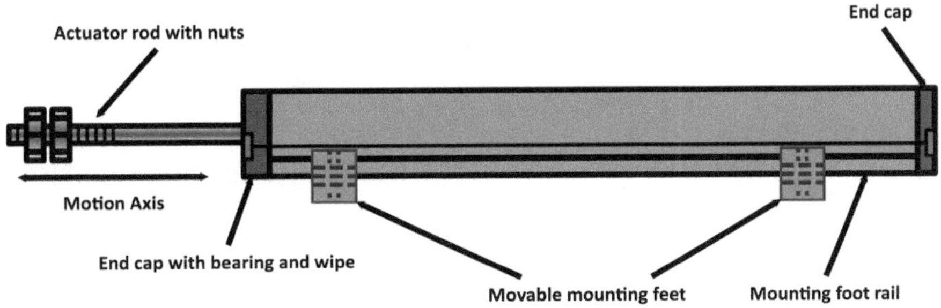

FIGURE 1.5 A linear potentiometer.

using a linear or rotary potentiometer as a sensor are their simplicity and that they often do not require signal conditioning. This sensing technology is also generally thought of as a low-cost solution for many linear and angular sensing applications, although manufacturing automation being implemented with other types of sensors is closing any cost gap that may still exist.

The disadvantage of a contact sensor is that there is a finite lifetime associated with the rubbing elements. Further explanation of the lifetime limitation and the optimization of operating life and reliability are presented in Sections 1.8, 2.19, and in Chapter 4.

In a contactless position sensor, the device making the conversion between the measurand and the sensor output incorporates no physical connection between the moving parts and the stationary parts of the sensor. The "connection" between the moving parts and the stationary parts of the sensor is typically provided through the use of inductive, capacitive, magnetic, or optical coupling. Examples of contactless position-sensing elements include the LVDT, inductive, optical, Hall effect, distributed impedance, magnetostrictive, magnetoresistive, and ultrasonic sensors. These are explained further in their respective chapters later in the book, but as an example, we consider the LVDT (Linear Variable Differential Transformer) here briefly.

An LVDT linear position sensor with core is shown in Figure 1.6. The core is attached to the movable member of the system being measured (the target).

The LVDT housing is attached to a stationary member of the system. As the core moves within the bore of the LVDT, there is no physical contact between the core and the remainder of the LVDT. AC power is applied to the LVDT primary coil. Inductive coupling between the LVDT primary and its secondary windings, through the magnetically permeable core, affords the noncontact linkage.

Contactless sensors are generally more complicated than linear or rotary potentiometers, and typically require signal conditioning electronics. The use of an LVDT requires signal conditioning electronics that comprise AC excitation, signal

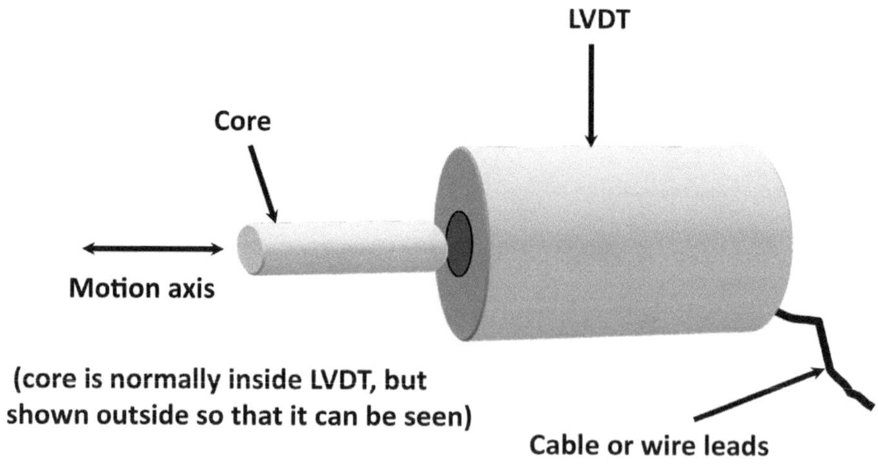

FIGURE 1.6 LVDT linear position sensor with magnetically permeable core.

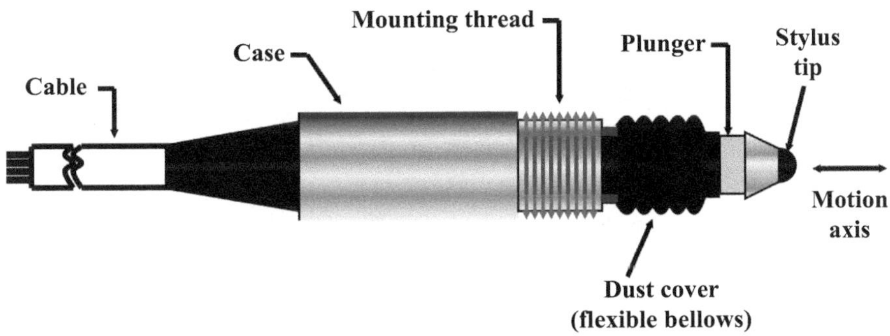

FIGURE 1.7 Contacting actuation in an LVDT gage head, with the LVDT coils mounted within the case.

demodulation, amplification, filtering, scaling, and an output amplifier or other circuit to provide the desired type of output signal. This is explained further in Chapter 8.

In addition to contactless operation within the sensor, a sensing system may utilize contactless *actuation* when there is no mechanical coupling between the sensing element and movable physical element (the target) of which the position is being measured. As an example of magnetic coupling, a permanent magnet can be mounted to a movable machine toolholder, and a magnetostrictive position sensor (as shown in Figure 1.3) can be mounted along the motion axis of the toolholder. The measurement of tool position is thus made without any mechanical contact between the tool holder and the sensing element.

Contactless actuation, obviously, does not utilize any rubbing parts that can wear out and reduce life or accuracy of the measurement. Conversely, contacting actuation may be used with an inherently contactless sensor, as when a toolholder presses the plunger of an LVDT gage head, for example. See Figure 1.7 for a pictorial representation of an LVDT gage head sensor, similar to several models designed by the author.

(Note: the spelling of *gage head*, as opposed to gauge head, has been the standard in the industry for many decades, and therefore gage head is the accepted spelling, and used in this book. This is similar to the accepted spelling of *strain gage*, another type of sensing element that is sometimes replaced by an LVDT gage head.)

In Figure 1.7, the LVDT coils are mounted within the case. A bearing or bushing is within the threaded area, and a return spring is usually contained within the dust cover. The stylus connects with an internal rod and the core. The core moves within the LVDT coils. Even though the LVDT itself operates as a contactless sensor, the contact actuation of the plunger leaves the system somewhat open to reduced life and varying accuracy due to wear. In this example, repeated rubbing of the gage head shaft against its bushings can eventually result in wear, possibly affecting performance through undesired lateral motion of the shaft, or increased operating force. To combat this possibility, LVDT gage heads are often constructed with high-quality ball bearings and a hardened stylus tip that serve to provide long life even with contact actuation.

1.5 LINEAR/ANGULAR CONFIGURATION

Linear and angular position sensors and transducers operate by utilizing any of a large number of technologies. Some of these being resistive, cable extension (also resistive), capacitive, inductive, LVDT, distributed impedance, Hall effect, magneto-resistive, magnetostrictive, optical, and ultrasonic. Each of these technologies can be utilized in the design of both linear and angular sensors (except that optical triangulation and ultrasonic sensors may not be readily adaptable for sensing rotation angle). For example, a resistive type of linear position sensor shown in Figure 1.5 operates in much the same way as a resistive angular position sensor constructed to measure an angular or rotary measurand. See Figure 1.8.

A resistive angular position sensor is also known as a rotary potentiometer. The rotational nature of the sensor dictates the addition of a rotating shaft to hold the wipers, and the resistive element is circular in shape. Other than that, the basic theory of operation is the same as with a resistive linear position sensor. The resistive element and electrical terminals are stationary with the housing. The wipers rotate with the shaft. A highly conductive layer (usually metallic) is placed in the same plane with, and around the resistive element. Some wipers rub against the resistive element, and some wipers rub against the conductive layer. Since the wipers are connected together, a voltage is picked up from the resistive element and placed onto the conductive layer. The conductive layer is then connected with the center terminal of the sensor. The two end terminals are connected with the two ends of the resistive element. A power supply voltage is connected with the two end terminals to apply a voltage across the resistive element.

1.6 POSITION, VELOCITY, ACCELERATION

In addition to measuring a position, it is sometimes necessary to measure how fast a position is changing (velocity), or how fast a velocity is changing (acceleration). Separate types of sensors are available to measure velocity and acceleration.

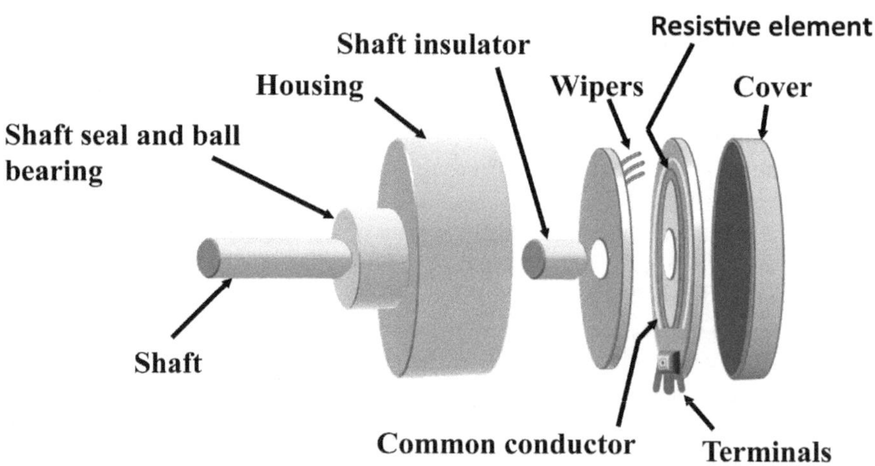

FIGURE 1.8 A resistive angular sensor (aka a rotary potentiometer).

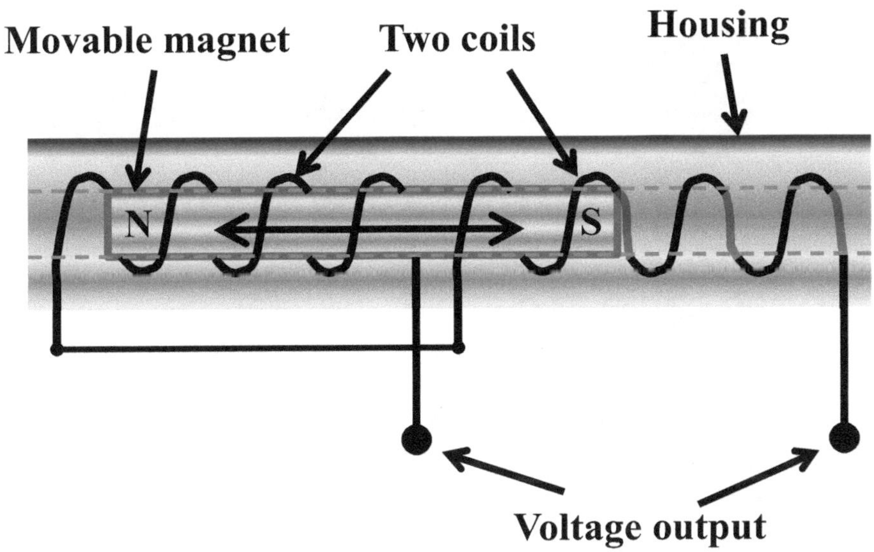

FIGURE 1.9 Velocity sensor with magnetic core that is movable within the bore of a pair of coils, all inside a nickel-iron alloy housing that provides magnetic shielding.

Figure 1.9 is a linear velocity sensor designed by the author using an inductive principle.

A rod magnet is movable within the bore of a sensing element. Two coils are arranged coaxially along the bore. Moving the magnet induces a voltage across each coil, according to Faraday's law:

$$e \propto -N d\Phi / dt \tag{1.1}$$

in which a voltage, e, is induced across a coil having a number of turns, N, of a magnitude proportional to a change in magnetic flux, Φ, with respect to time. This also means the voltage induced across a coil is proportional to the strength of the magnet, the number of coil turns, and the velocity of movement of the magnet.

If only one coil was used, the voltages induced from the north and south poles (ends) of the magnet would cancel each other out. Having two coils connected in series opposing polarity as shown allows the two voltages to add, producing an output voltage that has a linear relationship with the velocity of magnet motion.

As Figure 1.9 shows a magnetic induction velocity transducer, Figure 1.10 shows two types of accelerometers designed by the author: spring-mass and force-balance.

In the spring-mass type, a small mass (called the seismic mass) of ferromagnetic material is supported by a spring. The spring-mass sensor of Figure 1.10(a) was developed by the author and utilizes a ball housing inside which a spherical seismic mass (steel ball) is mounted on a tungsten spring wire suspended between two sensing coils. The two sensing coils were each wound onto bobbins having a hemispherical cavity on one side. A hole was laser-drilled through the steel ball to admit

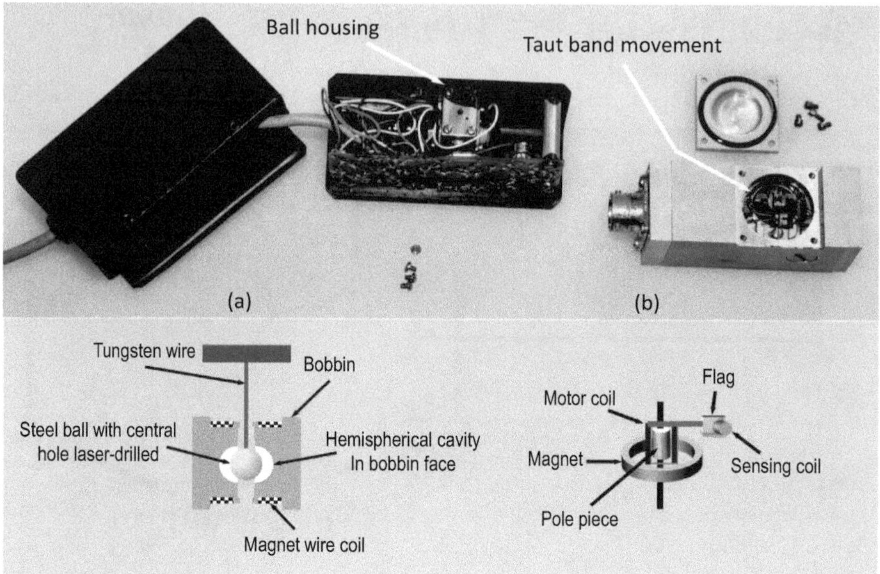

FIGURE 1.10 Accelerometers with covers removed: spring-mass type on the left (a), and force-balance type on the right (b). Sketch of sensing element shown below respective photo.

the tungsten wire. As the mass experiences acceleration, it moves against the spring force by an amount proportional to the amount of acceleration. The coils are driven by an AC excitation, a signal is demodulated, and an output signal is developed that is indicative of the magnitude and direction of the acceleration (such as +5 V to −5 VDC for an acceleration of +1 to −1 *g*). Accelerometers/inclinometers of the types of Figure 1.10 are usually filled with a light oil to dampen oscillation. A silicone oil of viscosity 50 or 100 cSt (centistokes) is common.

Similar spring-mass accelerometers are available that use a micromachined sensing element, providing for a very small sensor. They often use diffused semiconductor strain gage sensing elements to sense force or motion of the seismic mass. Accelerometers of ±1 *g* or less are sometimes called inclinometers, for indicating the angle of inclination from a horizontal attitude.

In a force-balance type of accelerometer, a seismic mass is suspended near a detector. Acceleration that tends to move the mass closer to or farther from the detector is counteracted by a force generated in response to an error signal from the detector. The force-balance accelerometer of Figure 1.10(b) uses a system similar to a taut band analog meter movement. A motor coil of a rectangular shape is held in place by "taut band", a flat spring suspended on an axis, or can be suspended by jeweled pivot points. The motor coil is free to rotate, and is returned to its resting position by the taut band acting as a torsion spring. Or, with a jeweled movement, returned by a light coil spring (coil spring not shown). A ferromagnetic pole piece is mounted within the motor coil, and a ring magnet is outside. A needle attached to the motor coil has a flag at one end, which forms the seismic mass. A sensing coil nearby detects the presence of the flag. The sensing coil is energized by an oscillator, the frequency of which depends upon the coil inductance. The oscillator typically operates at a frequency of about 1 MHz. Changes in the oscillator frequency, due to the proximity of the flag, are demodulated, amplified, and used to drive the motor coil, keeping the flag in the original position. *Force-balance* results as the motor force is adjusted to exactly balance against the acceleration force, so that the flag does not noticeably move with respect to the sensing coil. The amount of current in the motor coil needed to maintain the flag position is conditioned to produce an output signal indicative of the acceleration amplitude and direction (direction = + or –). A typical analog voltage output for an inclinometer could be 1.0 VDC for positive acceleration of 1 *g*, or –1.0 VDC for negative acceleration of –1 *g*.

The signals provided by velocity sensors and accelerometers can sometimes be replicated by a position or displacement sensor. Velocity is the first derivative of displacement, and acceleration is the second derivative. So, when position or displacement information is available to a microcontroller, the velocity and/or acceleration can be calculated. Alternatively, an analog differentiation circuit can be used, as shown in Figure 1.11.

With input voltage V_{IN}, capacitance C, and resistance R, the circuit of Figure 1.11 will provide an output voltage V_{OUT} according to the following formula:

$$V_{out} \propto -RCdV_{in}/dt \tag{1.2}$$

So, for example, if a position sensor has an output of 0–5 VDC, adding the circuit of Figure 1.11 could provide a velocity output of 0 V when there is zero velocity, 0 to –5 V for velocity in one direction, and 0 to +5 V for velocity in the other direction. Adding a second stage of the same circuit could provide an acceleration output.

Whether the velocity or acceleration signal output is obtained from a position or displacement sensor by using an analog circuit, or by calculation within a microcontroller, the signal integrity may be limited by the presence of ambient electrical noise.

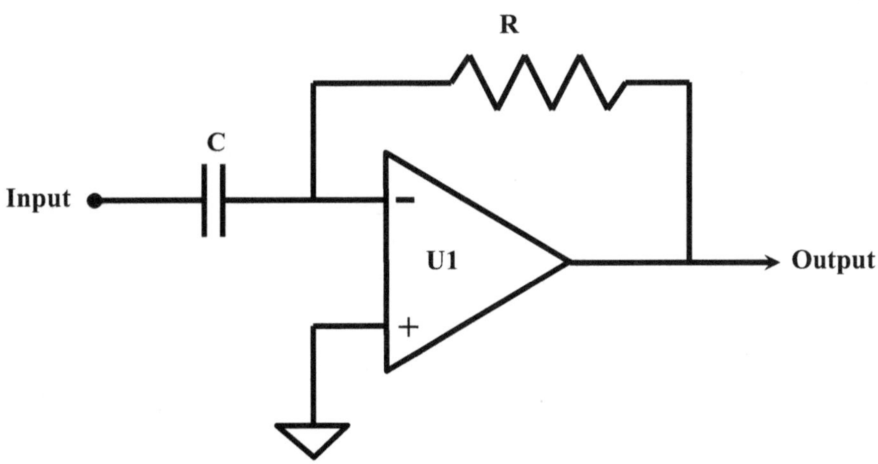

FIGURE 1.11 Analog differentiator circuit to provide velocity output from position input.

In such cases, it is sometimes better to use a velocity sensor or an accelerometer rather than deriving those functions from position or displacement sensor signals.

1.7 APPLICATION VERSUS SENSOR TECHNOLOGY

Linear and angular position sensors can be designed that are based on one or more of a wide variety of technologies, as mentioned previously and presented later in this book. When determining which sensor type to specify for use in a specific application, it may be important to match the technology of the sensor to the requirements of the application.

If the sensor will undergo continuous repetitive motion, as with constant vibration, contactless sensing and contactless actuation may be required in order to eliminate parts that could wear out. In this case, magnetic, inductive, distributed impedance, ultrasonic, or optical coupling to the sensor can be used, for example.

If it is desired to use the same linear position senor type for short strokes (tens of millimeters) as well as long strokes (several meters), then a sensor technology with this operating range capability may be required. Magnetostrictive or distributed impedance technology can be used in this case.

Advantages and disadvantages for each technology are listed in the respective chapters, but Table 1.1 provides general information on application suitability.

1.8 OPERATIONAL LIFETIME

The rated lifetime of the sensing element can be an important consideration in the application of a contact-type linear potentiometer in the presence of continuous vibration. A typical lifetime rating for a potentiometer is 20 million cycles. If the motion system has a constant dithering or vibration at 10 Hz, for example, this number of cycles can be accumulated at a small spot on the element within 2 months.

TABLE 1.1
Application Suitability of Various Sensor Technologies

Technology	Absolute	Noncontact	Lifetime	Resolution	Range	Stability
Resistive	Yes	No	Low	Medium	Medium	Medium
Cable Extension	Yes	No	Low	Low to Med.	High	Medium
Capacitive	Yes	Some models	High	Low to High	Low	Low
Inductive	Yes	Yes	High	Medium	Medium	Low
LVDT	Yes	Yes	High	High	Medium	Medium
Distributed Impedance	Yes	Yes	High	High	High	High
Hall Effect	Yes	Yes	High	High	Low	Low
Magnetoresistive	Yes	Yes	High	High	Low	Low
Magnetostrictive	Yes	Yes	High	High	High	High
Encoder	Some models	Some models	Medium	Low to High	Medium	High
Optical Triangulation	Yes	Yes	High	Low to Med.	Low	Medium
Ultrasonic	Yes	Yes	High	Low	High	Low

Many motion systems have two primary positions in which they operate over 90% of the time. The number of cycles of the example in each of these two positions is represented, per month, by Equation 1.3:

$$10 \text{ Hz} \times 2.59 \times 10^6 \text{ s/month} \times 50\% \times 90\%$$
$$\text{DC} = 11.6 \times 10^6 \text{ cycles/position/month} \quad (1.3)$$

where DC is duty cycle, and 50% accounts for the two positions.

This assumes that the two primary positions are used about equally. Accordingly, a contact-type resistive sensor (potentiometer) exposed to 10 Hz dithering in two positions can wear out within weeks. See Chapter 4 for more details on resistive sensing.

1.9 QUESTIONS FOR REVIEW

1. A distance measurement between the present position and a previous position of a given target is:

a. Displacement
b. Differential transformer
c. Acquisition
d. Double precision
e. Slow

2. The circuitry that is used to power a sensing element, filter and amplify its signal is called a:

a. Multiplier
b. Divider
c. Function circuit
d. Signal conditioner
e. Signal pattern

3. A position sensor of the type that indicates with respect to a constant, or reference, datum:

a. Incremental
b. Absolute
c. Integrating
d. Interval
e. Index

4. This type of analog circuit can provide a velocity output from a position input:

a. Gyrator
b. Peltier

c. Cyclotron
d. Kelvin
e. Differentiator

5. An unusable area near the end of the measuring range is called the:

a. Keep out area
b. Partition
c. No crossing zone
d. Twilight zone
e. Dead zone

6. Faraday's law shows that a voltage induced across a coil is proportional to a change in:

a. Capacitance
b. Elevation
c. Magnetic flux
d. Phlogiston
e. Light waves

7. An analog potentiometer usually has this many terminals:

a. Eight
b. Two
c. Two for power supply, plus four for signals
d. Three
e. Four

8. An LVDT utilizes this type of coupling between its primary and secondary coils:

a. Resistive
b. Inductive
c. Capacitive
d. Conductive
e. DC

REFERENCES

[1] E. Herceg, *Handbook of Measurement and Control*. Pennsauken: Schaevitz Engineering, 1976.
[2] J. Fraden, *Handbook of Modern Sensors*. New York: Springer-Verlag, 2010.

2 Specifications

2.1 ABOUT POSITION SENSOR SPECIFICATIONS

The measurement of a data point includes at least a numerical value, the unit of measure, and the maximum expected amount of error. For example, a linear position measurement could be 5.000 mm, ±0.1% FR (i.e. an error may be present of up to ±0.1% of full range). A sensor will have a set of specifications that define its expected performance, and may include the measuring range, zero and span adjustment, repeatability, nonlinearity, hysteresis, drift, temperature sensitivity, and other parameters.

The list of parameters that are important to specify when characterizing a position sensor may be somewhat different from those that would be important to specify, for example, a sensor for gas analysis. Compared with a gas sensor, the position sensor specification may have a similar requirement to list power supply voltage and current, operating temperature range, and nonlinearity; but there may be differences related to the specific measuring technique.

An angular position sensor specification should indicate, for example, whether the reading is absolute or incremental, whether it uses contact or contactless sensing and actuation, shaft diameter, and the amount of torque required to rotate the shaft, among other things. Conversely, a gas sensor specification would indicate what gas is to be detected, how well it ignores other interfering gases, if it measures gas by percent volume or partial pressure, and the shelf life and operating life (if it is an electrochemical type of gas sensor having a limited lifetime).

So, there exist a number of specifications that are important when describing the performance capability of a position sensor, and its suitability for use in a given application. These specifications are presented here.

2.2 MEASURING RANGE

The measurand, or the physical quantity being measured, must have a magnitude that is within the measuring range of the sensor in order for the sensor to provide an accurate reading. A position sensor can have a measuring range specified from zero to full scale, or it can be specified as a ± full-scale range (FSR). It is common with an LVDT, for example, to specify bipolar ranges, such as ± 100 mm FSR. In this case and with a ± 10 volts direct current (VDC) output specified, the output voltage would vary from −10 to +10 VDC for a measurand changing from −100 to +100 mm. In the center of travel, the output would be 0 V. Since the example sensor is specified over the range of −100 mm to +100 mm, the FSR is 200 mm. If the corresponding output range were ±10 VDC, then the full-range output (FRO) would span 20 VDC. These are the amounts used when other parameters are specified as a percent of FSR or FRO [1, pp 4–4]. For example, with an LVDT and signal conditioner specified for a

DOI: 10.1201/9781003368991-2

maximum zero shift of 1.0% per 100°C, an FSR of ±100 mm, and an FRO of ±10.0 VDC, then a 100°C temperature change can produce an error of 2.0 mm or 0.20 V. That is, 0.01 × 200 mm = 2.0 mm or 0.01 × 20.00 V = 0.20 V.

In a magnetostrictive position sensor, the sensing element measures a time period starting from one end, thus making an absolute, zero-based measurement. Even so, it is possible to produce a sensor having a bipolar range by adding an offset incorporated within the signal conditioning electronics; but the most common configuration is to have a range of zero to full scale (unipolar), with zero being located near one end of the sensor. An example of a unipolar range is an output of 0.0 VDC to +10.0 VDC, corresponding to an input position of 0.0 to 1.0 m.

2.3 ZERO, SPAN, AND FULL SCALE

In the field of sensor technology, the terms *zero*, *span*, and *full scale* have specific meanings, and are used to describe aspects of the measurand and/or the output of a sensor. On a unipolar scale, zero is the lowest reading, full scale is the maximum reading, and span is the difference between the full scale and zero readings. For example, a position sensor may have a measuring range of 0.0 to 1.0 m and produce an output of 4.0 to 20.0 mA. In this case, the input measurand has a zero of 0.0 m and a full scale of 1.0 m. The span is also 1.0 m. The output has an offset, however. The output has a zero of 4.0 mA, and a full scale of 20 mA. The output span is therefore 16.0 mA. So, 16.0 mA of output span represents, and is proportional to, 1.0 m of input measurand. The output sensitivity thus being 16.0 μA/mm. This output sensitivity means that, from any starting point in the measuring range, the output will change by 16.0 μA for each millimeter of position change.

Understanding the distinction among zero, span, and full scale is important when troubleshooting errors, since knowing whether the error is a zero shift or a span shift can indicate the error source. If, for example, a technician is temperature testing a position sensor with an output of 4 to 20 mA, corresponding to an input range of 0 to 100 mm, the technician would first mechanically fix the sensor at or near its zero position (≈ 0.0 mm). The output will be approximately 4 mA. As the temperature is varied in an environmental chamber, changes in the output would be recorded as temperature-induced *zero error*. Next, the sensor would be fixed at or near its full-scale position (≈ 100.0 mm). The output will be approximately 20 mA. After again changing the temperature over the same range, changes in the output would be recorded as temperature-induced *full-scale error*. Subtracting the zero errors from the full-scale errors would provide a data set indicating the temperature-induced *span errors*. By analyzing these errors, the source(s) of any temperature sensitivity problem can be categorized as being related either to zero or to span functions. Things that cause zero error are offset-related errors. They can be mechanical, such as thermal expansion of a mounting feature or actuator rod, or electrical, such as input voltage drift of an amplifier or resistance change in a voltage divider circuit that provides a zero offset.

Things that cause a span error are gain-related errors. They can also be mechanical, such as a changing spring rate, or electrical, such as change in a transistor gain, a resistance change in an amplifier feedback loop, or a capacitance change in a coupling capacitor for an alternating-current (AC) signal.

Full-scale error should not be investigated separately, since the error at full scale is the sum of zero and span errors.

Knowing this cause-and-effect link helps to guide one's efforts in the troubleshooting of sensor errors, as well as when designing a sensor or transducer to meet the specifications required in the product development stage.

When designing a sensor that has zero and span controls (either analog or digital), the controls are much easier to adjust when they are not interactive. This is especially true with an offset zero, such as with a 4 to 20 mA output. With non-interactive controls, adjusting the span control while the measurand is at zero or null should not affect the output current at all. Adjusting the zero control should affect the output reading by the same amount, at any output level within the operating range. Figure 2.1 is a circuit diagram of an amplifier that has interactive controls: it has the undesirable quality that adjusting the span control can also affect the output when the output signal is at zero. For example, if the input voltage is zero, and there is a current going through R_2 into the U_1 inverting input (such as may be needed to provide 4 mA at zero), changing the setting of span pot VR_1 will change the output voltage. The amount of change would be equal to the R_2 current multiplied by the change in VR_1 resistance. The input signal current (through R_1) and the zero adjustment current (through R_2) each enter operational amplifier U_1 at its inverting terminal. So, span control VR_1 affects the amplification of both the input signal and the zero signal. So, when adjusting zero and span, the user would need to employ several iterations of adjusting zero, then span, then zero, and so on, in order to get both zero and span set correctly. (Hint: if you need to calibrate a sensor having interactive controls, first adjust the span control to get the desired change in output in response to a specific change in input, then adjust the zero control.)

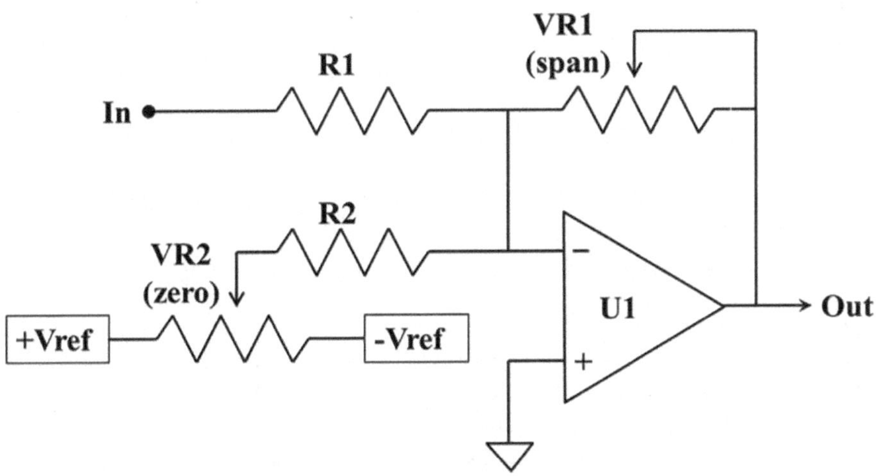

FIGURE 2.1 Amplifier circuit that has interactive zero and span controls.

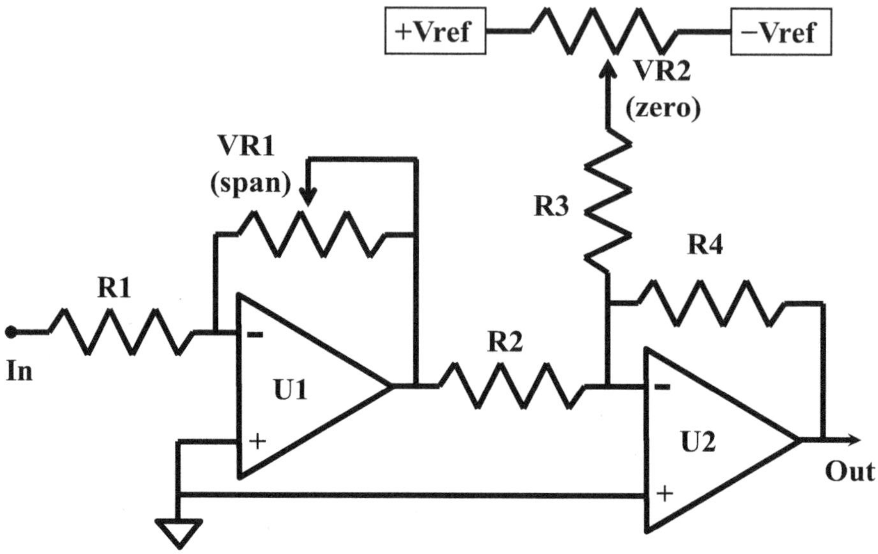

FIGURE 2.2 Amplifier circuit that has non-interactive zero and span controls.

The amplifier circuit has been improved in Figure 2.2 to provide non-interactive controls, so that adjusting of the span control does not affect the zero setting.

In Figure 2.2, gain is applied to the input signal, as scaled by VR_1. U_1 and U_2 each have a virtual ground (that is, non-inverting input connected to common, so it will also hold its inverting input at the same voltage as common, by changing the voltage at its output as needed). Since U_2 has a virtual ground, currents flowing into the inverting input will not change the voltage at the inverting input (inverting input has the "–" sign). Current flowing into the U_2 inverting input through R_2 represents the signal after scaling by VR_1. Current flowing into the U_2 inverting input through R_3 represents the zero offset, as adjusted by VR_2. So, changing the adjustment of VR_1 does not affect the zero offset provided by VR_2: this means that the zero and span controls are not interactive.

So for example, regarding the circuit of Figure 2.2, if the input signal is 0 V and the zero control (VR_2) is adjusted for a 1 V output signal, then adjusting the span control (VR_1) does not change the voltage of the output signal. Note: in order to achieve truly non-interactive zero and span controls, the op amps need to have low input offset voltage and input bias current. Low offset voltage can be achieved with a chopper-stabilized op amp, or one that is factory-trimmed for a low offset voltage.

The figures show analog circuits, but the same theory applies to digital calculations. In a digital system, the measurement readings are converted to digital representations by an analog to digital converter (ADC or A/D). A microcontroller (μC) then scales the data by making appropriate zero and span adjustments as needed to arrive at the desired output scaling. Care should be taken in the coding algorithm (as with analog circuit designs) so that offsets are first removed, then scaling factor applied, then the desired offset added back in, so that zero and span adjustments are not interactive.

2.4 REPEATABILITY

Repeatability is specified as the deviation that can be expected in consecutive measurements under the same conditions and using the same measuring instrument. To determine the repeatability of a sensor, the input parameter to the sensor is exercised over a set of conditions, the sensor output is recorded, then the exact same conditions are met again, and the sensor output is again recorded. The degree to which the same result can be expected in consecutive measurements is called repeatability. This is usually tested by maintaining fixed temperature, humidity, and other environmental conditions and then exercising the sensor by changing the measurand among fixed points. For example, while environmental conditions remain constant, the core of an LVDT may be exercised from zero, to full scale, to zero, then to half scale. A data point is taken at the last position. Then the movement of the core is repeated to full scale, to zero, then to half scale again. The second data point is taken at the last position. This is done many more times to obtain a set of data. The standard deviation of this data set is the repeatability that is listed in the sensor specification.

Here is an example of calculating the repeatability from a set of measurements, using measurements that were all taken of the same measurement point, under the same conditions (according to the method just described).

The measured data points are given in column 2 of Table 2.1.

The sample population of 10 data points are summed in the third column from the left. In column 4, the sum is divided by the number of points to find the mean. Column 5 takes the difference between the data of column 2 and the mean, providing the deviation of each point. Column 6 squares the differences (this removes any skewing effect of positive and negative differences, making them all positive). All of the squares are summed together in column 7. Since there are a limited number of samples, the sum is divided by 9 instead of 10 points, providing the sample variance (0.006). This use of $n - 1$ instead of n (for example, dividing by 9 instead of the 10 sample points in this case) accounts for the use of a relatively small random sample, instead of a total population. (This is known as *Bessel's correction*, and provides a better estimate of a total population standard deviation.) Then finally in the rightmost

TABLE 2.1
Calculating Repeatability

Point#	Measured	Sum	Sum/10	Diff	Squared	Sum	Sum/9	SQRT
1	9.950	100.110	10.011	–0.061	0.004	0.052	0.006	0.076
2	10.010			–0.001	0.000			
3	10.110			0.099	0.010			
4	10.050			0.039	0.002			
5	9.890			–0.121	0.015			
6	10.060			0.049	0.002			
7	9.930			–0.081	0.007			
8	9.970			–0.041	0.002			
9	10.020			0.009	0.000			
10	10.120			0.109	0.012			

column, the square root of the sample variance provides the sample standard deviation of 0.076. The repeatability of a sensor is usually specified as a percentage of FRO. So, if the previous data are from a sensor having a 0–10 VDC FRO, then the repeatability would be specified as ±0.76% FRO (which is 0.076/10 V × 100%).

It is theoretically possible to have a repeatability that has a smaller value than the resolution, by adding noise to the system and making a statistical analysis of the resulting set of data; but this is not usually helpful to someone using the sensor. So, the specified repeatability should not be smaller than the specified resolution. This assures that it is possible for a user to reproduce the specified level of performance. Repeatability can be the most important characteristic of a sensor if the receiving equipment is able to compensate for nonlinearity, temperature effects, calibration error, and other parameters. This is because repeatability is a sensor characteristic that cannot be compensated. Also, in many control systems, repeatability is more important than sensor accuracy because the system can often be programmed to provide the desired output in response to a given input from the sensor, as long as the signal received from the sensor is always the same for a given set of conditions.

2.5 NONLINEARITY

The set of output data obtained from a theoretically perfect (ideal) linear position sensor, when exercising it throughout the specified operating range and recording the output data versus input stroke, should form a straight line from the zero reading to the full-scale reading. In a real sensor, the data may not form a perfectly straight line, and the end points may not be exactly at the specified zero and full-scale points. This is depicted in Figure 2.3, somewhat exaggerated for clarity.

The maximum amount of difference between the sensor characteristic and an ideal characteristic at any point within the measuring range is defined as the maximum error. This error may sometimes be reported as a percent of FRO and called the percent accuracy, since it is often the largest of the sources of error. But instead, inaccuracy should be described as the individual components comprising it. And also, rather than report an accuracy of 0.5%, for example, it should be reported as an *inaccuracy*, or an error of 0.5%. (In that case, the accuracy would really be 99.5%.)

It is more appropriate to separately specify the major components comprising inaccuracy or error, since some applications may have a greater sensitivity to one particular error source as compared with the error sensitivities of another application. For instance, it may be most important that a hydraulic cylinder return to the exact point as before rather than that the exact position of the point be known (that is, repeatability may be the most important error to consider in this case). In a different application, it may be most important that the rotation angle of a shaft not change with changing temperature of the environment (so, insensitivity to variations in temperature may be the most important error to consider in this case).

The term "static error band" is properly used to indicate the sum of the effects of nonlinearity, repeatability, and hysteresis in position sensors. Environmental effects are typically reported separately. Nonlinearity itself, however, can be interpreted in several ways, as presented next. Hysteresis is then presented in the following section.

Typically, the most important characteristic of position sensor error is nonlinearity. A straight line is drawn that closely approximates the sensor characteristic. The

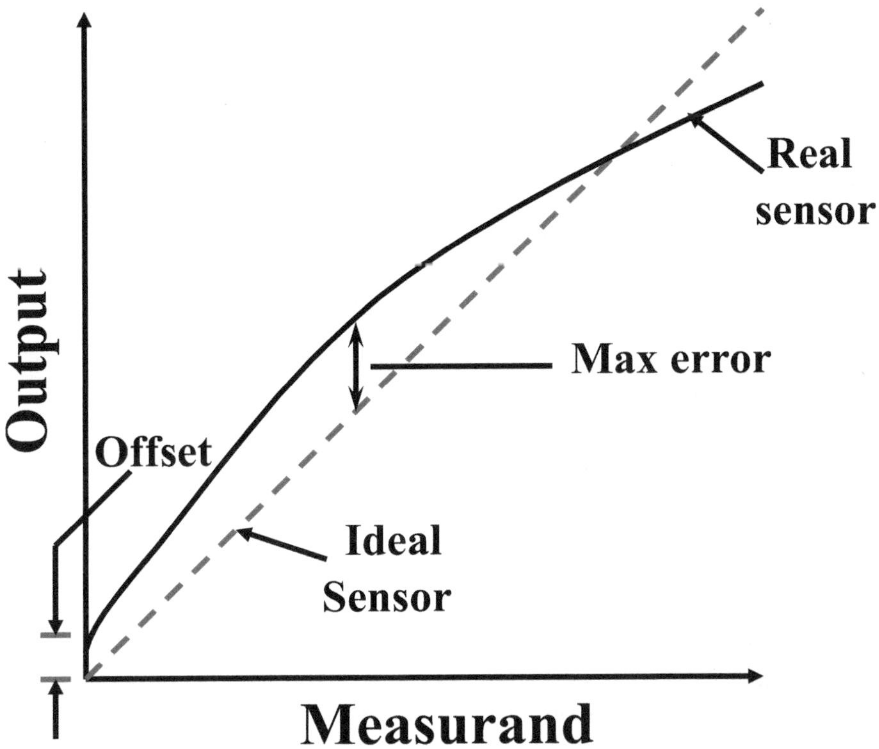

FIGURE 2.3 Error is measured by comparing an ideal characteristic (a straight line) with the real sensor characteristic.

difference between the straight line and the sensor characteristic is the nonlinearity, reported as a percentage of FRO or FSR. The *nonlinearity* error specification is often improperly referred to as the sensor "linearity".

For example, if the maximum linearity error (between the sensor characteristic and a straight line) is 0.5 mm and the FSR is 100.0 mm, then the nonlinearity is 0.5%. This sounds simple enough, but there are a number of ways to arrive at a "best" straight line that closely approximates the sensor characteristic, and to which the sensor output data will be compared.

2.5.1 Best Straight Line Nonlinearity

The best straight line (BSL) can also be called the *best-fit straight line* or *independent BSL*. When the BSL or the best-fit straight line is all that is named as the nonlinearity reference in the specification, or the independent BSL is named, it is not required that any specific point on the BSL be drawn through any specific data point of the sensor output characteristic. The BSL does not have to go through zero or full scale. The purpose is only to find a straight line that comes closest to matching all of the output data points of the sensor. The stated nonlinearity is then the maximum deviation of any data point from this straight line. A graphical representation of this is shown in Figure 2.4.

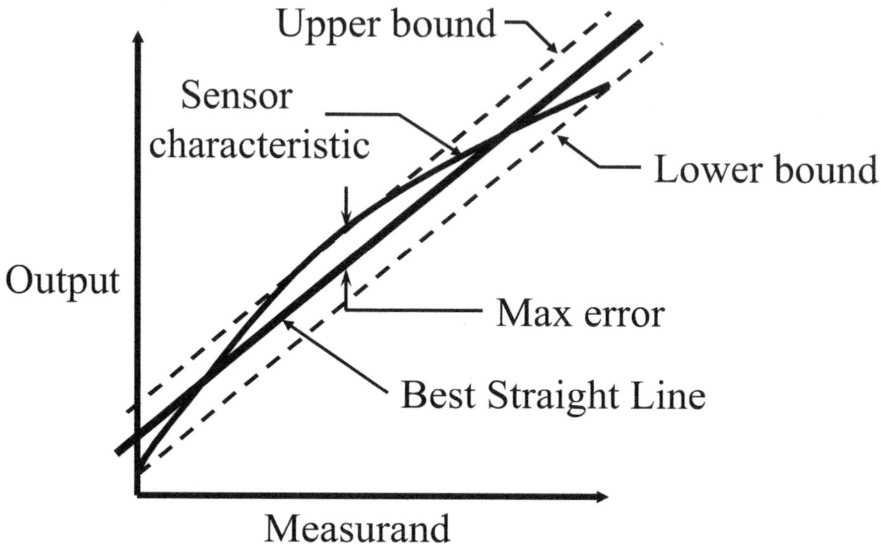

FIGURE 2.4 Finding the Best Straight Line (BSL) and maximum nonlinearity error.

Two parallel lines are placed on a graph of the sensor characteristic, one above and one below the line representing the sensor data. These are called the upper and lower bounds. The two lines should be brought as close together as possible, while encompassing all of the sensor data between them. They do not have to be parallel to the sensor data. A third straight line is then placed along the center between the two parallel lines. This third line is the BSL. The maximum deviation (error) between this line and the sensor data, expressed as a percent of FR, is the sensor BSL nonlinearity. This line can be defined in *Y*-intercept form as:

$$Y = mX + B \tag{2.1}$$

where m is the slope of the line and B is the *Y*-intercept. This means that m is the scaling factor (affected by the span control), and B is the zero offset (affected by the zero control).

One can visualize that half of the distance between the two parallel lines drawn on the graph (measured vertically) is the BSL nonlinearity, being the absolute value of the amplitude of the maximum deviation (Max error) of the output from a straight line. The method for calculating the BSL without using a graph, however, may not be evident at first glance. A practical way to find this line from the data is to first find the least-squares line through the data (explained here, after a few paragraphs), use this to derive a line equation in *Y*-intercept form (Eq. 2.1), and then use an iterative method with small changes in slope (m) and intercept (B) until a line equation is found that yields the minimum deviation from the sensor data. But the most common and effective way is to simply use the least-squares method, explained later in this chapter, and not continue with any further iteration or adjustment.

During sensor calibration, zero and span controls are adjusted so that this BSL coincides with the ideal line that was shown in Figure 2.3 (and was labeled as Ideal sensor). Any remaining difference between the BSL and the ideal line is calibration error. Calibration error can be listed as zero offset error and gain (or span) error.

2.5.2 Zero-Based Nonlinearity

When it is desired to ensure that the output indicates zero when the measurand is zero, a zero-based nonlinearity may be specified. This may be preferred when the indication of a negative position would not make sense and the equipment receiving the sensor signal cannot make the correction. In this case, one end of a straight line is set equal to the zero measurand/zero output point (in a graph, the origin), and the other (full scale) end of the line is moved up or down (changing the slope) until minimizing the maximum deviation of the sensor output data from the line (see Figure 2.5).

There will usually be one or more points on the sensor characteristic that fall above the straight line, as well as one or more points that fall below it. In an approximately uniformly curved characteristic as in Figure 2.5, there will be one maximum somewhere near the midpoint, and another near full scale (assuming that zero error

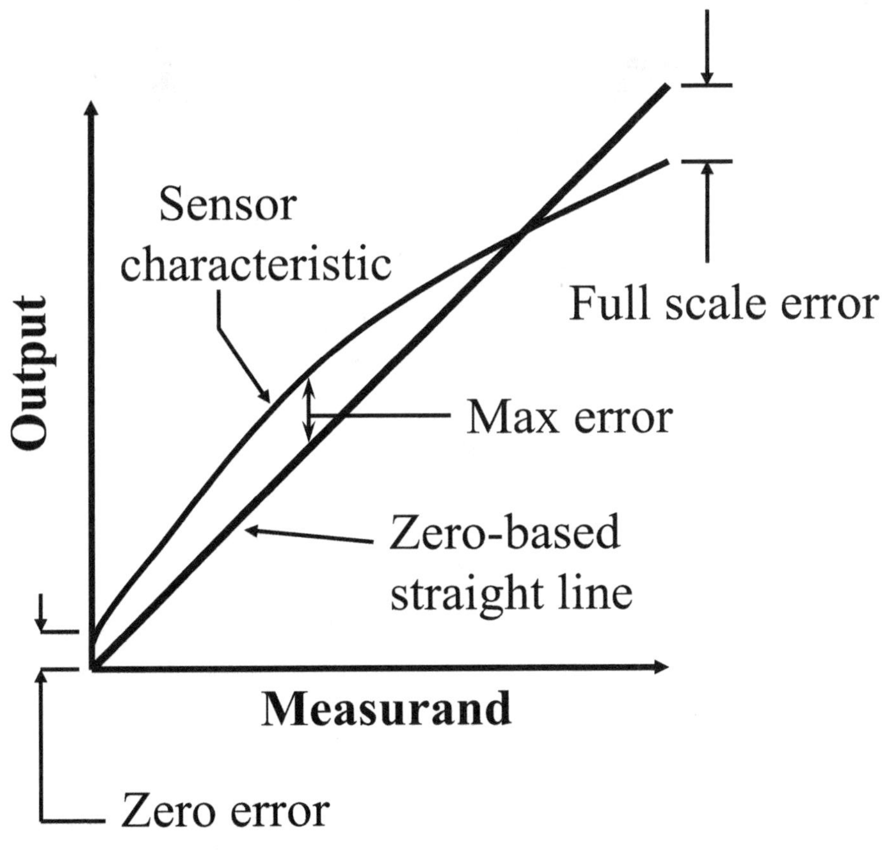

FIGURE 2.5 Zero-based nonlinearity.

has been minimized by adjustment of the zero control). These two error amounts should be approximately the same, if the straight line is properly placed.

2.5.3 End-Point Nonlinearity

A straight line can be drawn between the sensor outputs at zero measurand and at full scale (these two points are called the end points). The maximum deviation between this line and the sensor data is called the end-point nonlinearity (see Figure 2.6).

Sensor manufacturers prefer to specify nonlinearity according to one of the other methods, though, because the magnitude of the end-point nonlinearity is in the range of two times the number obtained by one of the other methods. End-point nonlinearity may be of interest to a user whose equipment does not have a means for correcting gain errors of the sensor.

2.5.4 Least-Squares Straight-Line Nonlinearity

Nonlinearity based on a least-squares regression (LSR) of the input data versus the output data is the most popular type of specification because it can be easily calculated. The disadvantage is that it can be very close to the optimum line, but the straight line found is not necessarily the absolute best one, since it is a statistical estimation. The degree to which the LSR line actually represents the "best" straight line depends on the number of data points taken and the nonuniformity or erratic nature of the data. The result will be less representative when the data do not follow a continuous smooth curve and when the number of data points is smaller. Still, it is the most popular method to find a BSL, since it is easy to implement mathematically.

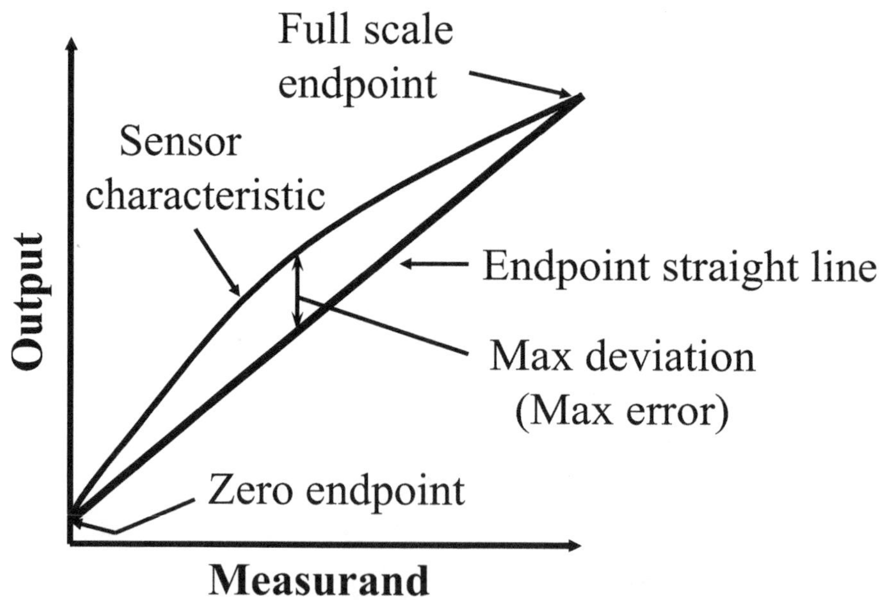

FIGURE 2.6 End-point nonlinearity.

If the LSR straight line is defined in the form $Y = mX + B$, then the slope m is found by:

$$m = \frac{\sum_{d=1}^{n} X_d Y_d}{\sum_{d=1}^{n} X_d^{\,2}}. \tag{2.2}$$

where X_d and Y_d are the data from the input measurand and sensor output, respectively, and n is the number of data points. Once the slope m is found, then the Y-intercept that yields the lowest overall deviation must be found. After that, the maximum deviation is reported as the least-squares nonlinearity. Although tedious to manually calculate on paper, it is easy to implement on a set of data using a calculator or spreadsheet program.

Using a scientific calculator, select the linear regression function. Enter the input measurand data consecutively as the set of values for the first variable of a two-variable array. Enter the corresponding sensor output data as the set of values for the second variable of the array. Select the calculate function.

In a spreadsheet program, select the linear regression analysis tool; this performs a linear regression using the least-squares method to fit a straight line through the data selected as input data columns in a spreadsheet. For example, in Excel, load the analysis tool pack. Then select the regression analysis tool. Make a spreadsheet with a first column of input measurand data versus a second column with the corresponding sensor output data over the range of sensor operation. Select the input measurand data (first column) as the input X range. Select the output data (second column) as the input Y range. Then calculate the slope of the LSR line using the SLOPE function. Find the Y-intercept using the INTERCEPT function. This will provide the slope and the Y-intercept of the LSR line (the slope and intercept are listed in third column of Table 2.2). Next, the data of the LSR line are calculated (fourth column, BSL), using the slope and Y-intercept applied to the X range (first column) according to the formula $Y = mX + b$. Then calculate the errors (fifth column) as the difference between the sensor output and the calculated LSR line (second and fifth columns). The percent error is calculated in the sixth column, as the errors of the fifth column divided by the FRO. The FRO in this example is 5 V, so each error of column 5 is divided by 5 V and then multiplied by 100 to get percent that is shown in column 6 (absolute values are shown). The maximum number from the sixth column is selected and reported in the seventh column as the least-squares nonlinearity (Max error %). See Table 2.2, and also refer to Figure 2.7, which is a graph of the data of Table 2.2. The 5.8% error shown would be very high for a position sensor, but was used so that the error could be seen in the graph.

After the spreadsheet functions are used to find the slope and intercept of a least-squares line, the least-squares nonlinearity error is specified as the maximum difference between the sensor data and least-squares BSL data, divided by the FRO. The slope and intercept constants for a particular sensor can be entered or downloaded into a computer or other equipment connected with the sensor, thereby allowing the

TABLE 2.2
Example of an Excel Spreadsheet with Reference Position versus Sensor Output. The Least-Squares Nonlinearity is 5.80% FRO

Ref. Pos. (cm)	Sensor Out (V)	BSL Calc.	BSL Data (V)	Error (V)	ABS Error %	Max Error %
0.015	0.051		0.341	–0.290	5.801	5.80
0.447	0.323	Slope = 0.519	0.565	–0.242	4.849	
0.808	0.563		0.753	–0.190	3.800	
1.210	0.854		0.962	–0.108	2.156	
1.588	1.106		1.158	–0.052	1.043	
1.938	1.362		1.340	–0.022	0.441	
2.313	1.567		1.535	0.032	0.645	
2.691	1.798		1.731	0.067	1.338	
3.042	1.979		1.913	0.066	1.311	
3.422	2.207		2.111	0.096	1.923	
3.794	2.425		2.304	0.121	2.419	
4.152	2.651		2.490	0.161	3.219	
4.527	2.873	Intercept = 0.333	2.685	0.188	3.763	
4.915	3.106		2.886	0.220	4.393	
5.297	3.279		3.085	0.194	3.884	
5.639	3.441		3.262	0.179	3.571	
6.008	3.599		3.454	0.145	2.897	
6.382	3.796		3.648	0.148	2.952	
6.709	3.920		3.818	0.102	2.035	
7.105	4.091		4.024	0.067	1.341	
7.480	4.232		4.219	0.013	0.265	
7.832	4.405		4.401	0.003	0.068	
8.181	4.504		4.583	–0.079	1.578	
8.566	4.640		4.783	–0.143	2.858	
8.935	4.772		4.975	–0.203	4.051	
9.299	4.922		5.164	–0.242	4.833	
9.505	4.996		5.271	–0.275	5.493	

computer or equipment to correct the sensor signal to improve overall system accuracy. One way in which nonlinearity may be corrected by using an algorithm in a computer or microcontroller is to implement a correction table and interpolation technique, as explained in AN942 from Microchip.com.

2.6 HYSTERESIS

Regarding the output signal of a position sensor, hysteresis is the variation between upscale and downscale approaches to the same position. More specifically, when a sensor is steadily indicating an increasing output (moving upscale), crossing through position *a* of the measurand, then reverses direction and steadily indicates a decreasing reading (moving downscale), again passing through position *a*, there will be a

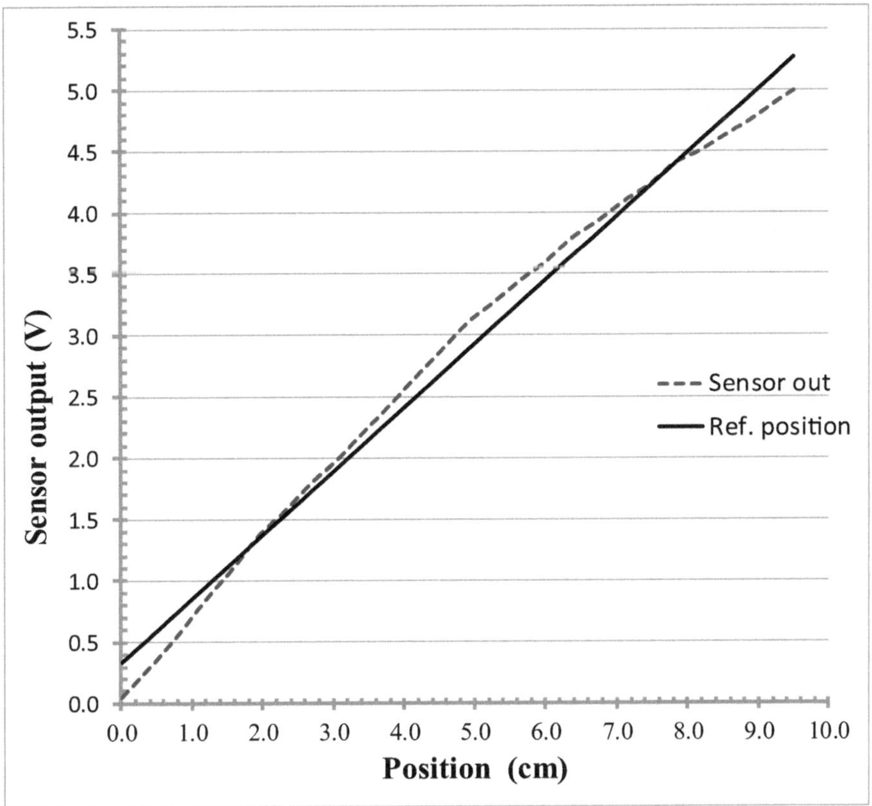

FIGURE 2.7 Graph of the sensor data and least-squares straight line of Table 2.2.

slight difference in the reading recorded for the increasing and decreasing approaches to position *a*. This characteristic is shown, exaggerated, in Figure 2.8. Position *a* is shown as a point on the lower curve, corresponding to an increasing measurand. A different output is shown for the same point *a* on the upper curve, corresponding to a decreasing measurand. The (vertical) difference on the chart between the two points is the maximum error due to hysteresis.

The specified parameter that is typically called hysteresis may include effects from mechanical backlash, the building of spring force before a wiper moves, magnetic remanence in a sensing element magnetic circuit, and/or plastic deformation of a sensing member, among others. These are typically reported only as an overall hysteresis error, and the individual elements are not reported separately.

In a position sensor based on the use of a magnetic field, for example, one cause of hysteresis is the magnetic remanence of the material that is being affected by the magnetic field. An initially non-magnetized material would first follow line (*a*) as shown in Figure 2.9 upon exposure to a magnetizing force. As the external field (magnetizing force) builds up, the magnetic material becomes magnetized. Then, when the magnetizing force is reduced, the remanence of that material causes some

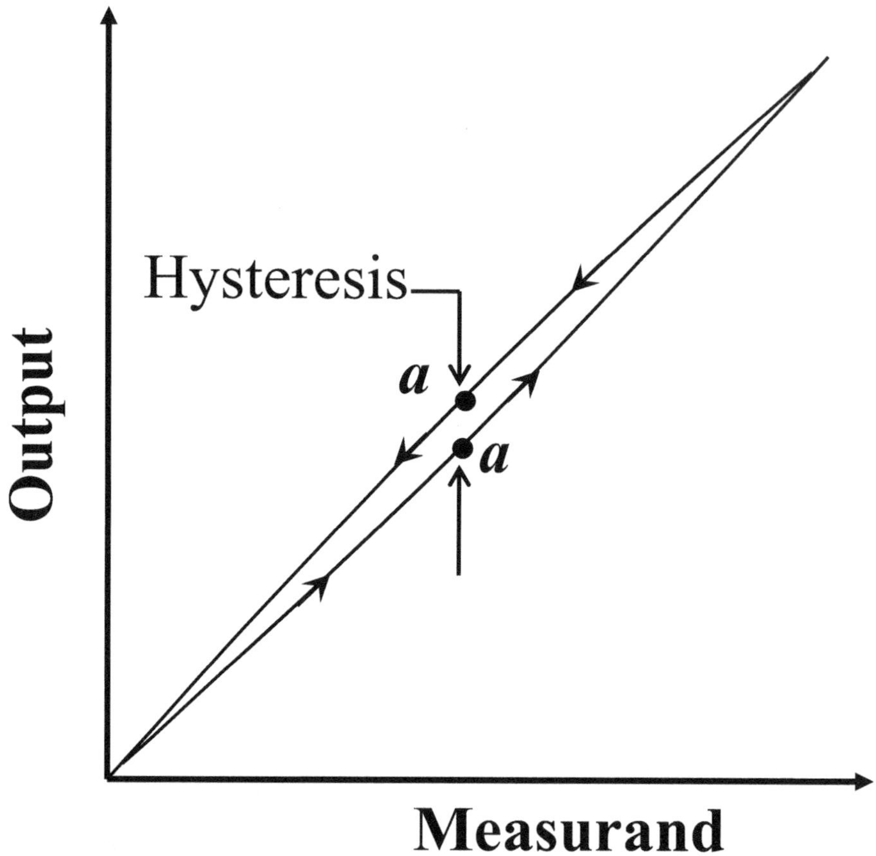

FIGURE 2.8 Hysteresis shown in a plot of measurand vs. sensor output.

of the magnetic field to remain in the magnetic material—it has become somewhat "magnetized". Thereafter, the field strength in the material would follow lines (*b*) and (*c*) when subjected to further reductions and increases in magnetizing force. A remanent field, that has not been accommodated in the sensor design, may result in an error in the sensor output signal. The measurement of the magnetic remanence of a material is the value of the flux density, *B*, retained with the magnetizing force, *H*, removed, after magnetizing the material to saturation [2, p 333].

Hysteresis in a potentiometric type of position sensor comes from other sources. The wiper may flex slightly down as it is being moved up, and then start to flex slightly up as it is being moved down. The lagging of the output reading with respect to the input motion may cause a difference between the upscale and downscale readings (see Figure 2.10).

The amount of flexing depends on the flexural strength of the wiper, the wiper spring force pressing it against the resistive element, and the surface friction of the wiper contact against the resistive element. There may also be backlash in the actuator that drives the wiper movement. Backlash, also called "play", is the amount of motion that is lost in a gear set, or other type of drive train, when moved in one direction and then in the opposite direction, due to gaps between the gear teeth or other

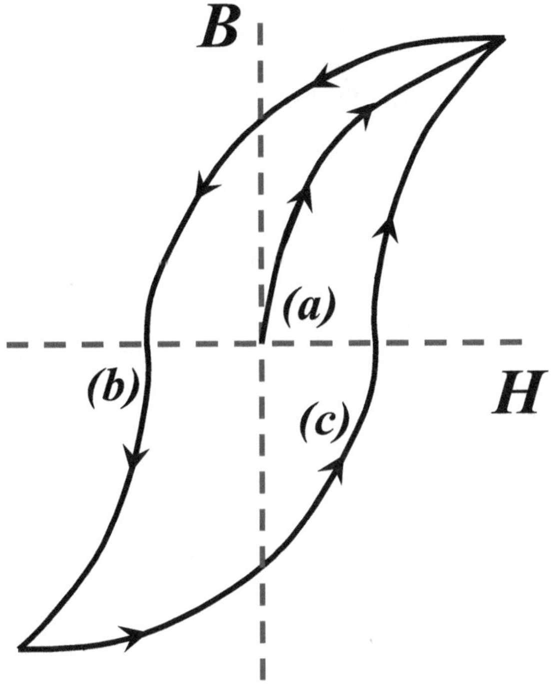

FIGURE 2.9 Remanent magnetic field in a magnetic material. Line (*a*) is an initial non-magnetized state.

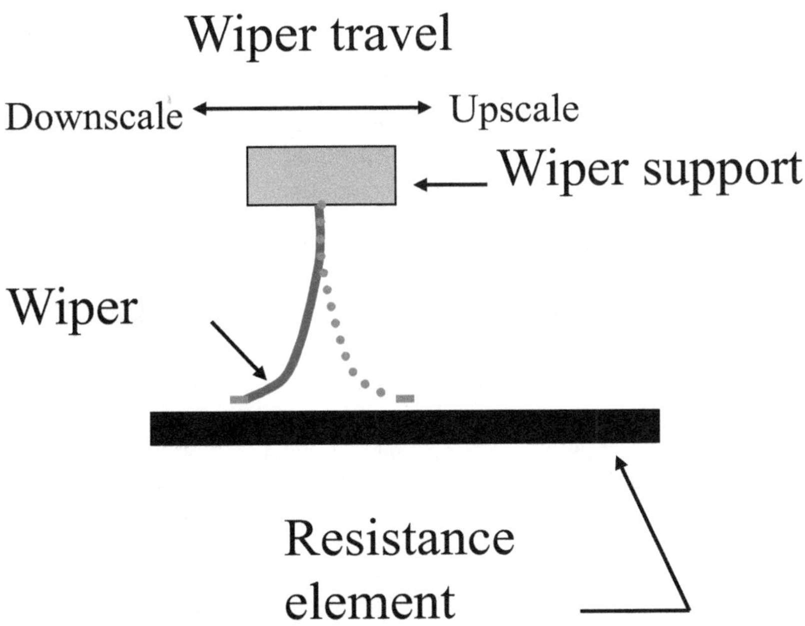

FIGURE 2.10 Wiper flexing may cause different upscale and downscale readings.

mating parts. Backlash is sometimes listed separately from other sources of hysteresis in screw- or gear-driven potentiometers, but has the same effect on performance in a given application as would hysteresis arising from other sources.

The accepted way to measure the hysteresis in the output of a position sensor is to first exercise the sensor throughout its full range in order to have a reproducible starting point. Then the position is smoothly varied to move the measurand starting from the zero reading, up to full scale, and then back to zero, while recording data at approximately uniformly spaced points along the range of the measurand. The upscale and downscale tracks are plotted. Then the maximum deviation between the two is noted. This deviation is reported as hysteresis and specified as a percent of FRO. Typically, the maximum error will be in the middle of the stroke. In an LVDT that travels in both a positive and negative direction, with respect to the null or zero position, the maximum hysteresis error is normally around the null point. Sometimes a bipolar range LVDT will have a unipolar hysteresis specification, as well as one for bipolar operation, or report a null hysteresis separately.

A related parameter of potentiometric sensors is friction error, due to the friction of the sliding wiper. This is usually included in the hysteresis spec as described previously, but not always. In a potentiometer application that will be accompanied by constant vibration, the effect of wiper friction may be greatly reduced. To accommodate this possibility, a hysteresis error with *friction-free* measuring is sometimes stated in the specification of a contact sensor (such as a pressure capsule mated with a contact potentiometer). During testing, a vibration is applied to overcome friction and allow the wiper to move to the friction-free point. This is also called *mechanical dithering*. (An interesting fact: an analogous electrical dithering can be used to average out quantizing error for increased resolution in some digital electronic circuits.)

2.7 CALIBRATED ACCURACY

A sensor exhibits a given performance including errors due to nonlinearity, hysteresis, temperature sensitivity, and so on; however, the actual performance in the application is also affected by the accuracy to which the best straight line through the sensor output data was calibrated to a known standard. Even if it is assumed that the calibration instruments are perfect, there can be a limit to how closely the sensor can be calibrated to match the standard (such as regarding settability of a calibration potentiometer, or the resolution of a digital adjustment). Differences in offset and slope between an ideal line and the best straight line through the sensor output data is called calibration error. See Figure 2.11.

As shown in Figure 2.11, calibration error of the best straight line through the sensor output data can have a zero calibration error (offset) and/or a span calibration error (a difference in slope), as compared with the ideal line. In the figure, best straight line *a* has an offset error, and BSL *b* has a span error. In a real application, there will be both zero and span error at the same time. Of course, in a real laboratory, the calibration standard instruments also have their own level of error, but it is typical to require that a calibration standard have an error at least ten times less than the error to be claimed for the device under test (DUT). So, in that case, errors in the calibration equipment should have a negligible effect on the reported data.

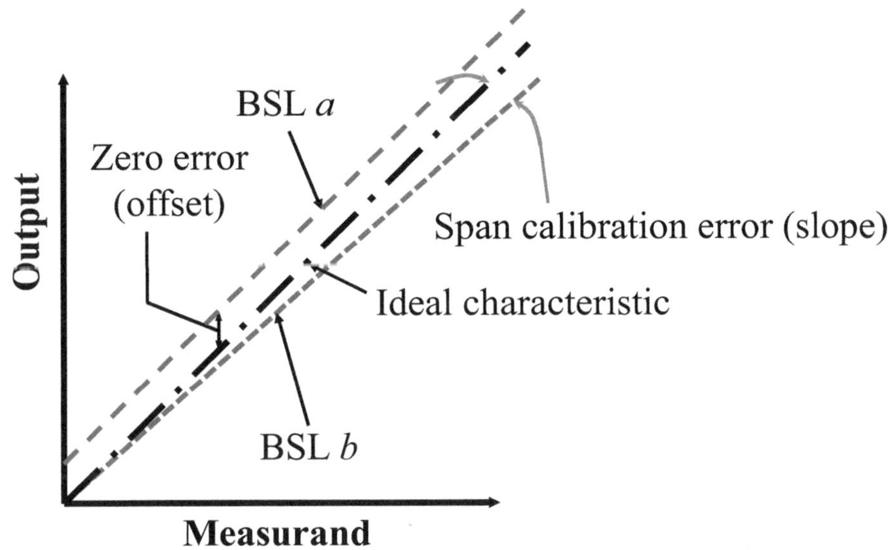

FIGURE 2.11 Calibration error.

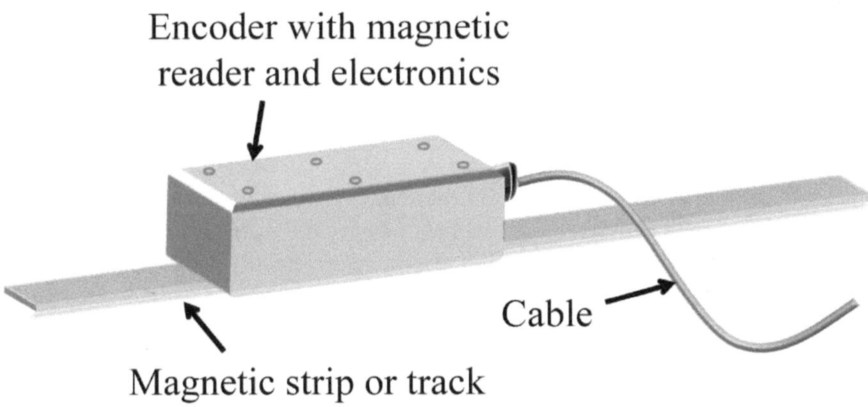

FIGURE 2.12 Linear encoder (magnetic strip type).

For a linear position sensor, length measurements having reference accuracy can be obtained from a linear encoder (Figure 2.12), a laser interferometer (Figure 2.13), or another sensing technique capable of accuracy sufficiently higher than that expected from the sensor being measured (that is, an error of at least ten times lower than that of the DUT).

With the linear encoder of Figure 2.12, a strip or track that has digital position information magnetically encoded onto its surface is stretched out along the path of the measuring axis of the DUT. Many other form factors of the magnetic strip and reader are possible, so this figure is shown as an example of one configuration. The encoder module is attached to the movable part of the test apparatus. As the encoder module (or read head) is moved, electrical pulses are sent to a computer or other

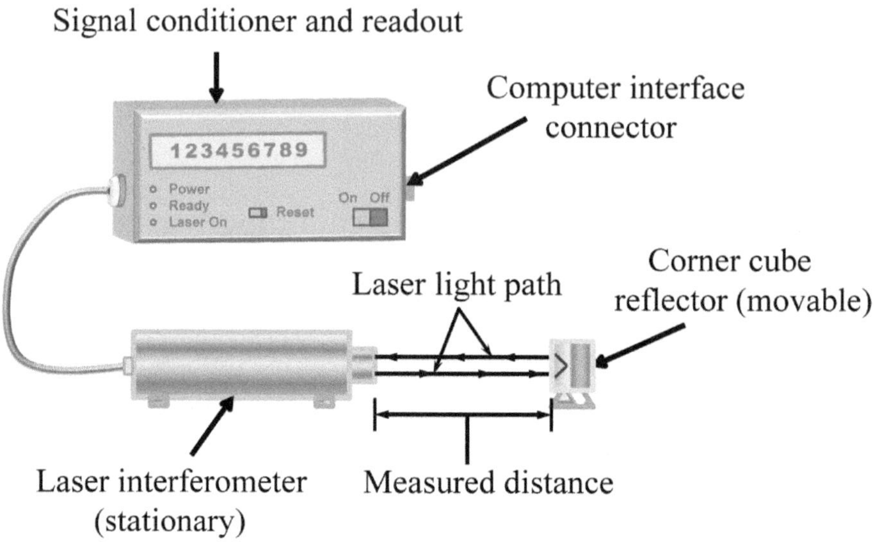

FIGURE 2.13 Laser interferometer.

interface device to indicate the exact amount of motion measured, usually in the format of square wave-in-quadrature (which is explained in Chapter 13).

An excellent measurement tool, for position sensors more than 1 or 2 meters long, is the laser interferometer. With the laser interferometer of Figure 2.13, a laser beam is sent from a stationary interferometer module, along the path of the measuring axis of the DUT, and reflected back from a corner cube reflector that is attached to the movable DUT fixture. The returning laser beam enters the interferometer, where it interacts with the original beam (through a mirror arrangement) such that an interference pattern is formed. The interference pattern comprises multiple fringes that occur at a regulator spacing. An optical sensor counts how many dark and light spaces pass, due to the fringes. The corner cube is attached to the movable part of the test apparatus. As the corner cube is moved, electrical pulses are sent to a computer or other interface device to indicate the exact amount of motion of the corner cube. This is an incremental measurement, so a count total represents the measured displacement. Measurement resolution is in the picometer range. As with a magnetic or optical encoder, the electrical pulses are usually in the format of square wave-in-quadrature.

Rotary encoders can be used for angular standards if they have sufficient resolution and sufficiently low errors that are ten times less than that of the device to be tested. In this case, as mentioned, the error of the reference device can be essentially ignored. Sometimes, though, this ratio of error is not practically available. When using a ratio of less than ten, an allowance should be made for this when evaluating the data.

Calibrated accuracy is the absolute accuracy (but actually a percent inaccuracy) of the individual sensor calibration and incorporates error of the standard used, as well as the ability of the calibration technique to produce a setting that matches the standard. For example, if the setting is made by turning a potentiometer adjustment, the operator tries to obtain a setting that results in a particular output reading. The operator will be able to achieve this to within some level of error. That error will become part of the calibrated accuracy specification, in addition to any allowance made due to the accuracy of the reference standard that was used. When evaluating the total error budget of an application, the calibrated accuracy must be included as well as the nonlinearity, hysteresis, temperature error, and so on.

2.8 DRIFT

Drift encompasses the changes in sensor output that occur over time even though there are no changes in the measurand or environmental conditions. The only variable when measuring drift correctly is the elapsed time. Some specifications may erroneously list "temperature drift", but that should be called temperature sensitivity. In a position sensor, no position change should be allowed during a measurement of drift. The sensor actuator is normally locked into a stationary position when drift testing. The test is run at constant temperature, constant humidity, constant power supply voltage, constant load impedance, and so on, while the sensor output is recorded. If the environmental temperature changes with time of day, for example, then readings may be compared from one day to the next when the environmental temperature happens to be the same.

Drift is reported in two components: short-term and long-term, and is expressed as a percent of full-range output (FRO). On a typical position sensor, short-term drift is that which occurs in less than 24 hours, unless another time period is specified. It is reported as error in percent of FRO per hour, and is common to be sometimes positive and sometimes negative regarding a given sensor. Long-term drift is specified in the same way, but the time period is usually per month. On some reference-grade equipment, long-term drift may have a time period of per year. Long-term drift tends to be in a given direction, either positive or negative, so the drift may accumulate over time. This can be due to normal aging of one or more components. A drift specification may sometimes be separated into its components of zero drift and span drift. Drift can sometimes be called aging. And sometimes the aging can be expected to be in a known direction. For example, as an incandescent lamp ages, it can be expected to produce less light over time at a predictable rate (because some of the tungsten filament evaporates with age, and increases filament resistance, resulting in less power and less lumens of light produced).

Sources of short-term drift include such things as instability in electronic circuits, and instability of mechanical components. Whereas long-term drift originates from changes in electrical and mechanical component characteristics, and from wear. For a short-term drift example: a sensor may have a normal operating cycle that causes some self-heating followed by cooling back down. Some components might be affected by the self-heating-induced temperature change and cause the output to constantly drift slightly up and down. A source of long-term drift might include an electrolytic capacitor, which can change capacitance value or equivalent series resistance (ESR) as the electrolyte dries with age. Operational amplifier offset voltage may have both short- and long-term drift. Mechanical components can undergo wear and/or fatigue, causing long-term drift. Identifying the type of drift experienced (short- or long-term) can give clues to the possible sources of the drift.

Sometimes drift can be reduced by changing the sensor design. If a bushing is experiencing wear within its expected lifetime, maybe the bushing material can be changed to one that is either tougher or harder (whichever may be suitable in the given instance) so that it will better withstand the conditions. If a spring experiences fatigue, maybe the material, its dimensions, or its processing can be changed. A small capacitance value aluminum electrolytic capacitor can be changed to a ceramic one. If an operational amplifier has too much offset voltage drift, a chopper-stabilized version can be used.

2.9 WHAT DOES ALL THIS ACCURACY STUFF MEAN TO ME?

An engineer tasked with implementing a position sensor into a control system must determine whether or not the control system will be capable of exhibiting the specified position accuracy when incorporating the feedback element (position sensor) that is planned to be used. Errors can be divided into the categories of either the static or dynamic type. Static errors in a sensor typically include nonlinearity, hysteresis, and repeatability. Dynamic errors include phase shift or amplitude variation due to the sensor frequency response, amplitude variation due to damping factor, and so on. Errors due to environmental conditions are normally reported separately and include errors from changes in temperature, humidity, moisture, vibration, shock, pressure, salt spray, radiation, and others.

The position sensor specification will probably not list an overall error level (or accuracy) that can be expected to include a combination of all of the static and dynamic errors (that is, performance over the temperature range, including nonlinearity, hysteresis, and so on). Rather, all of the specifications should be listed individually, and then it is up to the user to decide how to add up or otherwise choose to utilize the specified errors in determining suitability of the sensor for obtaining a desired system performance. Some sources of error may directly apply to the application being considered, and some may not. For example, the application requirements may include a wide ambient temperature operating range, but have only a slowly changing measurand. In this case, the temperature error will be important, but not the frequency response, phase shift, or damping factor. In another application, limitation of dynamic errors may be paramount; and in that case, it may be essential to investigate the individual effects of each dynamic error in detail.

The *static error band* of a position sensor normally includes nonlinearity, hysteresis, and repeatability. Since each of these can be in the same range of magnitude, the errors must be added up in some way and evaluated for their accumulated effect on the performance in the sensor application. The sum of the static errors is called the static error band. At a Christmas party of a former employer, the author (who enjoys rock music) witnessed a performance of a rock musical group of company employees that called themselves “The Static Error Band”. A good play on words, but their performance, not so much.

If all of the individual specifications were simply combined as an arithmetic sum, this could be used as the overall accuracy specification. Doing this, however, would not be realistic. It is not likely that all of the errors would each simultaneously be at their maximums, and at the same time, and for each to act in the worst case direction so their effects would add. Instead, some errors will be positive, and some errors will be negative. Some errors will be near maximum, others will be around average, some are likely to be lower than average. One way to statistically sum these error specifications in the design of industrial products is to use a root-sum-of-squares (RSS) estimation.

In the RSS method, each individual error percentage is squared, the results added together, then the square root of this sum is calculated.

$$e_{Sum} = \pm\sqrt{e_1^{\ 2} + e_2^{\ 2} + e_3^{\ 2} + \ldots e_n^{\ 2}} \tag{2.3}$$

or,

$$e_{Sum} = \sqrt{\sum_{j=1}^{n} e_j^{\ 2}} \tag{2.4}$$

where e_1, e_2, and so on are individual sources of error and n is the number of error sources. When using this method, it is assumed that each of the errors act independently and has an evenly symmetrical distribution. This simple RSS statistical sum solves for an interval of σ, accounting for approximately 63% of the specimens of the product, under the assumptions of the example. The standard deviation, σ, is also the square root of the dispersion.

For a sensor to be used in a high-reliability and critical application, it may be more reasonable to solve for a 3σ interval, assuring that nearly all specimens of the product (99.75%) will perform as calculated. Again assuming that the distribution is Gaussian, the maximum error for a 3σ interval is:

$$e_{Sum} = \sqrt{3 \cdot \sum_{j=1}^{n} e_j^{\ 2}} \tag{2.5}$$

For example, the important errors to be summed for a particular sensor may include hysteresis, nonlinearity, and repeatability. If the 1σ nonlinearity error is 0.21%, the hysteresis is 0.13%, and the repeatability is 0.05%, the combined error calculation using each method would be:

The simple sum would be 0.21 + 0.13 + 0.05 = 0.39%

The 1σ RSS would be sqrt $(0.21^2 + 0.13^2 + 0.05^2) = 0.25\%$

The 3σ RSS would be sqrt $(3(0.21^2 + 0.13^2 + 0.05^2)) = 0.44\%$

If the error specification is based on a 3σ interval, it may be listed as a guaranteed error, and the error over a 1σ interval may be listed as the "typical" error. But sometimes each guaranteed specification item is actually tested on every sensor specimen before shipping from the manufacturer.

2.10 TEMPERATURE EFFECTS

Experience has shown that the greatest contributor to the overall error budget in a measurement and control system is usually the temperature sensitivity of the sensor and associated circuitry (unless it is actually a temperature sensor). Characterizing the temperature sensitivity of a sensor, and implementing appropriate design changes in order to reduce its temperature sensitivity, are often a major part of the development program in the design of a new sensor.

Commercial products that will likely be used indoors have a typical temperature range specification of 0° to 70°C. Outdoor types and industrial sensors often have an operating temperature range of −40° to 85°C. Automotive sensors have several ranges, depending on where they will be mounted in the car. Engine compartment devices and those near other heat sources like the exhaust system or shock absorber orifices or valves range up to 125° or 150°C and sometimes higher. Sensors for use in the passenger compartment will not need to operate at such high temperatures, and may sometimes be rated for −40 to 70°C operation. The Automotive Electronics Council (AEC) is a United States-based organization originally established in 1993 by Chrysler, Ford, and Delco Electronics. The purpose of AEC is to set qualification standards for components in the automotive electronics industry. A stress test qualification standard called AEC-Q100 lists the following automotive operating temperature ranges:

Grade 0: −40 to 150°C

Grade 1: −40 to 125°C

Grade 2: −40 to 105°C

Grade 3: −40 to 85°C

Grade 4: −40 to 70°C

The author has designed sensors for continuous operation at temperatures of up to 210°C. In addition to the operating temperature range, there may be a storage temperature range, and there will be a temperature sensitivity specification while in the operating temperature range.

The storage temperature applies when the sensor is not required to operate, and is a survivability specification (any expected effect on calibration may also be noted). The operating temperature sensitivity specification, however, is sometimes the most important system performance specification.

On a linear position sensor, there should be a temperature sensitivity specification for zero and also one for span. Span is the difference between the zero reading and the full-scale reading (see Section 2.3).

Zero shift is due to thermal coefficient of expansion, the warping and shifting of mechanical components, as well as the changes in offset voltage of op amps, mismatching of temperature coefficients of resistor bridge circuits, and so on.

Span shift is due to changes in gain factor with temperature, which can be mechanical or electrical. To measure span shift, zero shift is first measured. Then full-scale shift is measured. The span shift is obtained by finding the difference between zero shift and full-scale shift.

When developing a new product and finding unwanted temperature sensitivity, it is important to separate temperature-induced errors into those associated with zero and those associated with span. This enables the engineer to have a clue pointing to the source of the errors, so the design can be optimized to reduce these errors. Zero shift can be avoided in the sensor mechanical design by selecting thermal expansion coefficients of the materials of construction, for example, and in the electrical design by selecting an amplifier with a low input voltage drift with temperature (or chopper-stabilized). Span shift might be avoided by using a Ni-Span C™ spring material in a pressure capsule, manganin resistance wire in an LVDT primary, or a resistor or capacitor with a controlled rate of sensitivity to temperature. Manganin wire is an alloy of copper, manganese, and nickel that can be used to reduce temperature sensitivity. Manganin magnet wire is sometimes used in winding the coils of an LVDT, especially the primary, because it has a low thermal coefficient of resistance, approximately +0.00002%/°C [3, p 75]. If it is required to reduce the zero and/or span shift still further, they can be compensated in hardware and/or software to some degree.

Because of all of this, it is important to complete an extensive set of temperature testing when developing a new sensor. This requires an environmental test chamber that can reach the desired temperatures and hold the temperature steady, and that is suitable to accommodate the sensor under test. An example of an environmental test chamber, in the author's lab, is shown in Figure 2.14.

For most efficient use, a temperature test chamber should have a window in the door, an interior light that can be turned on/off from the outside, an access port for wires or other support attachments, and the ability to have its temperature computer controlled. Better performance can be obtained if some insulating material is stuffed into the access port from the inside, as shown in Figure 2.14, as well as a rubber plug installed from the outside, as shown in Figure 2.15. The rubber plug (a 2 inch diameter one is shown) should have a hole in the center, and a slot cut to the center from the outside edge. Such a hole and slot in the rubber plug allows wiring to the DUT to be easily accommodated. If the test data (such as voltages, current, and/or digital data) can also be acquired by a computer, then a temperature test can proceed without the need for anyone to be present once the test is under way.

FIGURE 2.14 Environmental test chamber with heating and mechanical refrigeration.

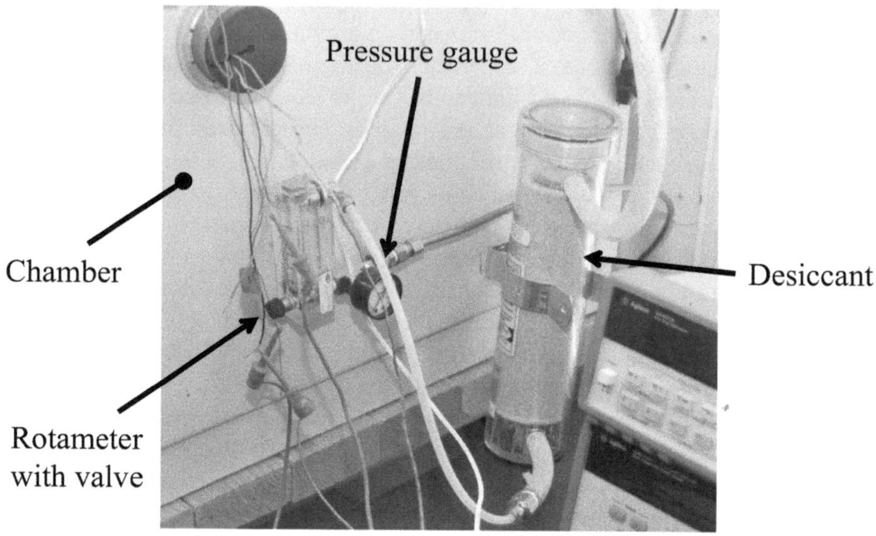

FIGURE 2.15 Pressure gauge, rotameter, and desiccant provide final drying near chamber.

It is usually important for the chamber to have the capability of going cold, as well as hot. The range of temperature capability should include at least −40 to 150°C to cover the automotive ranges. Some chambers use a liquid CO_2 cylinder to enable going cold (the liquid CO_2 expands to gas in a nozzle inside the chamber, taking on the latent heat of vaporization and thus greatly cooling the gas). In this case, it is important to use the type of cylinder having a bottom siphon so that liquid comes from the bottom of the tank. (If a top exit were used instead of a bottom siphon, the CO_2 would be gas instead of liquid, and would not cool well since the latent heat of vaporization would be expended in cooling of the CO_2 cylinder, instead of the environmental chamber.) When using liquid CO_2, the ambient temperature must remain below 31°C. The critical temperature of CO_2 is 31°C, and it will all be gas at any temperature higher than that, even when under pressure. On a factory floor without air conditioning some years ago, the environmental chambers were suddenly not getting cold. After the technicians and manufacturing engineer could not find the problem, they called for help. The author read the room temperature at 90°F, and pronounced "they won't be working again until the temperature cools down". And then went on to explain why.

But the most convenient method of cooling an environmental chamber is to avoid using liquid CO_2, and instead, use mechanical refrigeration as in the chamber shown in Figure 2.14. In this case, cooling is accomplished in the same way as a household freezer operates. Heating is done by using a resistance element as in a household oven. Chambers of up to about 3.5 cubic feet inside dimensions can be heated and cooled while operating on 115 VAC, normal household power in the USA. Larger than that will require 220 VAC power.

Sometimes a sensor under test is considered as having two areas of concern regarding temperature sensitivity: a sensing probe and an electronics module (which may be separated by an interconnecting cable, or may be directly connected, soldered, or welded together). To test the probe sensitivity to temperature separately from the electronics, the probe may be positioned inside the chamber while the electronics module is outside of it (with a plate of thermal insulation at the chamber wall, having a small hole for the probe to pass through). Or the electronics module may be inside the chamber and the probe outside if it is desired to test the electronics alone. Note that in Figure 2.14, as previously recommended, some fluffy white insulating material has been stuffed into the access port from the inside, and in Figure 2.15, a black rubber plug is inserted into the access port from the outside of the chamber. These steps are taken to reduce heat transfer, and to reduce humidity infiltration. The rubber stopper is slotted in order to allow wires to be inserted more easily from the side, rather than having to thread them through a hole in the stopper.

A compensating component, such as a resistor or capacitor, having a known sensitivity to temperature can be used to compensate for undesirable zero or span sensitivity of the sensor. A coil of nickel wire has been used inside a sensor by the author as a temperature compensating resistor, because nickel has a nearly linear temperature coefficient of resistance of approximately +0.0067 Ω/Ω/°C. Some thermistors are also suitable for use in temperature compensation. Such a temperature compensating resistor or thermistor can be used in the feedback loop of an op amp to adjust span sensitivity to temperature, or at a summing node to adjust zero. Some positive temperature coefficient (PTC) linear thermistors are based on nickel, and perform similarly to nickel wire. But negative temperature coefficient (NTC) thermistors have a negative temperature coefficient of resistance that is nonlinear. Accordingly, it is often necessary to add one or more standard resistors to tailor the temperature coefficient of an NTC thermistor, as necessary. This is usually called thermistor linearization, and can be found online, such as at TI, under application note SNOAA12.

Besides using both PTC and NTC thermistors, the author has used temperature compensating capacitors, such as one designated N1500, to compensate an oscillator frequency, or other capacitance-sensitive parameter; and has also used the change in diode forward voltage with temperature to provide temperature compensation in voltage-controlled circuits. A silicon diode has a predictable change in forward voltage of about −2.2 mV/°C, depending on the diode type and somewhat on the forward current.

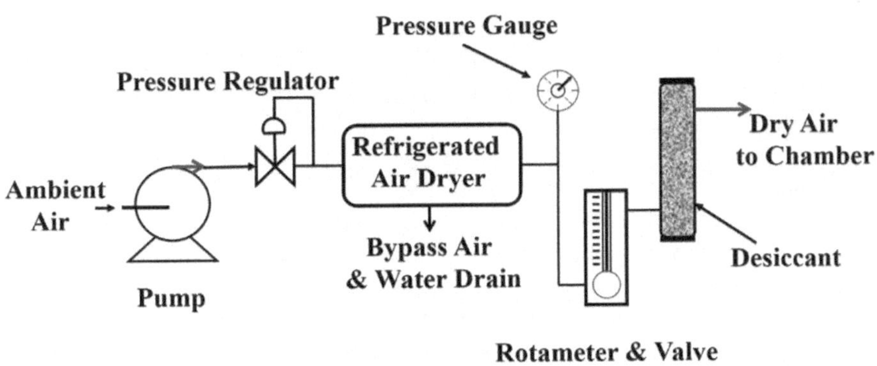

FIGURE 2.16 Diagram of complete air-drying system.

When the environmental chamber is going cold and there is an electronics module inside that is not sealed against moisture, condensing moisture may affect the performance of the electronics module. To avoid this, the chamber may be constantly purged with dry air. See Figures 2.15 and 2.16.

Figure 2.15 shows installation of a desiccant-filled cartridge at the outside of an environmental chamber in the author's lab. The chamber in the figure normally does not control the relative humidity, and so dry air is used to keep the humidity level below the dew point while the chamber goes cold. Of course, if the chamber has humidity control built in, then the humidity can be set appropriately.

In order to prolong the life of the desiccant, the air is pre-dried by a pump and refrigerated air dryer that are located in another room in order to reduce audible noise in the lab. (The author uses a refrigerated air dryer sold by Harbor Freight Tools, having a website at www.harborfreight.com.) See Figure 2.16 for a diagram of the complete air-drying system.

Ambient air is drawn into the pump, and brought up to a pressure of about 100 PSIG (pounds per square inch, gauge pressure). The refrigerated air dryer cools the pressurized air to about 2°C, causing some water to condense. A slight bypass stream of air exits the bottom of the refrigerated air dryer, carrying the condensed water with it. Then a pressure regulator (not shown) at the refrigerated air dryer reduces the pressure to 5 PSIG. As the main stream of air at 5 PSIG follows a copper tubing to the location of the environmental chamber, the air warms back up to ambient temperature. Next, the air flows through a rotameter (air flow indicator) with a control valve and a pressure gauge. The rotameter valve is adjusted to allow 2 lpm (liters per minute) of the dry air to flow through a desiccant column, and then into the chamber, via a dispersion nozzle. In the author's system, the dispersion nozzle is formed of a 1/4 inch diameter copper tube having a sealed terminal end and 20 small holes randomly positioned along a 30 cm length, the holes each being of approximately 1 mm diameter. The refrigerated air dryer is switched on and off by remote control, using a relay and key fob commonly available on internet sales sites.

When running a temperature test on a sensor, increasing up to a specified temperature may provide a slightly different sensor output measurement as compared with decreasing down to that same temperature. The effect of such thermal hysteresis on the data can be avoided by always changing temperature in the same direction during a test. For example, starting out at 25°C, and wanting to test from 100 to −40°C, the following schedule could be followed for temperatures of 100, 75, 50, 25, 0, −25, −40°C:

1. Ignore data at 25°C and increase temperature to 100°C.
2. Allow to stabilize at 100°C for 1.5 hours, and then take data.
3. Change to next lower temperature and allow to stabilize 45 minutes, then take data.
4. Repeat step three until reaching −40°C.
5. Return to 25°C (but do not take data there again because this would be increasing temperature).

The same 45 minutes is allowed for the last change even though for a fewer number of degrees change (from −25 to −40°C) because the chamber may take a longer time to reach the lower temperature.

Following this procedure will provide data that are most suitable for analysis and compensation, because all of the data are taken with decreasing temperature. Also, when making sure to always start at the hottest temperature, any humidity that may have been present in the sensing element or electronics can evaporate before the start of the test. The result is that any changes due to water freezing when going cold afterward will be far less likely, and will be certainly eliminated if using the recommended air-drying system.

2.11 RESPONSE TIME

Response time, of course, is the amount of time elapsed between the application of a change in the measurand at the sensor input and the resulting indication of that change in the sensor output. This simple explanation begs for more detail, though, when one is trying to account for the actual differences in timing between changes in the measurand and the sensor output signal. Thus, the total response time in a fully damped system may be further divided into a "lag time" before start of response, a "time constant" based on natural frequency and damping, and a "stabilization" or settling time while the final reading is being approached. See Figure 2.17, in which the input measurand makes a step change at time t_0.

The lag time is that time which passes between the start of a change in the measurand to the start of a change in the sensor output ($t_1 - t_0$ in Figure 2.17). This lag time can be due to propagation delay in electronic components, or the equivalent in mechanical, pneumatic, or other types of components.

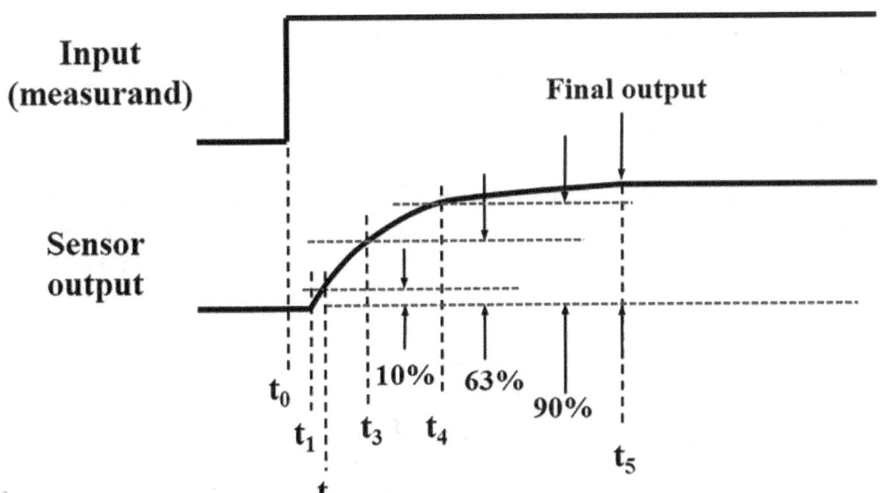

FIGURE 2.17 Input signal and output signal vs. time, showing lag time, time constant, and stabilization.

The "time constant" or main component of response time is usually based on the natural frequency of the sensing element, the maximum time between samples of a sampling type sensor, or the frequency of an electronic filter somewhere in the signal path. Coupled with a damping factor, this comprises most of the specified response time. It is usually specified, with a step input, as the time between the start of response until reaching 63% of the final response ($t_3 - t_1$ in Figure 2.17).

Final output is typically the level that would be indicated after waiting the lag time, plus five time constants for stabilization of the output after a step change in the measurand. The time from application of a step input until reaching the final output level is $t_5 - t_0$ in Figure 2.17. The amount of change in output after waiting one time constant to five time constants is usually determined by a combination of mechanical damping and electronic filtering.

Alternatively, sometimes response time is stated as the time between 10% and 90% of the final output response to a step input ($t_4 - t_2$ in Figure 2.17). This is less specific than listing lag time, time constant, and final reading. If only the 10% to 90% response time is specified, testing may be required for verification that the performance is suitable for response time-critical applications.

The response time information given so far is based on the output amplitude as a percentage of the expected final amplitude. In real-time feedback systems, it may also be important to look at the phase lag in addition to the amplitude variation. The combination could be specified, for example as –3 dB at 1 kHz with 10° phase lag. This would mean that, with the measurand varying as a sine wave of frequency 1 kHz, the sensor output voltage would be 0.707 of the theoretical output, and delayed by 10° (of the sine wave) as compared to the measurand.

Note that attenuation of –3 dB means that *voltage* amplitude is reduced to 0.707 of its value before attenuation, since voltage amplitude is a scalar quantity. Attenuation of –3 dB means that *power* is reduced to 50% of its value before attenuation, since power is a vector quantity, ($0.707 \times 0.707 \approx 0.500$).

Most position sensors incorporate a low-pass filter somewhere in the signal path before producing an output voltage, output current, or digital output. The filter may be implemented in hardware, software, or both hardware and software. When both, the hardware filter can be used to ensure that the software filter will not be fooled (for example, an anti-alias filter), and then the software filter provides the desired amount and type of filtering. A digital filter may be implemented in such a way that the amount and type of filtering can be easily adjustable by digital commands. But if such adjustability is implemented in hardware, it requires changes in values of resistors and/or capacitors. Even if an adjustable filter is implemented with changes in hardware component values, this adjustability can be accomplished through the utilization of digital switches, and thus the adjustment can be carried out via digital commands.

Most hardware filters utilize resistors and capacitors (usually, inductors are avoided, since they are often more expensive, may be larger, have wider accuracy tolerance, and may produce unwanted magnetic fields). Each RC pair forms a delay element. A simple filter may utilize one delay element, and is called a first order filter. A filter having two delay elements is a second order filter, and so on. Higher-order filters offer a sharper cutoff rate and therefore can have a calculated operating frequency (f_0) that is closer to the natural frequency of the sensor system, thus having a shorter settling time than could be had when using a lower order filter. This can appear as a faster response time or as a reduced error for a given frequency of variation in the measurand. The circuit may be configured in various ways, and the delay element values calculated accordingly, such that numerous filtering characteristics maybe achieved. The most commonly used filter characteristics are Butterworth, Bessel, Chebyshev, and Elliptic (Figure 2.18). (Some filter circuits will be shown in Figure 2.20.)

A Butterworth filter solution (named for Stephen Butterworth) is most often used, providing a maximally flat amplitude response, and having the steepest roll off that may be achieved while not introducing a passband ripple. Roll-off is a measurement of the rate at which the level falls as the signal frequency exceeds the corner frequency of the filter design. A Bessel solution (named for Friedrich Bessel) does not have as sharp a roll-off as the Butterworth, but has the advantage of linear phase response, so that the wave shape of a filtered signal is preserved.

A Chebyshev filter solution (named for Pafnuty Chebyshev) offers a faster roll-off, as can be seen in Figure 2.18, but at the expense of adding a signal amplitude ripple in the passband (or in the stopband), and it has a nonlinear phase response. The amplitude of the ripple is proportional to the steepness of the roll-off. Such ripple is normally not desired in a position sensor and can give excessively higher error when velocity or acceleration signals are derived from the position signal.

An elliptic filter solution (also called a Cauer filter, for Wilhelm Cauer) has ripple both in the passband and the stopband. It has the steepest roll-off, but due to the ripple, is not often used with position sensors. More information about filter circuits can be found in Johnson and Hilburn [4]. Also, there are many apps available online for

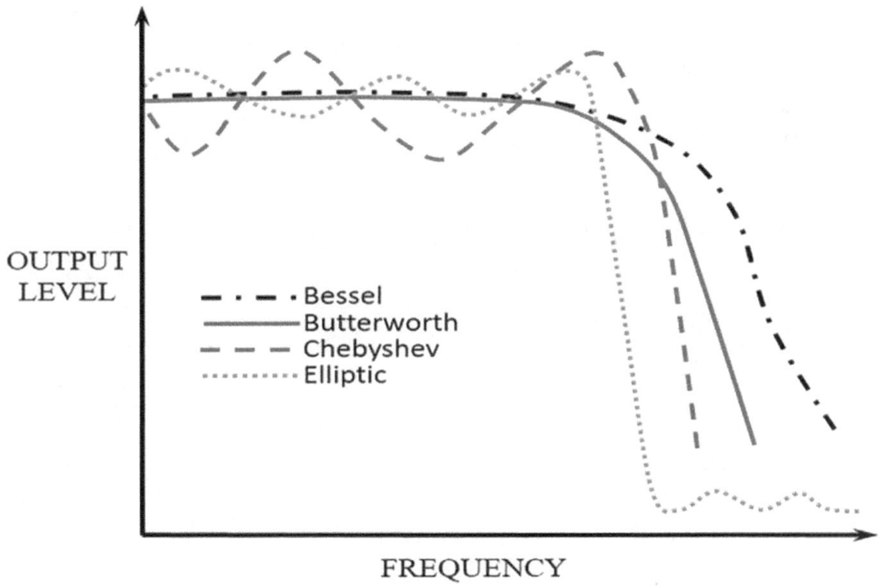

FIGURE 2.18 Commonly used filter characteristic types: Butterworth, Bessel, Chebyshev, Elliptic.

determining the component values of various filter circuit topologies (high pass, low pass, band pass) having the selected filter response (Butterworth, Bessel, and so on).

2.12 DAMPING

Some form of signal damping is often implemented in both the mechanical and electronic elements of a sensor. Without any damping, a sensor mechanical system may oscillate continuously at its natural or resonant frequency. With slight damping, it may oscillate for a short time after upset, until it settles down. Critical damping provides the fastest approach to a final value (or steady state value) of the sensor output after an upset (or step input of the measurand) without overshoot past the final value. See Figure 2.19.

In the figure, the ordinate is the amount of output overshoot, so that when fully settled, the output will come to its final value. A positive upset of value 1.0 is introduced at time zero. Over time, the output falls back toward its final value, when overshoot comes to 0.0. When damping is more than critical, it is called overdamping (or the sensor is overdamped), and the sensor output approaches its final value more slowly (such as along the 2X critical damping line). When damping is less than critical, it is called underdamping, and the sensor output approaches its final value more quickly, but has some overshoot past the final value. Damping ratio is the ratio between a given damping amount and critical damping. So, with critical damping, the damping ratio is 1. If underdamped, for example, the damping ratio might be 0.5; if overdamped, the damping ratio might be 2.0.

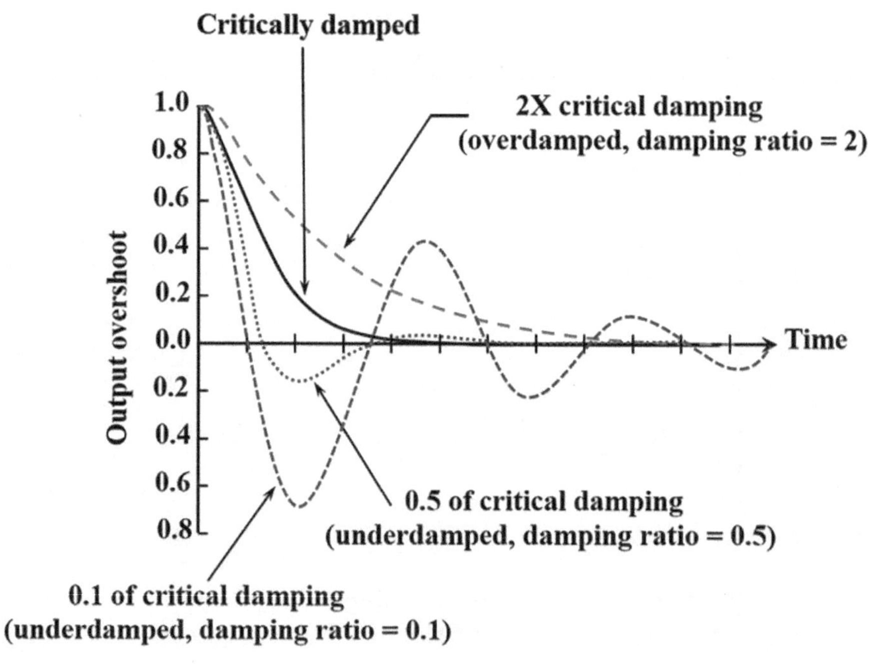

FIGURE 2.19 Sensor output stabilization versus damping.

Mechanical damping is often provided by adding silicone oil (such as polydimethylsiloxane, or PDMS) to a chamber containing the moving parts. Care must be taken to avoid bubbles of air in the oil that could cause erratic behavior. The author uses a method in which the device to be oil-filled is subjected to vacuum. Dissolved air is removed from the oil to be used by exposing it to vacuum. Then the oil is allowed to flow into the device while still under vacuum. After the device is filled, then air at normal pressure can be returned.

Oil of various viscosities may be utilized in order to obtain the desired damping ratio, considering that a more viscous ("thicker") oil will provide a greater amount of damping. The viscosity of a fluid is an indication of its resistance to deformation by shear or tensile stress. The kinematic viscosity of a fluid is usually measured in the unit of centistokes (cSt). A Stoke (St) is equivalent to 1 centimeter^2 per second, which is large, so cSt is used, and is one-hundredth of a Stoke. Water at 20°C has a kinematic viscosity of about 1 cSt. Silicone oil for sensor damping, such as Dow Corning DC200, will usually be in the viscosity range of 10 to 500 cSt when oil-filling a position-sensing element, but silicone oil is available in a very wide range of viscosities from 0.65 to 20,000,000 cSt. Dynamic viscosity is measured in centipoise (cP). To convert centipoise to centistokes, divide centipoise by the density of the fluid. The density of most hydrocarbons will be in a range from 0.85 to 0.90, so the centipoise value will be lower than the centistokes value by about 10 to 15%.

Mechanical damping may also be implemented through friction of rubbing surfaces, or by allowing a movable part to have contact with various elastomeric materials. Polynorbornene (also called by the tradename Norsorex™) is an elastomeric material having a very high loss, and tends to absorb or dampen kinetic energy. But many other elastomeric materials are also suitable for use in damping, depending on the amount of damping needed, physical space allowed, operating temperature range, and so on.

Some sensors have input transducers that provide a continuous DC measurement, such as resistive elements. Other sensor types may go through a cycle that produces a signal pulse, and then another cycle must take place before the next signal pulse is obtained, as with a magnetostrictive sensor. Such cyclic sensors need filtering in order to provide a continuous output signal, such as an output of 2.0 volts, with a 0 to 5 VDC output range, or an output of 16.0 mA with a 4 to 20 mA output range. A sample and hold circuit (S/H) or equivalent software function may be used instead of, or in conjunction with, a filter. Filtering is also used in many sensors in order to smooth out noise that may be generated from nearby electromagnetic energy, or other noise sources.

In electronic circuits, damping is typically provided by use of a low-pass filter circuit. See some examples in Figure 2.20.

The filter circuit of Figure 2.20(a) forms a single-pole low-pass filter with R_1, C_1. The input signal sees a moderately high impedance looking into the filter (R_1, C_1), so the circuit supplying the signal is not overly loaded down. The non-inverting input of op amp U_1 provides a relatively high impedance to the R_1/C_1 filter, so the filtering performance is not affected by the op amp. U_1 also provides a low impedance output looking back from the circuit that will follow. The double-pole low-pass filter circuit

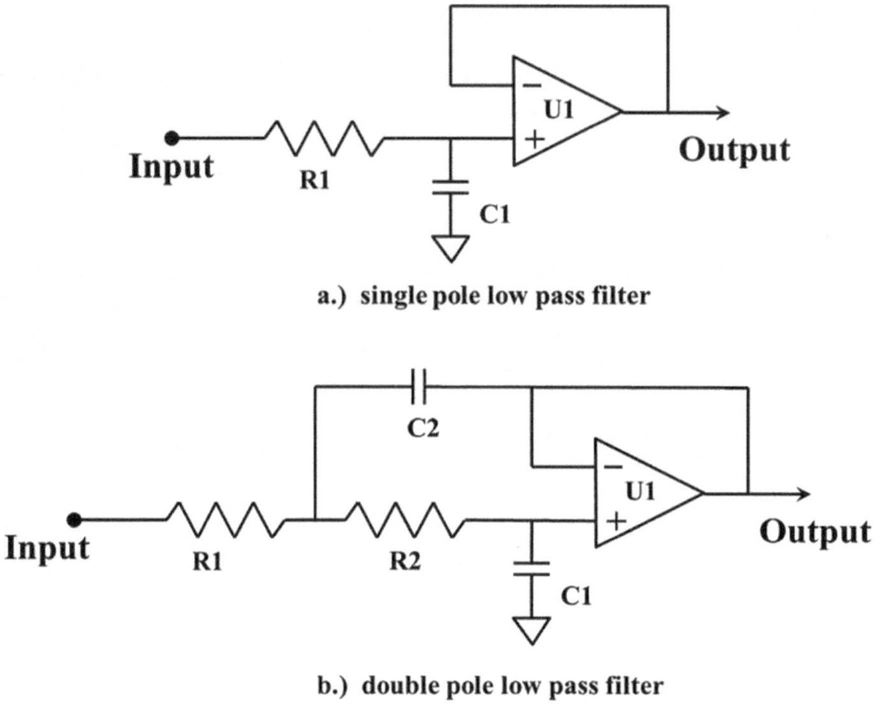

FIGURE 2.20 Low-pass filters: (a) single-pole; (b) double-pole.

of Figure 2.20(b) has similar impedance-matching qualities, but adds another filter pole, resulting in a steeper roll-off.

Rather than calculating the component values from scratch, which can be an exhausting task, the values may be obtained by using one of the several filter calculation programs available online. Three of these are: WEBENCH from Texas Instruments (www.TI.com), Analog Filter Wizard from Analog Devices (www.Analog.com), and FilterLab by Microchip (www.Microchip.com). In these programs, you can select the number of poles, the filter frequency, filter type (low pass, high pass, or band pass), amount of voltage gain, and filter response. The filter response (roll-off) can usually be selected among Butterworth (maximally flat), Bessel (constant delay), or Chebyshev (maximum roll-off). The Butterworth or Bessel are most common. Chebyshev provides the maximum roll-off, but adds variation in signal amplitude (ripple) in the passband. So, Chebyshev may be useful when there is only one specific frequency that is being filtered out (such as when filtering a fixed-frequency square wave to produce a sine wave).

Alternatively, Johnson and Hilburn [4] provides charts and tables for easy calculation of component values, and also shows many different circuit configurations. In lieu of, or in addition to, implementing one or more analog filter circuits, digital filtering of the output signal can be accomplished within the program of a microcontroller (μC), after the signal data have been digitized and entered into the μC. There are many algorithms available to provide various types of digital filtering, but a simple type that works well for many sensors, and often utilized by the author, is the median filter. A median filter works on a set of data, for example: five measurements. In the example of using five measurements, the measurements are held in a memory stack, and this procedure is followed:

1. The five data are arranged from highest to lowest.
2. The middle one is selected, and used as the output.
3. One new measurement datum is pushed into the stack, and the oldest measurement datum is pushed out.
4. Repeat from step one.

Step one can be accomplished through a bubble sort. A bubble sort works by repeatedly swapping adjacent numbers if they are in the wrong order, until the ordering is complete.

The implementation of such a median filter method assures that any high or low flyer (that is, a datum outside of the range of the other data) that may be caused by system or environmental noise is thrown out. The number of measurements used in the memory stack is usually at least five, and may be more, depending on how noisy the signal might be.

2.13 CROSS SENSITIVITY

If an angular sensor is used to indicate the rotation angle of a shaft, the sensor should not respond to a radial or axial force applied to the shaft if the rotation angle has not changed. The degree to which a position sensor responds to inputs other than the desired input to be measured is called cross sensitivity or cross-axis sensitivity.

In Figure 2.21(a), a Hall effect sensing element is disposed within the housing such that it will measure rotation of the shaft, and should be insensitive to cross-axis inputs. A cross-axis input would include a sideways force on the shaft, or a force pushing the shaft inward or pulling it outward. In Figure 2.21(b), An LVDT sensor measures linear motion of its core, and should be insensitive to cross-axis inputs, such

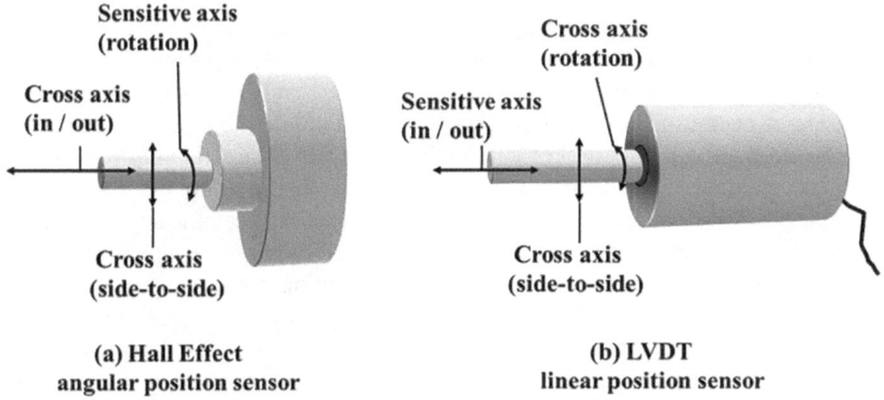

FIGURE 2.21 Cross axes in (a) a Hall effect angular or rotary sensor; (b) an LVDT linear sensor.

as a sideways motion of the core, or twisting of the core. (Note: in Figure 2.21(b), the core is shown outside of the LVDT for clarity, but would normally be fully within the LVDT housing.)

Cross sensitivity is often not specified for position sensors, but should be kept in mind when considering the best way to implement a particular technology. Some examples are the following:

Magnetostrictive linear position sensor—a ring or c-shaped position magnet provides less cross sensitivity (lateral, moving the magnet side-to-side) than does a single bar or rod magnet. A mechanism that moves the position magnet along the probe length should keep the magnet distance to the probe relatively constant for the best cross-sensitivity rejection. A sensor housing with a captive slide provides the least possibility of cross sensitivity for a given magnet configuration, because the slide mechanism maintains the position magnet at a fixed distance from the waveguide.

LVDT or *Inductive*—when the bore is substantially larger in diameter than the core, there may be some sensitivity to moving the core side to side within the bore. A bore-liner may be used, or other means for keeping the core centered in the bore.

Capacitive position sensor—it is important that parallel measuring plates maintain their alignment throughout the stroke range. With a set of parallel plates, providing overlap on one of the plates can reduce cross sensitivity.

Hall effect, magnetoresistive—it is important that the position magnet(s) maintain their alignment with the sensing element(s) throughout the stroke range.

2.14 SHOCK AND VIBRATION

Most position sensors will undergo some level of shock and vibration exposure during normal use, and so shock and vibration ratings should be included in the sensor specification. This specification is used to qualify the sensor for two purposes: how will the normal shock and vibration levels affect accuracy, and how long will the sensor last under higher levels of shock and vibration before failure.

The installation components and mounting technique can make a difference in the performance and survivability of a sensor in response to a given vibration input; thus increasing or decreasing the susceptibility to shock and vibration. Mounts made from a dampening type of elastomeric compound may improve the reliability of a position sensor, for example, by reducing the peak amplitude of the vibration; but may also reduce the sensor accuracy by allowing motion to occur between the sensor mounting flange, or other reference point on the sensor, and the rigid sensor mount (the elastomeric material being inserted between the sensor mounting flange and the rigid mount).

Sometime during the development of a new position sensor design, a sample should be mounted, in the intended way, into a vibration test fixture. The performance and survivability are tested over an extended time at several discrete frequencies and amplitudes. A frequency scan is also performed to find any tendencies of the sensor to resonate at particular frequencies. If there are any resonant frequencies found, the design can be modified to remove or dampen the undesired resonance. Alternatively, the sensor can be tested for an extended time at each resonant frequency discovered, to make sure that the proper performance and reliability are maintained. A typical vibration machine setup is shown in Figure 2.22. Note that the vibration system comprises a signal source, power amplifier, forcer, sensor mounting fixture, and an accelerometer. The signal source will have variable frequency and amplitude, and may also have selections of other waveforms in addition to sine wave, but sine wave is typically used. A forcer is the device that actually moves the DUT. The forcer shown in the figure is a voice-coil type, a larger and high-power version of the type used in a loudspeaker. Cooling of the forcer is also normally required. The forcer of Figure 2.22 utilizes a cooling fan, but some may need a cool water supply. An accelerometer is mounted to the test fixture to measure the amount of vibration being produced (some accelerometers require their own driver, to energize the accelerometer and to condition its signal). The vibration level and frequency are confirmed by mounting one or more accelerometers to the sensor or mounting fixture, wired to a driver/readout device.

The vibration testing conducted in the development lab is done to make sure that the design does not contain any major flaws. An additional vibration test is usually conducted at a certified testing lab to make sure that industry standards are met. At the same time, mechanical shock tests are performed. It is not common for a development lab to have shock testing equipment on site. Shock testing can be done by allowing the sensor and fixture to fall along a track a pre-determined distance into a calibrated stopping area (see Figure 2.23), or a moving mass can be slammed into the sensor mounting fixture.

In the tester of Figure 2.23, a device under test (such as a sensor) would be mounted to the test carriage. The carriage is raised to a desired height, and then allowed to fall

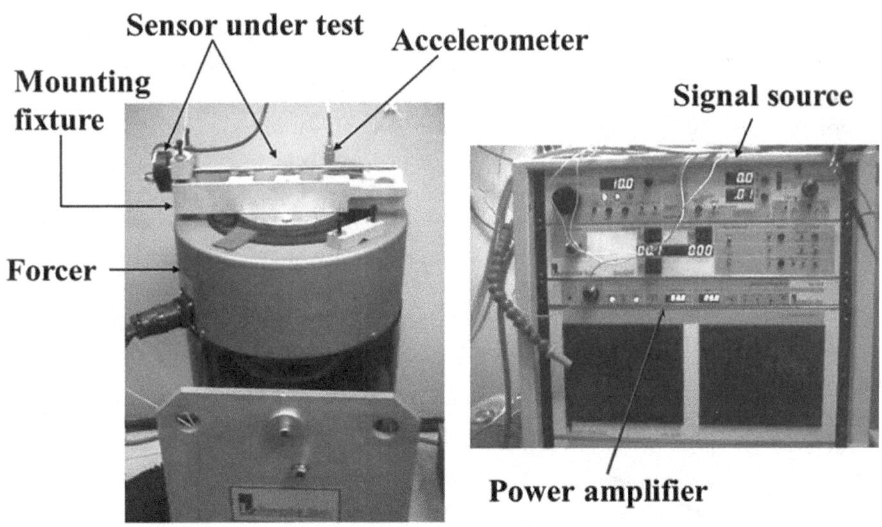

(a) (b)

FIGURE 2.22 (a) Sensor mounted to a vibration forcer by using a mounting fixture; (b) source and amplifier.

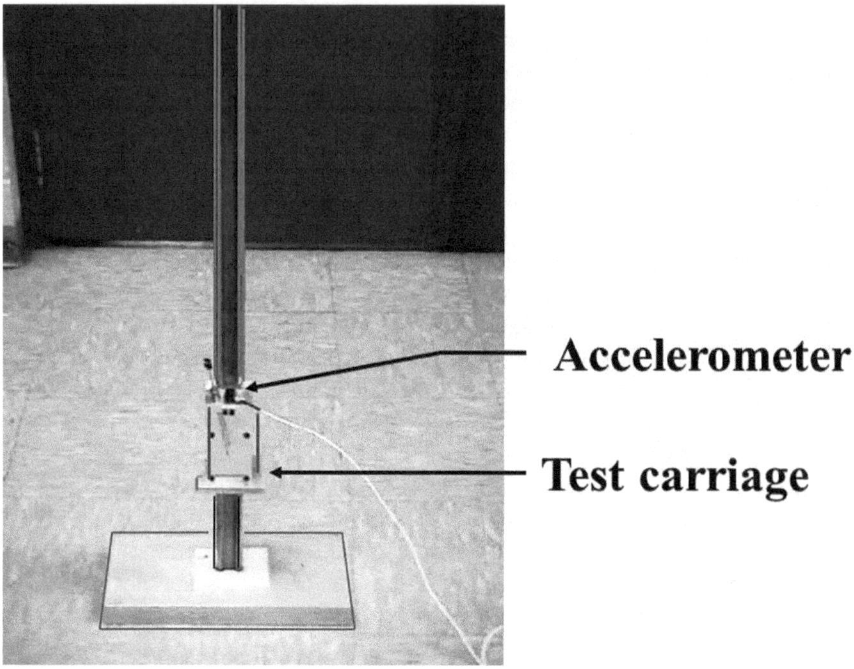

FIGURE 2.23 A simple type of shock tester, with accelerometer attached.

until the carriage slams into the damping pad (aka—the calibrated stopping area). The pad has a hardness and energy loss so that the carriage will not bounce after contact, and so the desired shape and duration of the shock pulse is obtained. An accelerometer that is mounted to the test carriage provides measurement of the shock applied to the test carriage. The accelerometer output can be monitored on an oscilloscope having storage capability, so that the shock amplitude, duration, and wave shape can be observed and recorded.

The author has also visited a facility where larger equipment was submerged in water and explosive charges were used to produce the shock. A related test is the drop test. This can usually be performed in the development lab. The sensor is dropped from a pre-determined height (usually desktop height) onto a concrete floor. It is dropped at least six times, once on each normal face (top, bottom, four sides). In addition, if there is a particular angle in which the sensor could be expected to be more sensitive, dropping at this angle is also tested.

In addition to the standard shock and vibration testing that will be listed in the product specification, it is prudent for a sensor manufacturer to perform additional tests to weed out any potential problems before they can cause failures in the field. This is sometimes called highly accelerated life testing (HALT). This testing method subjects a product to a series of high stress environments, that may include: vibration, rapid thermal changes, humidity, and electrical stress.

The vibration source is normally random, and applied along several axes. This can be accomplished by mounting many solenoids at different angles to the test fixture, and constantly energizing the solenoids with pulses at random sequences. The thermal part of the test may have extended periods at the min and max temperatures, plus rapid changes in temperature. Electrical stress can include voltage pulses and running at or slightly above the rated power supply voltage.

A HALT can cause a sensor to fail in days instead of the months that often are required in other types of life testing (such as elevated temperature only). Upon failure, the sensor is analyzed to find out what failed. After re-design, it can be re-tested, and this can be repeated until the sensor reliability has been sufficiently improved. A properly executed HALT program can help to assure the success of a sensor product, and ultimately greatly reduce the rate of field failures and warranty support cost.

2.15 ELECTROMAGNETIC COMPATIBILITY

The Electromagnetic Compatibility (EMC) of a sensor is a rating of how well it will operate in, and not substantially contribute to, an environment of electromagnetic radiation. When electromagnetic radiation is emitted from one device so that it affects the performance of another device, it is called Electromagnetic Interference (EMI). EMC is divided into two areas: emission and susceptibility. A well-designed sensor will not emit sufficient levels (emission) of electromagnetic energy to disturb the performance of other devices, nor will it be affected (susceptibility) by EMI emitted from other devices at specified allowable levels. In either case (emission or susceptibility) a non-conforming sensor can usually be adjusted to conform by incorporating suitable countermeasures. Such countermeasures may commonly include the addition of parallel capacitance, series inductance, shielding, isolation, or changing the grounding circuit. Industrial product specifications also include ratings for resistance to ESD (electrostatic discharge), EFT (Electrical Fast Transient, or "burst"), and lightning surges, and these three are presented in Section 2.16.

FCC (Federal Communications Commission) part 15 addresses EMI to some degree, but the standard to meet is generally considered to be the EMC portion of the requirements for the Conformité Européenne (CE) Mark. Since 1996, all products that are liable to cause, or be affected by, EMI must have the CE Mark if they will be sold in or into any of the European Union (EU) countries. This includes all electrical and electronic products.

For the purposes of compliance testing, electromagnetic energy may either be conducted or be radiated through space. When conducted, that means energy comes into the device through the connection leads or through the case. The EMC standards for the CE Mark come from the International Electrotechnical Commission (IEC). These are part of the IEC 61000 series of standards.

Circuits to provide the protection needed to meet EMC requirements include those contained within the integrated circuits, as well as additional components mounted to the PCB (printed circuit board), housing, and/or connectors. Integrated circuits that make external connections via the sensor input and output leads must be able to withstand the EMI, ESD, EFT, and surge, or components must be added to ensure compatibility. Back-to-back fast Zener diodes can be used to limit peak voltages. Back-to-back connected Zener diodes normally means that two Zener diodes are connected in series with their anodes connected together, and called a bidirectional Zener. One of the two leads goes to common or ground, and the other lead goes to the conductor to be protected. (Connecting the cathodes together instead, would also work.) Multiple devices contained within one package are also available. Sometimes, additional impedance is required in series with the Zeners to limit peak current, thus avoiding damage to the Zeners when a high-energy voltage spike is encountered. The impedance can be a resistor, or may be an inductor to aid in reducing the passage of higher-frequency interference, while not adding an unwanted voltage drop for DC signals. An inductor is often formed by adding a ferrite bead to a wire lead coming into the PCB, or may be an inductor component mounted to the sensor's PCB. A parallel connection of a ceramic capacitor (typically in the range of 0.1 to 10 nF) can be added, and serves to shunt high frequency energy to ground. These shunt capacitors can be connected between the sensor lead and the circuit common, but are frequently connected between the sensor lead and the case. So, for example, the power supply input connection of the sensor may incorporate circuit traces on the printed circuit board to accommodate a small series resistance, but more often a small series inductance. After the resistor or inductor, a bidirectional Zener diode might be connected to circuit common. And a small capacitance may be connected across the Zener.

Sensors with high frequency signals appearing in their sensing elements, microcontrollers, or communication circuits need special attention to avoid the "leaking" out or radiating of this energy and causing EMI (emission). Switching type power supply regulators also produce voltage spikes that can be radiated, or conducted to the circuitry being supplied. The amplitude of the conducted voltage spikes can be reduced by ramping the current at the switch points instead of switching sharply (some integrated circuits are available that provide this function internally). This also reduces the radiated EMI. Another way to reduce conducted EMI is to use a higher switching frequency (more than 1 MHz) so that smaller filter components can be used to attenuate the spikes. Radiated EMI can be spread across a wide frequency range (called spread spectrum) by constantly varying the switching frequency. This limits the energy in any given frequency, making it more likely to pass the EMC tests.

Testing to make sure that a sensor being developed will meet the EMC requirements usually takes place in two steps: in-house laboratory testing, and third party lab testing. For most engineering companies, an anechoic chamber, Faraday cage, and a full set of EMC testing equipment to guarantee meeting the latest industry specifications are not cost effective (see emcfastpass.com/anechoic-chamber-guide for information about anechoic chambers and Faraday cages). Instead, most development labs have some basic equipment that can be used to get a relative indication of whether or not the sensor will pass the "real" compliance test, by comparing results with other equipment that have already passed. This lab equipment can also be used to troubleshoot a new product design for EMI-related problems.

EMI is coupled from a source (the culprit) to the affected device (victim) by conduction or radiation. In a conducted EMI test, the test signal is applied to the sensor power and signal leads, and also to the case. For a radiated EMI test, the source is connected with an antenna.

After initial testing in the development lab, the sensor to be tested is sent to a third party test lab (third party means: not the developer, not the customer, but a third party that is certified to perform the testing). The third party test lab supplies a test report that can be used to prove to a customer that the sensor was tested and meets the specified requirements for EMC.

2.16 HIGH VOLTAGE PULSE PROTECTION

Under some conditions in the field, a sensor may be exposed to high voltage pulses that are above the maximum specified power supply voltage. As long as these voltage pulses have a peak voltage and time duration within certain limits, a sensor may be designed to withstand the high voltage without damage. The IEC system level test standard according to IEC 6100-4 comprises several parts, including IEC 6100-4-2 for ESD (electrostatic discharge), IEC 6100-4-4 for EFT (electrical fast transient), and IEC 6100-4-5 for surge. ESD testing is meant to assure that a sensor passing this test will not be harmed when subjected to a discharge of electrostatic energy up to a specified level.

ESD is a high voltage discharge similar to that experienced when one walks across a carpet and touches a doorknob. Voltages of more than 14 kV are possible. ESD sensitivity is normally tested by applying a specified voltage while using a simulation circuit (such as a small value coupling capacitor) that simulates a given category of test model. There are three main test model categories for ESD testing: the human body model (HBM), machine model (MM), and charged device model (CDM).

The HBM simulates a discharge that might occur when a human finger touches an electronic device; that is, when a finger of a human who has been electrostatically charged touches the device under test (DUT). The MM simulates what happens when a metal implement (such as a screwdriver, other hand tool, production fixture, etc.) that is already charged comes into contact with the device under test. The CDM simulates what happens when the DUT is already charged, and then it comes into contact with an electrically conductive material. This could happen, for example, when a device is sliding on a conveyer belt, picking up a static charge, and then touches a metal fixture. The results from the HBM and MM models of ESD tests are usually very similar, and so an ESD test program including only HBM and CDM testing is often deemed sufficient (omitting any separate MM testing). There are various testing levels, or classes. The designated test voltage and other details for a given test class can be found on the JEDEC website at www.jedec.org.

During an ESD test, a particular current profile is utilized according to the type of model being applied, as generally depicted in Figure 2.24.

The amount of time is relatively small, and will be given in nanoseconds, such as around 2 ns for CDM and around 10 ns for HBM. A typical test circuit for HBM

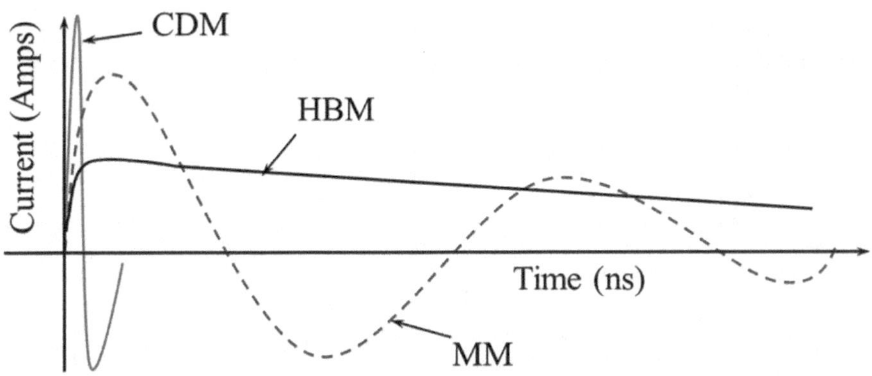

FIGURE 2.24 Example of current profiles for different ESD models.

could include something like discharging a 100 pF capacitor through a 1.5 k ohms resistor, with the capacitor having been charged to a voltage of 1 kV to 8 kV. For MM, the capacitance could be 200 pF, charged to 100 to 400 V, with 0 ohms resistance. And for CDM, charging the device at 500 to 1,000 V, and discharging it to ground through an inductance of 1 nH and 10 ohms series resistance.

2.16.1 EFT Immunity

The EFT test (electrical fast transient, commonly called the burst test) is described in the IEC (International Electrotechnical Commission) document IEC 6100-4-4, and simulates the burst of high voltage/high range of frequencies that can occur when a set of relay contacts open. Arcing between the contacts as they open can produce high voltage spikes over a range of frequencies, especially when the closed contacts were passing a substantial amount of current through an inductive load at the instant when the contacts opened (such as when switching off a motor).

The EFT test does not use direct electrical connection to the DUT, but uses a capacitive clamp to couple the energy to its cable wires. Accordingly, taking care to design the sensor to have appropriately sized cable with good shielding can reduce the amount of energy that is delivered to the DUT. The burst test is applied to all signal, power, and ground/earth wires. A burst comprises a series of test pulses produced by a burst generator in a sequence having a specified duration. An individual test pulse reaches a specified peak, and then attenuates to 50% of its peak value in a specified amount of time (for example, less than 100 ns). The next pulse usually occurs 1 μs later. A typical burst of pulses will last for 15 ms. A burst period is the time from the start of one burst to the start of the next burst, and is 300 ms. Additional bursts are generated until a total of 10 seconds have elapsed. That is followed by a rest period of 10 seconds. The 20 seconds total forms one test cycle. See Figure 2.25.

A complete EFT test includes six test cycles. The EFT test is characterized by low energy content, but having fast rise times and a high level of repetition. In contrast with the few cycles completed during an ESD test, the EFT test includes 3 million pulses within 110 seconds.

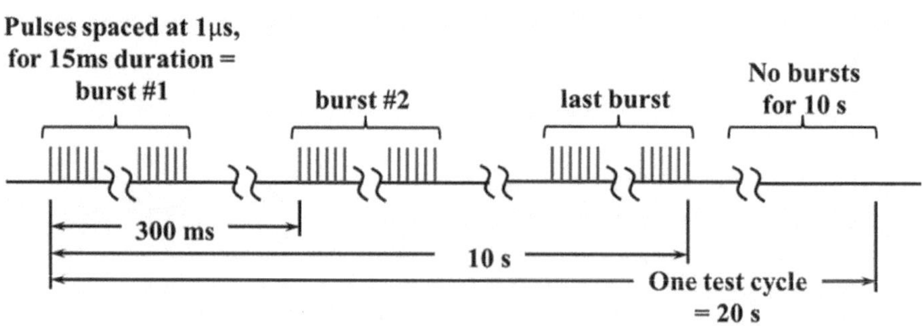

FIGURE 2.25 Burst test: one burst cycle lasts 10 seconds. One test cycle lasts 20 seconds. Test duration is 6 cycles, for 110 seconds total from start of first burst to end of final burst.

2.16.2 Surge Immunity

The surge immunity test is commonly called the lightning test, and depicted in Figure 2.26. Lightning-induced surges do not simulate direct lightning strikes (as these can typically melt or vaporize portions of the equipment, with an average lightning strike able to deliver about 30 kA). Rather, a lightning surge test simulates the increased voltage levels that may be induced into conductors when a lightning strike occurs in the general area. Nearby strikes induce high voltage peaks for shorter durations. Farther away strikes produce more moderate voltages for a longer duration. A lightning surge test can typically have a peak voltage in the range of 600 volts and a duration of 1.0 ms, to 5,000 volts, and a duration of 10 μs. Durations of longer than 1 ms, for the moderate voltages, can be simulated by multiple strikes. The waveform of the applied test voltage is a dual exponential one. There is an exponential rise time to the peak voltage, followed by an exponential decay back to zero over a longer time period. Inductance in the test circuit may allow the voltage and/or current to go below zero (negative voltage or current), and is called undershoot (not shown in the figure). A maximum of 30% undershoot is allowed in the test, so that the voltage or current should not go below zero by more than 30% of the peak value. The "surge immunity test" is described in IEC 61000-4-5.

The test equipment operates by first charging a capacitor. Then a switch closes so that the energy stored in the charged capacitor is delivered through a wave-shaping resistance and inductance (RL) circuit to the DUT. Component values for the wave-shaping circuit are calculated so that a defined voltage surge would be provided to an open circuit, and a defined current surge would be provided into a short circuit.

Protection against a lightning-induced surge may include a larger energy-handling device than required for the other protections already mentioned. The voltage level

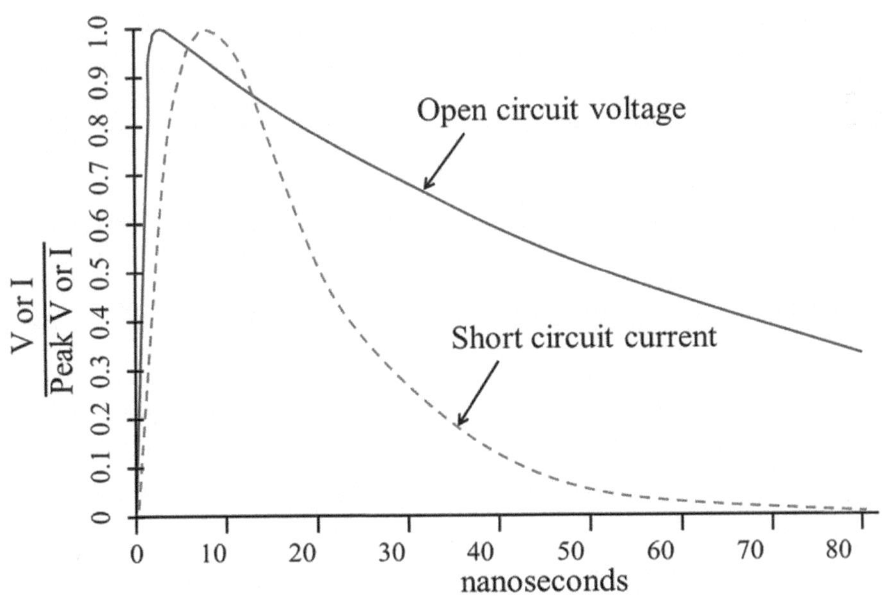

FIGURE 2.26 A pulse in accordance with the surge immunity test.

may not be as high as with ESD, but the energy level may be higher with lightning surge than with ESD because of the longer time period during which the voltage may be present when due to lightning surge. In addition to the use of Zener diodes, lightning surge protection may include the use of spark gaps or MOVs (Metal Oxide Varistor) to handle the higher energy. A spark gap can be used to protect a Zener against a higher energy level. Although MOVs are less expensive than fast, high surge current Zeners, MOVs can be degraded after handling multiple surges.

2.17 POWER REQUIREMENTS

Older industrial equipment often required ±15 VDC power. This was needed in order to operate the analog amplifiers within the sensors, controllers, etc., and to provide the ±10 VDC signal output that was sometimes used. With the availability of lower voltage and lower power amplifiers beginning in the mid-1970s, however, the requirement for the negative supply voltage was reduced. By the 1990s, the negative power supply voltage was rarely used. It has since become the standard for basic sensors to be connected into either +15 VDC or +24 VDC power systems, the negative side of the power supply being called "common" (and often grounded at some point in the system).

In order to enable a sensor to operate with either a +15 VDC or a +24 VDC power supply, it should be designed for operation from 13.5 to 26.5 VDC. This is derived (approximately) as 15 volts minus 10%, and 24 volts + 10%. The additional 10% on both ends of the supply voltage range allows for variation in calibration from one power supply to the next, changes in the supply voltage with line and load variation, and it allows for long-term drift of the power supply voltage.

Mobile equipment, on the other hand, usually operates from a nominal 12 or 24 volt battery system, and some automotive systems use even higher voltages, such as 42 VDC. Since a mobile environment has a charging system, a high starting load, and possible disconnection of the battery (which is otherwise normally providing a load to the charging system), a much wider range of power supply voltage must be accommodated than just the rated voltage of the battery. For example, in a 12 volt automotive system, the battery voltage during cranking could be 8 volts, or lower, while normal operation of a sensor with this supply voltage is still expected. When the battery is being charged at normal cruise conditions, its voltage can be more than 14 volts. Under conditions of *load dump* (which is when the charging system is operating normally, then the battery connection is abruptly broken), the system voltage can peak at several times the battery nominal voltage. So, a complete specification of possible power supply voltages and durations is needed for the proper design of a sensor that will be used in a mobile application. Sensors that are primarily designed for industrial use, but may often be used in 12 VDC mobile equipment are sometimes specified with a power supply voltage range of 8 to 36 VDC, and sometimes wider than that.

Common to all sensors, no matter which power supply type is specified, is the need for adequate circuit design to provide protection against reverse polarity of the power supply connections, overvoltage, and shorting of the output leads to power or ground. Figure 2.27 shows some of the methods used to protect against input and output overvoltages and shorts.

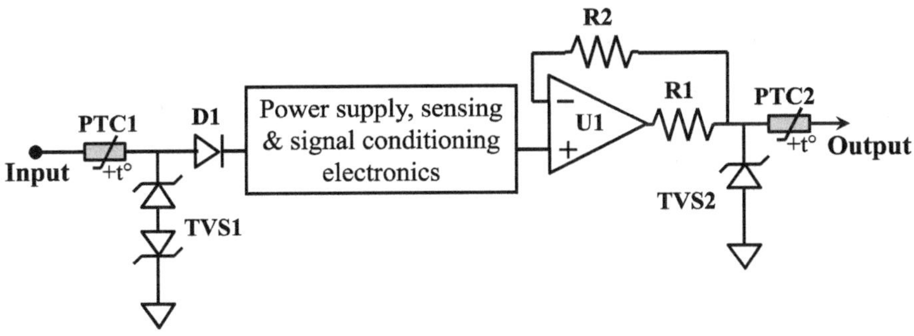

FIGURE 2.27 Protection of power input and signal output lines of a sensor.

In the figure, PTC1 and PTC2 are positive temperature coefficient thermistors that have a switch-mode action. At normal operating temperature and current, they have a very low resistance, such that the resistance can be ignored in the particular application. When passing a current that is higher than the rated switching current, they heat up abruptly and open-circuit to a resistance of a few thousand ohms or higher. When cooled down and the current is at a specified recovery level that is lower than the trip current, they switch back to a low resistance. So, they operate as a fuse that will reset when normal conditions return. They are sometimes called a resettable fuse. A one-time use fuse is rarely designed into a sensor, since the sensor would fail to operate after such a fuse is blown.

In the figure, TVS1 is a transient voltage suppressor, configured as two Zener diodes connected back to back. If TVS1 has a 28 V rating, for example, it will conduct whenever the input voltage is above +28 V or below −28 V. If TVS1 conducts a high current due to an excessive input voltage, then PTC1 opens up in order to protect TVS1 and any interconnecting printed circuit traces. D_1 will not conduct when the power supply voltage is connected backwards, providing reverse-polarity protection. For example, if −15 VDC was connected, no current would flow. It is important to place D_1 to the right of TVS1 (as shown in the figure), so that D_1 does not need to have a high current rating to handle transient suppression.

TVS2 is a single Zener transient voltage protector that protects the sensor electronics from damage if the sensor output is connected to any voltage outside the safe range of the output circuit. If the output range is 0 to 10 VDC, for example, one might use a 12 V TVS Zener. Then it will conduct if the output is connected to anything above +12 V, and will also conduct at about −0.7 V when inadvertently connected to a negative voltage (0.7 V is the forward bias voltage of one silicon diode). PTC2 limits any current that TVS2 might conduct. Sometimes a TVS Zener having a voltage high enough to avoid affecting the sensor output might have a clamping voltage that is too high to protect the output op amp when conducting a large current. For example, the 12 volt rated TVS2 might have a voltage of 16 V when considering the tolerance, temperature coefficient, and conduction current. And maybe a current of over 20 mA could flow into the amplifier U_1 and damage it. R_1 is included to limit the current and protect U_1 in that case. R_2 would probably already be part of the amplifying circuit, and protects the path back to the inverting input.

In addition to protection against reverse polarity and overvoltage, many mobile applications require protection against any combination of mis-wiring of the input and output leads, including the shorting of any of them to power or ground potentials. Sometimes it is possible to limit the added expense of overvoltage protection in a sensor for an automotive application by specifying that it be connected to the 5 volt regulated power available from the onboard controller or ECU (Engine Control Unit), rather than the 12 or 24 volt battery supply.

2.18 INTRINSIC SAFETY, EXPLOSION PROOFING, AND PURGING

Sensors of various types must often be able to operate in areas that may intermittently or continuously experience a hazardous (flammable or explosive) atmosphere. Intrinsic safety is a method used to prevent combustion or explosion by removing the likelihood of the presence of ignition energy. Explosion proof (also called flameproof) is a method used to contain an explosion, should it occur. Purging is a method that removes hazardous gas from the area surrounding the electrical equipment, so that combustion will not occur even if an ignition source happens to be present. Intrinsically safe, explosion proof, and purging are the three main explosion prevention methods, and will be described here, although some other methods are also used, such as encapsulation or sand emersion.

For the purposes of fire and explosion prevention, a hazardous atmosphere is one that contains flammable or explosive materials that can be ignited by a sufficient energy source when an oxidizer is present in a quantity capable of supporting combustion. The flammable or explosive material may be flammable gas or vapor (such as methane gas or alcohol vapor), flammable dust (such as coal dust or wheat flour), or flammable fibers or flyings (such as textile fibers). A *flash point* is the lowest temperature at which a given flammable liquid generates vapors at a rate high enough to support combustion, in the presence of sufficient oxidizer and an ignition source. Liquids having a low flash point temperature, like gasoline, generate flammable vapor at normal temperatures; so vapors are therefore in the category of flammable gas or vapor. In North America, the classifications for hazardous locations, and specified in the US by the NEC (National Electrical Code), are divided into three classes: Class I (gas or vapor), Class II (dust), and Class III (fibers or flyings). In addition to the class, hazardous locations are separated into two divisions.

A *Division 1* location is one where the hazardous condition may exist under normal operating conditions, such as in the vicinity of a process vessel having a hatch that can be opened during normal processing. A *Division 2* location is one where the hazardous condition can only exist under abnormal conditions, such as during repair, an accidental breach of a seal, or failure of a purge system. Areas other than Divisions 1 or 2 are labeled as nonhazardous.

The flammable or explosive materials of Class I and Class II locations are further broken down into groups, based on the minimum ignition energy of the most easily ignitable mixture of the material with air. For example, Class I Groups A through D include the following gases: A—acetylene, B—hydrogen, C—ethylene, and D—methane. As may be obvious, they are arranged from A being the most easily

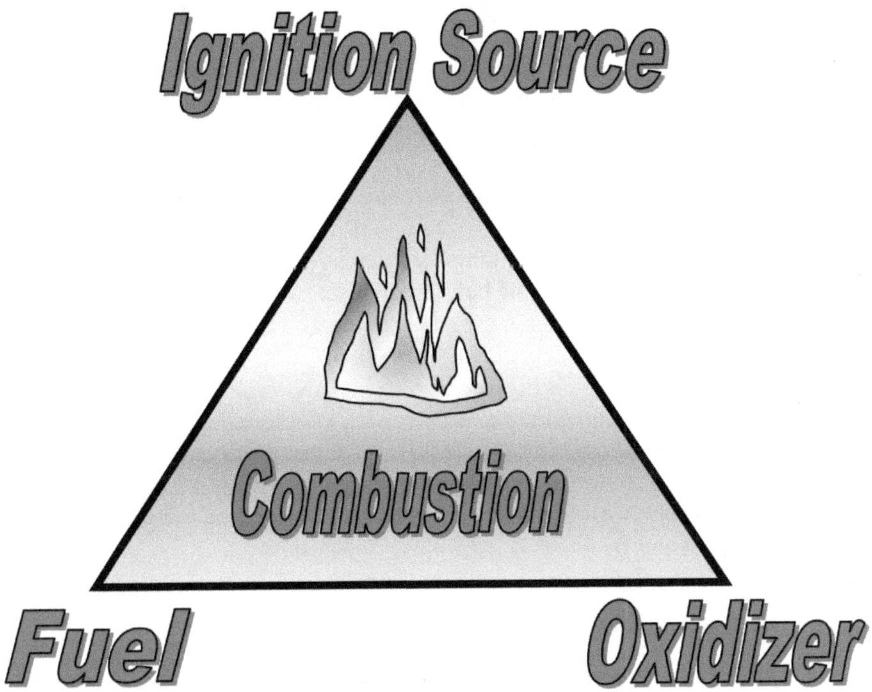

FIGURE 2.28 The Combustion Triangle.

ignitable, through D being less easily ignited. Many other gases and vapors are included in each group, but the ones listed are representative. Likewise, examples of Class II (dust) Groups include: E—aluminum dust, F—coal dust, and G—grain dust. Other types of dust and powders are included in each group, but the examples listed here are representative.

The elements required in order to initiate combustion are shown in the combustion triangle, Figure 2.28.

As depicted in the combustion triangle, fuel, oxidizer, and an ignition source are all required to be present before combustion can occur. If any one of these elements is absent, fire or explosion is prevented. The starting point for evaluation is the fuel. When a fuel source is not present, then safety precautions against fire or explosion are not normally considered. A certain minimum amount of fuel must be available in order to support combustion. With a gas or vapor, this minimum concentration is called the LEL (Lower Explosive Limit). If the fuel concentration is below the LEL for the type of fuel present, then combustion cannot take place. Of note is that in some special cases in closed containers, combustion can be prevented by keeping the fuel concentration above the UEL (Upper Explosive Limit). Above the UEL, the fuel to air ratio is higher than that capable of supporting combustion (i.e. there is enough fuel, but not enough oxidizer present), so combustion will not be initiated, even in the presence of an ignition source.

If fuel is available in a concentration between the LEL and UEL, then one needs to consider removing the oxidizer and/or ignition source. Even with fuel present, combustion can be prevented by limiting the oxygen level to below the minimum concentration needed for combustion of that particular fuel or fuel mixture. Such reduction of the oxygen concentration is called "inerting" (as opposed to purging, which is reduction of fuel concentration). Inerting is normally accomplished by forcing nitrogen (a relatively inert gas) into the container to be protected, in order to displace enough air to reduce the oxygen level below that needed to support combustion. Alternatively, another inerting gas may be used instead of nitrogen if this gas contains very little or no oxygen. When fuel and oxygen may both be present, combustion can be prevented by making sure that sufficient energy to cause ignition is not available. This is the principle behind intrinsic safety (explained later).

The oxidizer being considered is usually provided by air. Atmospheric air at zero relative humidity is approximately 20.94% oxygen (O_2) and 78.09% nitrogen (N_2), the remainder being argon (Ar), carbon dioxide (CO_2), and other gases. These percentages by volume are independent of the atmospheric pressure, but the partial pressure due to each gas (expressed in absolute pressure) is proportional to the barometric pressure. The ignition source could be an electric spark from a loose wire or from opening a set of relay contacts, a hotspot on an electrical component, or an electrostatic discharge, among others.

2.18.1 An Inerting System

The author invented and developed an automatic Inerting Control System (ICS) in the early 1980s that became a standard for fire prevention in the chemical, pharmaceutical, and paint mixing industries. This system measured the oxygen concentration within, and controlled the flow of nitrogen into, a process vessel by using a closed loop measuring and control system (see Figure 2.29). The oxygen concentration was measured by a fuel-cell type of oxygen sensor.

In the ICS, a gas sample is withdrawn from the process tank by a pump within the sample conditioning module. The sample conditioning module also included a particulate filter, a cooled coalescing filter, and a flow switch, assuring a particulate and condensate free flow of sample gas to the oxygen sensor. The clean sample gas assured a long lifetime of the sensor. The ICS opens a solenoid valve to admit nitrogen to the controlled vessel whenever the oxygen concentration is above a set level, such as 4% oxygen for some solvents. The setpoint is adjustable depending on the LEL of flammable gas present, and is set sufficiently below the LEL to assure effectiveness. The solenoid was normally open when unpowered, so the system was fail-safe: inert gas (nitrogen) would flow if the control system lost power. In case the solenoid valve could have an open coil or open circuit, the current driving the solenoid was monitored. If the valve should be closed, but no solenoid current flowed, then an alarm sounded. The alarm is also activated if the flow switch indicates a loss of sample gas flow.

2.18.2 Intrinsic Safety

In England in 1913, a gas explosion in a coal mine caused the loss of many lives [5, p 5]. It was believed that the ignition was caused by a system for signaling the

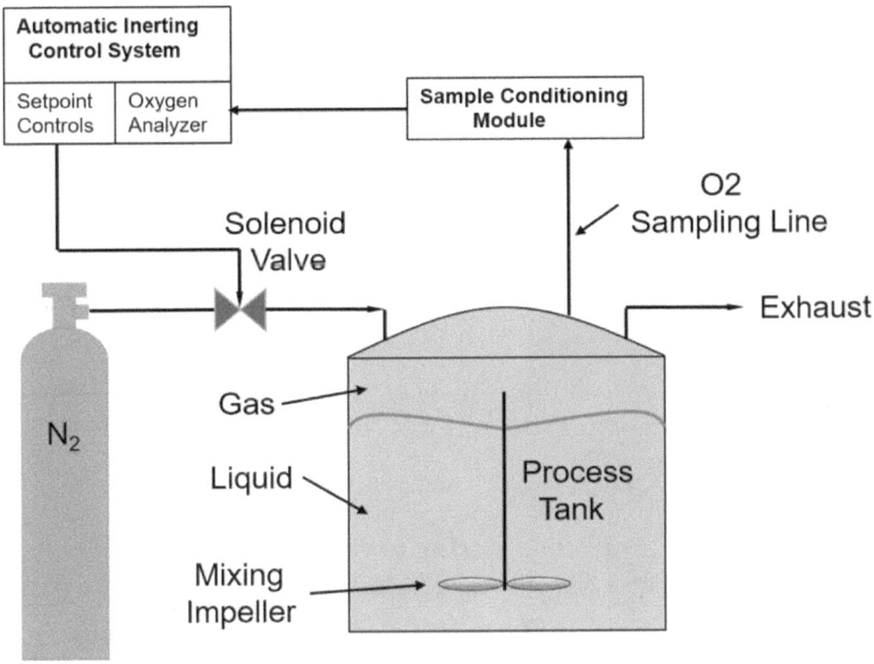

FIGURE 2.29 Automatic Inerting Control System for a liquid process tank.

surface crew that a coal car was ready to be brought up. The signaling circuit was energized by shorting two contacts with a shovel. After a lot of research and work, it was determined that this problem could be avoided by limiting the energy available for a spark. This was the birth of the idea of intrinsic safety.

When a device is rated as *intrinsically safe* (IS), its operating voltage, current, energy storage capability, and temperature are low enough that it is incapable of igniting the specified atmosphere group when used according to the recommended connection diagram. An IS-rated device is designed so that it does not have any (temperature) hotspots, has a low enough outside case temperature for the rated atmospheric gases, and does not store energy in excess of the prescribed level. The IS device is rated to operate in conjunction with one or more types or models of IS barriers. An IS system normally includes a series device called an *IS barrier* installed between the nonhazardous and hazardous areas (see Figure 2.30).

The barrier device is designed that voltage and current levels available to the protected device in a hazardous area are below that required for ignition of the specified types of atmosphere. The barrier device can be located so that the hazardous connection is in the hazardous area and the nonhazardous connection is in the nonhazardous area, as shown in the figure, or it can be mounted fully in the nonhazardous area if positive means are employed so that wires from the hazardous area can only come into contact with the appropriate terminals of the barrier and can contact no other electrical conductors.

As shown in Figure 2.31, a barrier has connection means near one end for wiring to the sensor (protected device) that will be located in the hazardous area, and

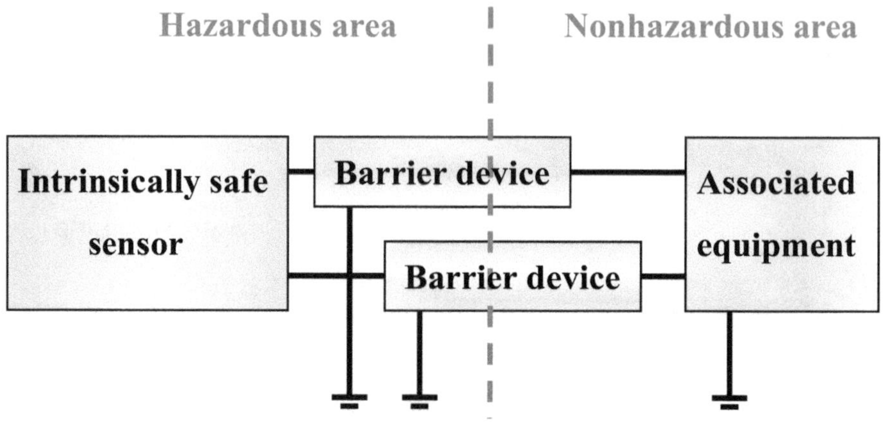

FIGURE 2.30 Intrinsically safe (IS) barrier devices installed between a sensor located in a hazardous area, and associated equipment located in a nonhazardous area.

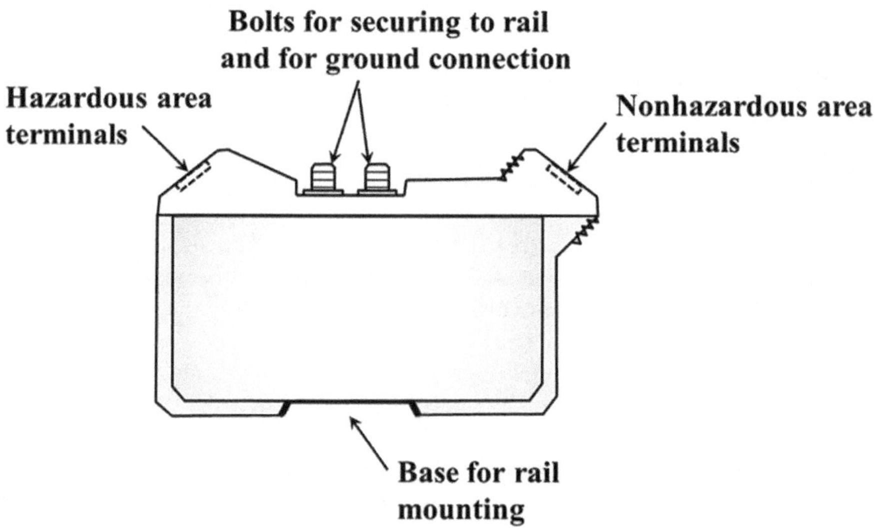

FIGURE 2.31 A typical intrinsic safety passive Zener barrier (about 100 mm long and 13 mm thick).

connection means near its opposite end for wiring to the associated equipment that is located in the nonhazardous area.

The passive Zener barrier of Figure 2.31 would usually have a plastic shell, and may be filled with potting material inside. The housing is often approximately rectangular as shown, and about 13 mm thick. By limiting the maximum voltage and maximum current, a barrier device makes sure that any energy passed into the hazardous area is at such a low level that it is insufficient to ignite the hazardous atmosphere.

IS barriers can be passive or active. The schematic of a single channel passive IS barrier device (also called "Zener barrier") is shown in Figure 2.32.

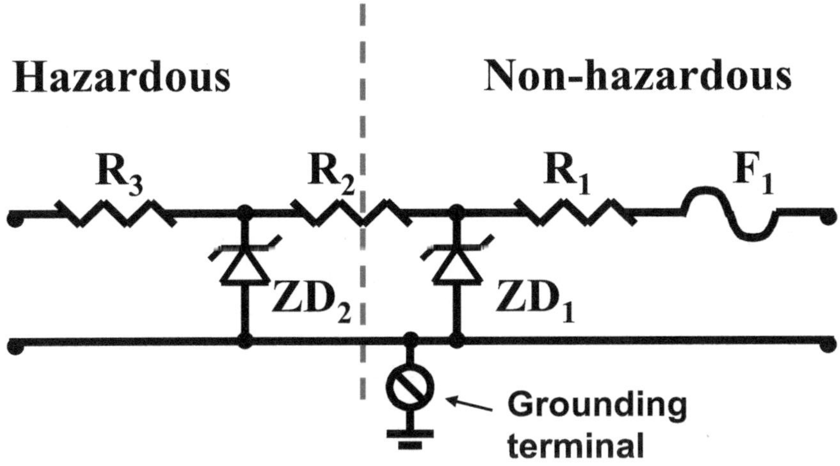

FIGURE 2.32 Schematic of an IS single channel passive Zener barrier.

This is the simplest type of IS barrier and is shown to give the reader a basic understanding of the theory. It includes a first Zener diode, ZD_1, to limit the voltage that can be applied at the nonhazardous side. A second Zener diode, ZD_2, is included for redundancy in case the first one fails. A resistor, R_2, is connected between the two Zener diodes so that it is possible to test the proper function of each diode. A resistor, R_1, is in series on the nonhazardous side to limit the current in an overvoltage condition originating from that side, with R_1 protecting the Zener and limiting the temperature. A fuse, F_1, is also in series on the nonhazardous side to protect the Zener, ZD_1, in case of a gross overvoltage from the equipment on the nonhazardous side, such as an inadvertent connection of line power (115 or 250 VAC) instead of the signal level of voltage for which the usual system is designed (less than 24 VDC). In such a case of high overvoltage, R_1 limits the fault current to ensure that ZD_1 will not fail during the short time it will take F_1 to open up. The fuse must be encapsulated to prevent it from becoming an ignition source, and in fact, the complete IS barrier device is typically encapsulated.

A resistor, R_3, is in series on the hazardous side in order to limit the current to a safe level, at the voltage rating of ZD_2 and ZD_1, for the group rating of the expected flammable materials. So, for example, if the Zener diodes are rated at 6 V, and maximum allowed current is 20 mA, then R_3 could be 300 ohms. So, the sensor would have to be able to operate normally at less than 6 V with 300 ohms in series. Barrier devices are normally designed to be mounted on a metal plate or DIN rail (DIN is a European standard—Deutsches Institut fur Normung). In a passive barrier, it is required that the ground connection be wired to the appropriate IS ground point of the system with a resistance of 1 ohm or less. This is sometimes a difficult task to accomplish reliably, and is one reason why active barriers are popular.

An active barrier includes electronic circuitry to perform more complicated functions or provide added convenience, in addition to providing intrinsic safety. Active barriers often resemble the general appearance of the passive barrier shown in

Figure 2.30, but may not have a need for a ground connection. An active barrier can provide galvanic isolation between the hazardous and nonhazardous parts of the system by using oscillator and demodulator circuits with inductive, capacitive, or optical coupling. This removes the need for a less than 1 ohm Zener barrier ground connection. Active barriers are also available for interfacing with switch inputs, annunciator outputs, or conversion between voltage and current signals, and communication with various digital protocols, such as HART. Each barrier device, whether passive or active, is labeled with the safety ratings, approval body, and limitations.

A simple device (called *simple apparatus*) itself may not be required to carry an IS approval, if it is used with an IS barrier approved for such use. For the purpose of IS requirements, a simple apparatus is one that does not include any appreciable energy source, storage capability, or hot surfaces. This usually means not greater than 10 μH inductance, 5 ηF capacitance, 1.2 volts potential difference, 100 milliamps current, 25 mW of power, or 20 μJ of energy.

This may include switches and simple contacts, LEDs, thermocouples, RTDs, photocells, and non-inductive resistors, thermistors, or potentiometers. Most other electrical devices must have approval from an accepted approval agency before they can be used in an approved IS system. In North America, the primary approval agencies are UL (Underwriters Laboratories), FM (Factory Mutual), and CSA (Canadian Standards Association), but many smaller agencies are also registered. The IS standards among these agencies are similar and are UL 913, FM 3610, and CSA 22.2, respectively. There is some acceptance of these agencies worldwide, but typically, an approval for the country of use is required. Many European countries are members of CENELEC (in English—the European Committee for Electrotechnical Standardization). There are also many other affiliated countries worldwide. CENELEC members have, by regulation, agreed to accept each other's approval systems. Even so, it is common for a customer to require the approval of his or her own country's agency, PTB in Germany, for example.

Outside of North America, the classifications are slightly different from those already described. The North American Division 2 is somewhat similar to the European Zone 2 (but not exactly identical), Division 1 is similarly compared with Zone 1. There is a Zone 0 in Europe, which is assigned to areas where explosive mixtures are expected to be continuously present or present for long periods. In North America, the continuous presence of a hazardous condition is included within Division 1.

2.18.3 Explosion Proof

In an explosion-proof housing, it is assumed that hazardous gases are likely to enter into the electronic equipment enclosure. It is also assumed that it may be possible for the contained device or wiring to cause an ignition. The housing is designed to be capable to withstand the resulting explosion or increase in pressure due to the ignition. The enclosure also has a pressure relief path, that vents the pressurized gas while cooling the gas to a temperature below the ignition temperature of the external atmosphere. See Figure 2.33.

The pressure relief path is called a flame path. It is usually designed to include an area of parallel metal surfaces or a threaded area such that a gas fueled flame passing

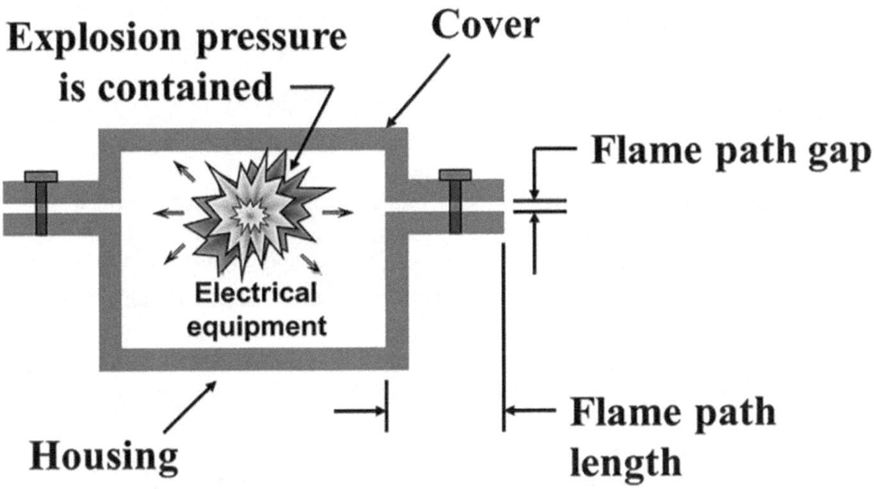

FIGURE 2.33 Explosion-proof housing and flame path.

along the path will be sufficiently cooled before escaping the housing. A sintered metal plug is sometimes used instead for this purpose. After sufficient cooling, the gas will no longer be burning and will be cool enough that it cannot ignite the atmosphere external to the explosion-proof housing. The parallel metal surfaces, which naturally have a relatively high thermal conductivity, are designed to have a small gap between them, called the interstice or flame path gap. The flame path length is made long enough to provide for the needed cooling time. The flame path also provides the means to allow the high gas pressure (from the combustion) to escape. This is important so that the service person does not potentially risk injury due to removing the cover from a pressurized vessel for maintenance or repair of the equipment contained within. The flame path allows equalization of the pressure inside the enclosure with atmospheric pressure. So that, after combustion or an explosion takes place inside the explosion-proof enclosure, the pressure subsides quickly. The equipment inside the enclosure must be designed so that a contained explosion will not compromise the enclosure, such as by throwing a piece of metal against a viewing window with excessive force (some explosion-proof enclosures incorporate a viewing window, not shown in Figure 2.22). Also, since it would not be desirable to have your equipment routinely exploding within the housing, the equipment within the explosion-proof housing is designed to avoid ignition sources. The explosion-proof housing is there just in case a hazardous condition occurs.

An explosion-proof housing will have one or more threaded holes (or ports) though one or more of its walls. Electrical wiring passes through these holes to the equipment within the explosion-proof housing. The wires are contained within a conduit, and screwed into a pour-seal that is mounted into the port. A pour-seal is used at each one of these port connections to provide sealing that is strong enough to contain pressure that may be generated by an internal explosion. Once assembled, a sealing compound is mixed with water and then poured into the pour-seal. When the sealing compound cures into a hard plug, that port is sealed. Some more expensive fast-curing types are available that do not require mixing with water.

2.18.4 Purging

A third alternative for fire prevention, in addition to, or instead of IS and explosion proof, is the use of a non-flammable gas to displace any flammable gas that may be present. This method is called purging. In Europe, the standard is EN 50016 (EN is "European Norm"). In the USA, the standard is NFPA 496 (NFPA is the National Fire Protection Agency). The NFPA specifies three types of purge for use with installations having various likelihoods for the presence of a hazardous gas: types X, Y, and Z. A type Z purge system reduces the enclosure hazard rating from Division 2 to nonhazardous. A type Y purge reduces the rating from Division 1 to Division 2, and a type X purge reduces the rating from Division 1 to nonhazardous. Types Z and X are the most popular, because the type Y reduces from Division 1 to Division 2, and then Division 2 standards must still be met. Whereas with type Z or X, the classification within the purged enclosure is then nonhazardous, and therefore many restrictions on electronics design are not required. Schematics of typical type Z and X purge systems are shown in Figure 2.34.

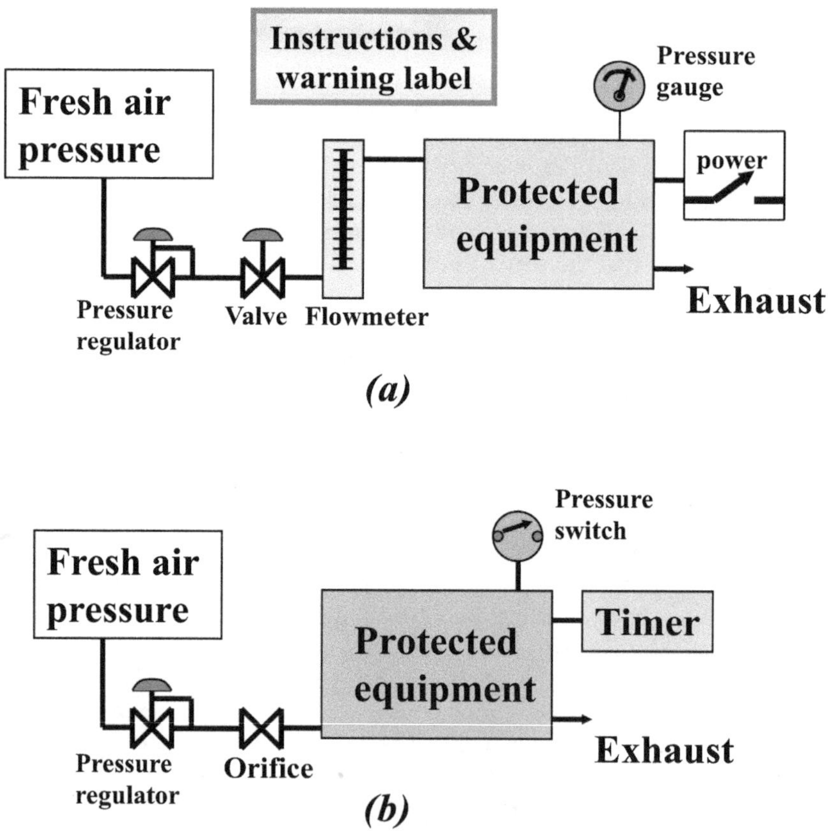

FIGURE 2.34 Purging systems, type Z (a), and type X (b).

The type Z purge is simpler, because it does not have an intrinsically safe timer or a pressure switch, but requires some attention by the operator. That's why it's only for use with Division 2 installations, since there is a lower potential for danger than with Division 1. In a type Z purge, an operator applies the purge gas at a specified flow rate (by adjusting the valve while observing the flowmeter) for a pre-determined amount of time before powering the system up by manually operating the power switch. This assures that the potential hazardous gas has been sufficiently displaced by fresh air (or another non-flammable gas) before the presence of an ignition source is possible. The enclosure also has a pressure gauge, so that a slight indicated positive pressure will assure that the enclosure is being purged and no flammable gas is entering (that is, if there are leaks in the enclosure, the purge gas is going out through the leaks, and flammable gas is not coming in through the leaks). If the purge is lost, the operator notices this on the pressure gauge or flowmeter and turns off the power. For a type Z purge, a label must be placed at the location of the protected equipment, listing a warning and the operating instructions.

In a type X purge, an automatic system is used to allow purge gas to flow for a pre-determined time while maintaining a minimum positive pressure inside the purged enclosure before the purge-protected equipment is energized. After the specified purging conditions have been continuously met for the specified time, then power is automatically applied to the equipment that is protected within the purged cabinet. If the positive pressure in the enclosure is lost at any time, the power to the protected equipment is switched off. In any purge system, the electrical equipment is contained within a reasonably sealed enclosure. It does not have to be airtight, but must be able to hold the required amount of positive pressure. During the initial purging time, the equipment is not yet protected, so that any electrical equipment operating that time (e.g. timer, pressure switch, solenoid valve, and so on) must have another means of protection. This protection is normally by means of encapsulation, explosion proof, or intrinsic safety. Figure 2.35 is an illustration of an X-type purging system designed by the author. In this example, the protected equipment (an oxygen analyzer and inerting system) is contained within a movable cart on wheels, so it can be rolled up for use at the job site.

In the figure, an oxygen analyzer and electrical controls are mounted within an explosion-proof housing mounted on top of the cart. The sample gas conditioner and purge equipment are contained within a purged enclosure mounted under the top of the cart.

Electrical equipment can be protected by any one of the described methods, but sometimes a system may be protected by two or more methods used together. An example would be using an explosion-proof or purged housing to contain intrinsic safety barriers within a hazardous area.

2.19 RELIABILITY

The probability that a product will perform correctly and be free from failure is called reliability. Intrinsic reliability of a product is a function of the quality of design and indicates the reliability that is theoretically possible. Achieved reliability may be different, and is usually lower than the intrinsic reliability, due to factors such as unexpected conditions during use (environmentally or customer induced), a

FIGURE 2.35 A tank inerting system designed to be portable by incorporating into a movable cart, with an X-type purging system for the lower cabinet. Cabinet door shown open, to show space where the sample gas conditioner and purge equipment can be mounted (but not operable while door is open).

temporary lapse in quality of manufacture, lack of maintenance, and so on. Achieved reliability is sometimes called operational reliability, and its measurement in the field is called demonstrated reliability.

Reliability is generally quantified by reporting a mean time between failures (MTBF). This is a statistical representation of the reliability of a product, and relates to the estimated service life of a product until it requires repair or replacement. Although mean time to failure (MTTF) should be used for non-repairable products, MTBF is typically used both for repairable and non-repairable ones. MTBF is expressed in terms of hours. For example, MTBF of a simple sensor could be 250,000 hours, while the MTBF of another more complex device could be

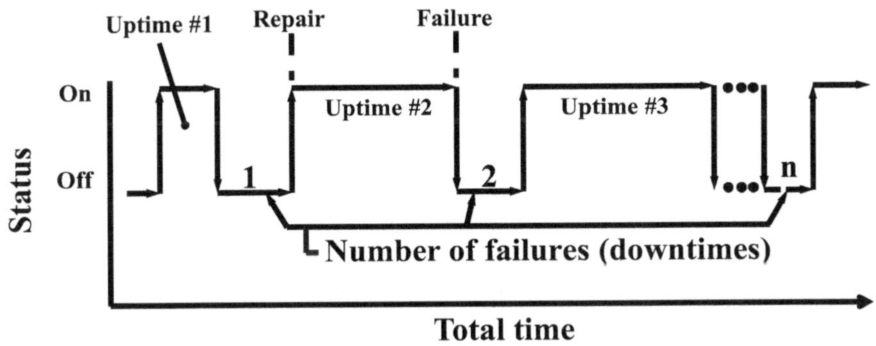

FIGURE 2.36 MTBF is the arithmetic mean value of operational time (uptime) between two failures.

150,000 hours. Contrary to intuition, the MTBF number is not very useful to predict how long a particular example of a product will last before it fails. If a product has an MTBF of 250,000 hours, it does not mean that one can expect an individual example of that product to last that long. The probability that a particular example of a product would last a given amount of time can be calculated using failure probability theory.

To determine the probability that a device will perform without failure over a specified time period, a statistical formula is used, but first, a base of information is needed.

The MTBF is the arithmetic mean value of operational time (uptime) between two failures (unplanned downtimes), and should not be taken as a guaranteed time before first failure. See Figure 2.36. Uptimes and downtimes are shown, with respect to total time.

Since MTBF is the arithmetic mean of the operational times, the MTBF of a device as shown in the figure can be expressed as:

$$\text{MTBF} = (\text{uptime } 1 + \text{ uptime } 2 + \ldots \text{ uptime n}) / (\text{n}) \tag{2.6}$$

The MTBF is meant to be applied over a long period of time to a statistically large sample. If a product has a failure rate that remains constant throughout its life, then the MTBF is the inverse of the failure rate. This is routinely assumed, although often it is not actually the case.

There are two ways by which to arrive at a number representing MTBF: calculated and demonstrated. When calculating reliability, the device is broken down into all of its component parts. Each component part, such as a solder joint, weld, connection pin, resistor, transistor, etc. is assigned an individual failure rate. Then all of the failure rate numbers are added up to arrive at a total failure rate. The inverse of the total failure rate is the MTBF.

One method to calculate the individual failure rates, and the total, is described in MIL Standard 217 (check for the latest revision). Another one is the Telcordia/Bellcore SR-332 standard (again, check for latest revision). The MIL standard was developed, of course, with military applications in mind. The Bellcore standard was developed for the telecommunications industry and is a better predictor of the actual measured performance of products in commercial use. Both standards are in common use for industrial products. Software products are available to make calculations according to these standards easier, including Crimson Quality (www.crimsonquality.com), Isograph (www.isograph.com), Reliasoft (www.reliasoft.com), Item Software (www.itemsoft.com), T-Cubed Systems (www.t-cubed.com), and others.

Calculating an MTBF from estimated failure rates may be the only way to predict reliability before a new product has been in the field. Conversely, if a large number of the product has been installed and operating in the field for a sufficiently long time, an MTBF can be calculated from the field reliability data. This data comes from the internal reports of the manufacturer regarding products returned to the factory by the end users, for reasons related to reliability. Since this is based on actual performance results, it is called "demonstrated" MTBF.

To determine the probability that a given device will perform without failure over a given time period, *T*, the following formula is used:

$$R(T) = e^{-T/MTBF)} \tag{2.7}$$

where *R(T)* is the reliability over period *T*, and *e* is the natural log (approximately 2.71828). For example, if a sensor has an MTBF of 250,000 hours (about 28 years), then a particular example of the sensor has an 84% chance of operating for a period *T*, of 44,000 hours, or about 5 years. $R(T) = e^{-44,000/250,000}$ = 0.0839, or 84%. This may sound like a high rate of reliability. As mentioned earlier, however, a total failure rate of a system comprising several devices can be expressed as the sum of individual rates (and, consider failure rate as the reciprocal of MTBF). So if three of those sensors are installed on a single machine, then there is only a 28% chance (84%/3) that all three will last for five years before at least one of them will need repair.

Another important aspect of reliability is the extent to which the sensor is fit for use at the time when it is actually needed. This is called availability, and is expressed as:

$$\frac{Uptime}{Uptime + Downtime} \tag{2.8}$$

or, alternatively, it can be expressed as:

$$\frac{MTBF}{MTBF + MTTR} \tag{2.9}$$

where MTTR is the Mean Time to Repair. If a product never failed, then the availability would be 100%.

In order to provide a high reliability when designing a new sensor, it may be important to identify some possible sources of failure so that steps can be taken to

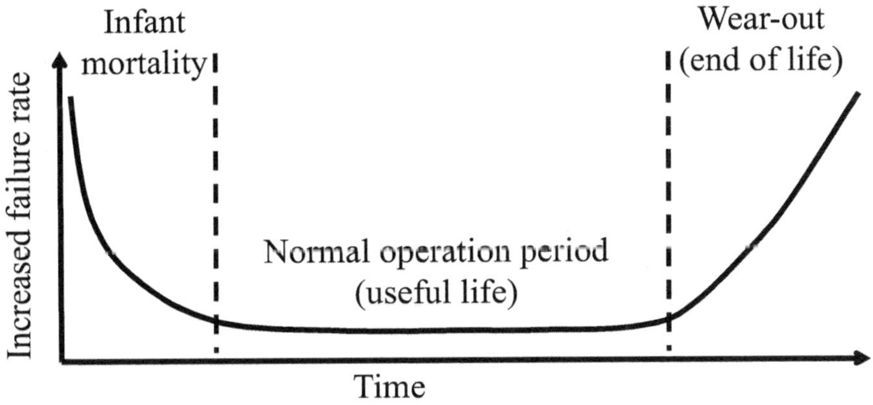

FIGURE 2.37 Bathtub curve includes infant mortality, useful life, and wear-out.

prevent the same. Failure modes can generally be divided into three stages: infant mortality, useful life, and wear-out. The lifetime of a population of products is often represented by the bathtub curve, as depicted in Figure 2.37.

The bathtub curve is not indicative of the failure rate of a single device, but depicts the relative failure rate over time of an entire population of a given device. Failures that occur during infant mortality are the result of defects and mistakes, and every practical step should have been taken in the design phase to prevent this (see FMEA next). Considering the low customer satisfaction level that may result, infant mortality is unacceptable. Sometimes infant mortality can be addressed by burn-in, or operating the device for some short amount of time, sometimes at an elevated temperature, before shipping. But it is surely desirable to correct the design as needed to avoid the problem before the product is shipped to a customer. Once the weak examples are weeded out, the infant mortality rate steadily decreases until a steady state is reached. This period of a relatively low and steady failure rate is the normal operation period, or useful life. Normal life failures are usually assumed to be due to random occurrences, but failed examples should be routinely analyzed in order to determine the cause and improve the design when possible. As some of the components start to wear out, the failure rate increases, and this begins the end-of-life period, during which some better examples may still continue to operate, but more and more units continue to fail over the extended time.

One method for improving the reliability of a sensor during the development phase is to utilize a failure mode effects and analysis (FMEA) procedure. This procedure may include a process FMEA (PFMEA) as well as a design FMEA (DFMEA). See a simplified version of a design FMEA in Figure 2.38.

The first step in developing an FMEA is to list the things that might likely go wrong. In the figure, that is the first column on the left, and here it is called function. Other column titles may be more appropriate for the FMEA of a given project. The second column in the example describes what could go wrong with that particular function. Then, the effects that failure would have on the product. The severity could be low if the effect is cosmetic, but may affect the customer's opinion of the product. It may be

Function	Potential failure	Potential effects	Severity	Potential causes	Occurrence	Current controls	Action
power supply	heats up and transistor Q1 fails	no signal output	high	insufficient heat sink size	2	test voltage	increase heat sink area by 25%
output circuit	long term short to + supply heats up transient supp. diode	possible failure of ESD protection	medium	no current limit for transient suppression diode	1	n/a	add a PTC thermistor between output and diode
protective surface	paint gets scratched	cosmetic defect	low	insufficient curing	2	inspect	cure paint with heat

FIGURE 2.38 Basic example of a design FMEA (DFMEA).

medium if the product will still function, but with some noticeable difference from full performance. The severity is high if the product function is seriously compromised, or is no longer usable. Potential causes should be identified and entered on the chart. Occurrence can be rated, for example, on a scale of 1 to 10, with 1 representing a low occurrence rate (failure of maybe 0.1% or less of product shipped), to 10 (failure of maybe 2% or more of products shipped). The controls that are presently in place that are supposed to prevent this failure should be listed. Action is listed that will remove the likelihood of this failure to occur. The action could include adding a checklist to make sure that an important step is not forgotten, adding a calibration or inspection step, requiring a supplier to certify an important specification on each example of a critical component, a change in design or process, and so on.

This FMEA example is very much simplified in order to present the general idea. Actual FMEAs may have many more columns and rows. Several suppliers offer software to simplify the implementation of FMEA charts. Some of them include: PTC WinDChill (www.ptc.com), Byteworx (www.byteworx.com), IHS (www.ihs.com), and many others.

2.20 QUESTIONS FOR REVIEW

1. **The first step in developing an FMEA is to list the things that might likely:**

    a. Increase
    b. Decrease
    c. Be adjustable
    d. Oscillate
    e. Go wrong

2. **A device rated for intrinsic safety must be designed so that it does not have any:**

    a. Hot spots
    b. Epoxy

c. Metal parts
d. Inductance
e. LEDs

3. The difference between the full scale and zero of a position sensor is called:

a. Span
b. Offset
c. T/C
d. The juice
e. Monotonic

4. Depicts the relative failure rate over time of an entire population of a given device:

a. Infant mortality
b. FIFO
c. Bathtub curve
d. MTBF
e. MTTF

5. Position sensor nonlinearity, repeatability, and hysteresis are included in the term:

a. Resolution
b. Turn-down
c. I.S.
d. Static error band
e. Linpeatysis

6. The three parts of the combustion triangle are fuel, ignition source, and:

a. Heat
b. Oxidizer
c. Carburetion
d. Vapors
e. Liquid

7. This wire is designed to have a very low thermal coefficient of resistance:

a. Copper
b. Cobalt
c. Manganin
d. Tungsten
e. Unobtainium

8. A series of pulses produced during an electrical fast transient test is called a:

a. Static discharge
b. Burst
c. Surge
d. Lightning test
e. Human body model

REFERENCES

[1] E. Herceg, *Handbook of Measurement and Control*. Pennsauken, NJ: Schaevitz Engineering, 1976.
[2] P. Neelakanta, *Handbook of Electromagnetic Materials*. Boca Raton, FL: CRC Press, 1995.
[3] J. J. Carr, *Sensors and Circuits*. Upper Saddle River, NJ: Prentice Hall, 1993.
[4] D. E. Johnson and J. L. Hilburn, *Rapid Practical Designs of Active Filters*. New York: Wiley, 1975.
[5] Elcon Instruments, *Introduction to Intrinsic Safety*. Annapolis, MD: Elcon, 1989.

3 Sensor Outputs and Communication Protocols

3.1 ANALOG OUTPUT TYPES

Sensor outputs are supplied in many variations of analog and digital formats.

Popular *analog* outputs include 0 to 10 VDC, ±10 VDC, 0 to 5 VDC, 4 to 20 mA (occasionally 1 to 5 mA), 10% to 90% ratiometric, 5% to 95% ratiometric, frequency, timed pulse, and pulse width modulation (PWM).

Timed pulse and PWM are often called digital outputs because they are suitable to be directly interfaced with digital circuits but are, in fact, analog signals because they can have a fully continuous nature with no quantization. Timed pulses and PWM do, however, usually provide their signal at the same voltage levels as digital signals (typically, where 0 VDC represents a *low* logic level and a *high* logic level is represented as +5 VDC or +3.3 VDC), or alternatively, at voltages and impedances according to various differential signal standards.

Voltage output circuits, including 0 to 10 VDC, +/– 10 VDC, 0 to 5 VDC, 10% to 90% ratiometric, and 5% to 95% ratiometric are either operated into a high impedance circuit or may have a load resistor within the customer's application circuit (see Figure 3.1). The sensor manufacturer provides a minimum load resistance specification. If the customer applies a load resistance lower than this amount, the output performance can be degraded, or the output circuit inside the sensor could overheat.

Figure 3.1 shows the output circuit of a typical position sensor, having an op amp driving a load resistor R_L, and also a protection device ZD_1. Protection against damage from ESD (electrostatic discharge) is required for a reliable voltage output circuit, and also with other analog output types (ESD protection is explained in Chapter 2). A Zener diode, ZD_1, is included in the figure to protect the output of the amplifier against high voltage spikes. Such Zeners are manufactured specifically for this purpose, and are able to handle much heavier current pulses than a standard type of signal Zener. They are called transient voltage suppressors (TSV). The output circuit of the figure is shown as unipolar, such as 0 to 10 volts DC. At positive voltages below the maximum output voltage of the amplifier (for example, below 10 volts), the TSV does not conduct. ZD_1 in this circuit will conduct current at any voltage above its Zener voltage (such as at 12 volts and above), preventing the voltage from going any higher and damaging the amplifier. ZD_1 will also conduct current at any output voltage more negative than about −0.6 to −0.7 volt (because that's the forward conduction voltage of a silicon diode or Zener). Some analog outputs deliver a current to the load instead of a voltage.

DOI: 10.1201/9781003368991-3

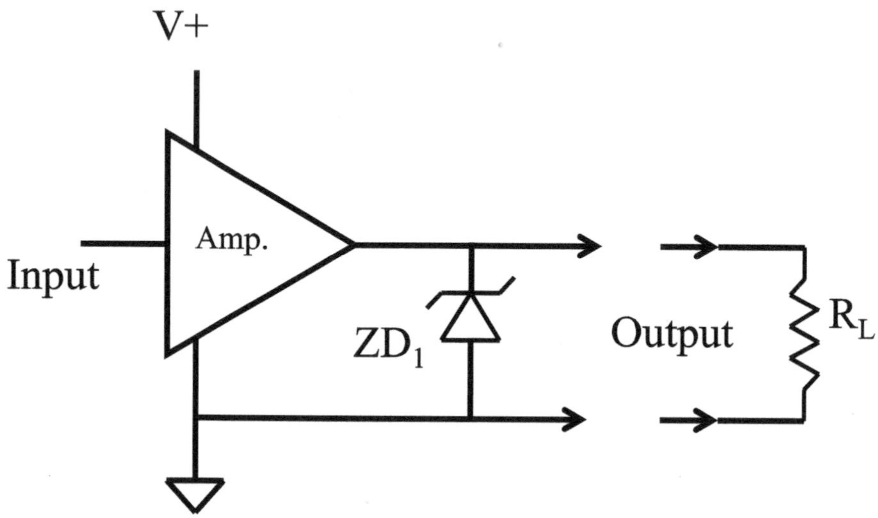

FIGURE 3.1 Voltage output sensor with load resistor.

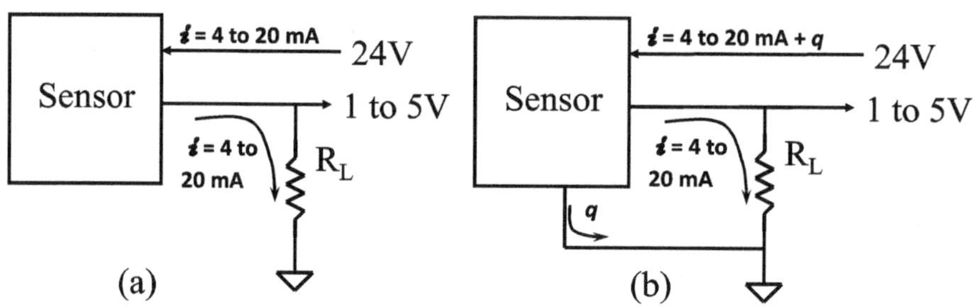

FIGURE 3.2 4 to 20 mA current output sensors: (a) two-wire (or loop-powered); (b) three-wire, having a separate positive power supply voltage.

Current output sensors, including 4 to 20 mA and 1 to 5 mA output types, have output circuits that are operated into a relatively low resistance load circuit near the receiving equipment (R_L, usually somewhere in the range of 20 to 1,200 ohms), or a precision load resistor may be installed within the user's application (such as a data acquisition card or system) to convert the 4 to 20 mA current into a voltage. See Figure 3.2. Sensors with a current output can be operated with a separate power supply input (3 or 4 wires) or the power and signal can both be incorporated on one pair of wires (2 wires). A two-wire sensor is shown in Figure 3.2(a), and a three-wire sensor in Figure 3.2(b). In the three-wire connection, quiescent current q is drawn as the sensor's normal operating current, and is not part of the 4 to 20 mA output current. With a four-wire connection, there would be two common wires: one for power return and one for the 4 to 20 mA common. In the figures, the voltage measurement of 1 to 5 V is made with a high impedance measurement circuit, so that negligible current flows into the voltage measuring circuit.

The single pair type of Figure 3.2(a) is called a loop-powered transmitter, two-wire transmitter, or transmitter. The two wires supply power, and also the signal. One advantage of a loop-powered transmitter is that only two wires need to be run from the data acquisition equipment to the sensor, usually a twisted pair, saving installation cost. Another advantage is that the resistance of the wire does not affect the signal level, because the signal is a current. The use of only two wires is especially important with intrinsically safe systems, where each non-grounded conductor going into a hazardous area must be protected by a safety barrier device (see the description of intrinsic safety in Section 2.18).

Most often, a 4 to 20 mA sensor is operated with a precision 250 ohms load resistor (such as one rated at ≤ 0.1% tolerance, ≥ 0.5 watt power rating, and ≤ 20 ppm/°C temperature sensitivity) at the far end of its cable (or within a data acquisition card) in order to convert the 4 to 20 mA of current into 1 to 5 VDC (because 4 to 20 mA × 250 ohms = 1 to 5 volts). The data acquisition system (DAQ) or other receiving equipment is usually located at the far end of the cable away from the sensor. Even though a current of 20 mA will cause only 0.1 W to be dissipated from a 250 ohms resistor, a 0.5 W resistor may be specified in order to limit self-heating of the resistor, and thereby reducing thermally induced inaccuracy that could have been caused by any temperature sensitivity of the resistor. The connection of a load resistor at the receiving equipment to convert the output to 1 to 5 volts still preserves the main advantage of using a current loop: voltage drops along the length of the cable are ignored. The sensor manufacturer provides a maximum load resistance specification that may be based on the power supply voltage to the sensor. For example, if the sensor power supply voltage is 15 volts, then the maximum resistance possible for generating a current of 20 mA could be 750 ohms (15 volts/0.20 amps = 750 ohms). But there is usually some voltage headroom required for operation of the current controlling circuitry (and for operating the sensor, in a two-wire system), so the maximum load resistance with a 15 volts supply would usually be something less than 750 ohms (maybe 600 ohms). With a 15 V supply, and a load resistance of 600 ohms, the voltage drop across the load resistance would be 12 V at 20 mA. This allows 3 V for operating the sensor (15 V – 12 V). With a 24 volt supply voltage, a typical sensor can usually drive a load resistor of up to 1,000 ohms, leaving 4 V for sensor operation. (Since the loop current can be as low as 4 mA, the internal circuitry of such a sensor must be able to operate with 4 V at less than 4 mA.) The sensor manufacturer may supply a chart showing a safe operating area that indicates the range of load resistances that can be used with a given power supply voltage. See Figure 3.3.

In the figure, a load resistance of 0 ohms is allowed for all power supply voltages from 13.5 to 32 volts. At 13.5 volts, the maximum load resistance is a little less than 400 ohms. At 32 volts, the maximum load resistance is a little more than 1,200 ohms. Some sensors may not be able to internally dissipate all of the power when the voltage is at maximum and the load resistance is zero ohms. In that case, the bottom line of the safe operating area chart will so indicate by changing its slope (slanting upwards, going above 0 ohms) at some voltage.

A recommended load resistance is typically 250 ohms, but a load resistance may be any amount within the safe operating area as shown in the figure. If the user applies a load resistance higher than the specified maximum, the output performance

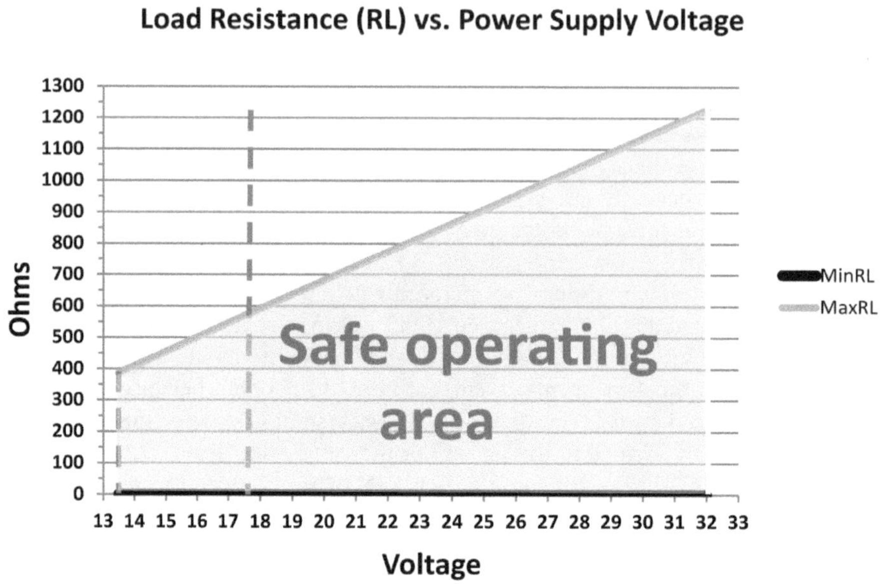

FIGURE 3.3 Safe operating area chart.

can be degraded, due to the lack of a sufficiently high voltage to drive the 20 mA of full-scale output current. If the safe operating area does not go to zero ohms (for example, having a minimum of 20 ohms), using less than that amount of load resistance could cause the sensor to heat up and result in failure. But as a general rule, a sensor should have been designed so that a load resistance of zero ohms is allowable at least at the lower end of its operating voltage range.

With a loop-powered transmitter, the transmitter (in this case, a position sensor) is powered over the same pair of wires as is used to indicate the sensor signal. The minimum signal level of 4 mA is sufficient to operate the sensor, and indicates the minimum reading of the measurand (zero, or zero scale). A larger measurand is indicated in the sensor output by the sensor circuit drawing a greater amount of current, so that full scale (FS) is indicated by a loop current of 20 mA. This means that a low-power circuit design and transducer must be utilized, so that the total power supply current drawn can be less than 4 mA. With a zero of 4 mA, and a full scale of 20 mA, the difference of 16 mA is called span. So, 0 to 16 mA of span is indicative of 0 to 100% of the measurand.

Ratiometric output sensors have an output voltage that varies as a percentage of the power supply voltage to indicate the value of the measurand. For example, with a 10% to 90% ratiometric sensor having a 5.00 volt power supply, a zero measurand input will produce an output voltage of 0.50 VDC, or 10% of the 5.00 VDC power supply. The output will vary from 0.50 to 4.50 VDC for a measurand change of zero to full scale. The advantage is that a voltage reference is not required in the sensor or the customer's receiving circuit. Voltage references are expensive and have their own error specification, adding to the total system error. Since the sensor circuit and the monitoring circuit of a ratiometric system both refer to the same power supply voltage, there is no additional error due to variations in voltage reference. But some users incorrectly assume that the output is just 0.5 to 4.5 VDC, and forget about the ratiometric part. That will result in apparent measurement error. For example, if the power supply happens to be supplying power at 4.950 volts (instead of 5.000 volts), the correct output scale would be 0.495 to 4.455 volts, not 0.500 to 4.500 volts.

So, when making laboratory measurements of the output voltage of a ratiometric sensor, a power supply voltage reading must be taken together with each output voltage reading. This is because the lab power supply may have small short-term variations and the test meter will otherwise read this as sensor error.

Note that some ratiometric sensors utilize a range of 5 to 95%, instead of 10 to 90%. Accordingly, the zero and full-scale voltages in this case are 0.25 to 4.75 V with a 5.00 V supply.

The output types presented so far have been analog. In addition, there are many digital formats in wide commercial and industrial use, so a few common ones are presented in the next section.

3.2 DIGITAL OUTPUT TYPES

Some popular digital protocols, suitable for use in communicating with sensors, include Serial Synchronous Interface (SSI), Controller Area Network (CANbus), DeviceNet, PROFIBUS (application Profile), Highway Addressable Remote Transducer (HART), and EtherNet Industrial Protocol (EtherNet/IP). Each of these is described in their own fairly complex manual of hardware and software interface, so explanations are presented here with the intent to provide sufficient information for a basic understanding. When designing a sensor having a digital communication protocol, it will be necessary for the engineer to review the entire hardware and software specification of the desired protocol.

3.3 SSI

3.3.1 Introduction

SSI was developed as a serial interface technique for use with absolute encoders in order to transfer data from a sensor to a controller. It was developed in 1984 by Stegmann, an encoder manufacturing company in Germany. This interface option is available on many controllers and PLCs (Programmable Logic Controller). Absolute encoders generally produce internally a parallel output in Gray code (see the chapter on linear encoders). The SSI protocol allows serial communication of that parallel data by using a very simple format. This technique is also available with other absolute position sensor technologies, such as magnetostrictive linear position sensors and distributed impedance sensors. An SSI connection system comprises two power, two clock, and two data lines (wires). The clock and data lines are each twisted pairs, and operate according to RS-422/RS-485 standards. The data lines connect a shift register in the sensor to a shift register in the user's PLC or other application circuit, through suitable voltage level and impedance matching circuitry. The user's application circuit sends clock pulses on the clock lines to shift the data *out of the register* located within the sensor, and *into the register* located in the application circuit. The register length is usually 24 or 25 bits, but can be less, depending on the type of sensor. The range of clock rates that can be used is specified for the sensor and varies with the cable length. A data transfer rate of up to 1.8 MHz is possible with a 15 meter cable. Longer cables, up to 300 meters, can be used if the clock rate is limited to 100 kHz (see Figure 3.5). After all of the data bits are transferred, a synchronization period follows. The state of the data line remains logic "high", and no data are transferred during the synchronization period. The duration of the synchronization period is longer than the period of the clock frequency. Synchronization is then possible by knowing that the next pulse on the clock lines after the synchronization period will be pulse number one of the data frame (see Figure 3.6). The first *high to low* transition is the signal for the sensor to latch its data (i.e. to cease taking new measurements and freeze the data that is in the sensor's output register). The next *low to high* transition signals transmission of the start bit. Then the following sequence of low to high transitions shifts out the remaining bits of data. After the data are transferred, the clock line remains high for at least the minimum synchronization period. The cycle is repeated at a rate set according to the internal update rate of the sensor and the requirement of the application for new data. Differential driver/receiver circuits and termination resistors are used to limit the possibility of electrical interference. This is not a bus connection, but is a one-to-one connection between the sensor and controller or other user application circuit. Data flows one way from the sensor to the application (user) circuit.

SSI is a good protocol for communication of absolute sensor position data in industrial applications for several reasons:

1. The differential inputs and differential outputs (RS 422/485) are relatively immune to corruption by electrical noise in the environment.
2. It is relatively simple to implement, with low component count, in the sensor and in the receiving equipment.
3. Data transmission between the sensor and the controller are synchronized by the controller, which makes it easier for the controller to utilize the signal.
4. Only six wires needed for transfer of fine resolution data (two for power, two for clock, two for data, see Figure 3.4).
5. Data transfer rate can be up to 10 MHz, depending on the cable length.

In many SSI sensors, the data can be specified as either binary or Gray code, with Gray code being more popular. The data comprise either 24 or 25 bits, configured during manufacture of the sensor. Such a large number of bits means that a high resolution can be presented, while still having a relatively wide full-scale range.

6. Optical coupling provides galvanic isolation, which eliminates communication errors due to ground loops.

3.3.2 SSI Hardware Configuration

Since digital data will be transferred during SSI communications, the sensor position data must be in digital form. The position data are resident in a set of parallel

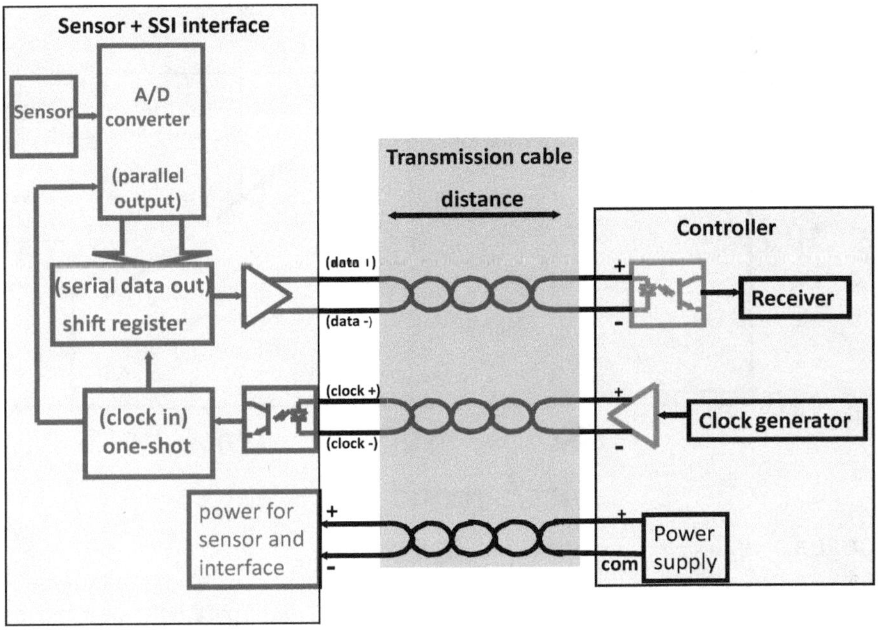

FIGURE 3.4 SSI hardware functional diagram. The hardware connection utilizes optical isolation and RS 422/485 parameters.

registers within the sensor. Having up to 25 bits of data capability, high resolution is possible. So, for example, with a resolution of one micron, the full-scale range can be more than 33 million times that (25 bits), which comes out to over 33 meters!

The hardware connection is shown in Figure 3.4. The clock and data transmissions utilize RS 422/485 parameters. This is usually implemented by designing with an RS 422/485 interface type of integrated circuit. The clock and data receivers are optically coupled, to provide galvanic isolation.

SSI communication provides synchronized data transfer. A set of clock pulses (that is, a clock pulse train) from the clock generator (in the controller or other equipment acting as the leader), is used to serially clock out the data from a shift register located in the sensor. Parallel data had been previously stored into the shift register from an A/D converter (or microcontroller) in the sensor. For each positive-going transition of the clock pulse train, one bit of position data is transferred from the shift register in the sensor to the receiver in the controller (or other receiving device). The shift register data in the sensor are latched with the start of the first clock pulse (the falling edge), and the data set remains latched for the duration of the clock pulse train (as detailed later in the data configuration).

The balanced transmission line provides good noise immunity, and twisted pairs are usually sufficient. But even greater noise immunity can be achieved with the use of shielded twisted pairs. Maximum transmission rates are determined by the cable length, according to the chart of Figure 3.5. (Note: there is also an RS485 specification for intrinsic safety, called RS485-IS, that limits drive current and voltage, and

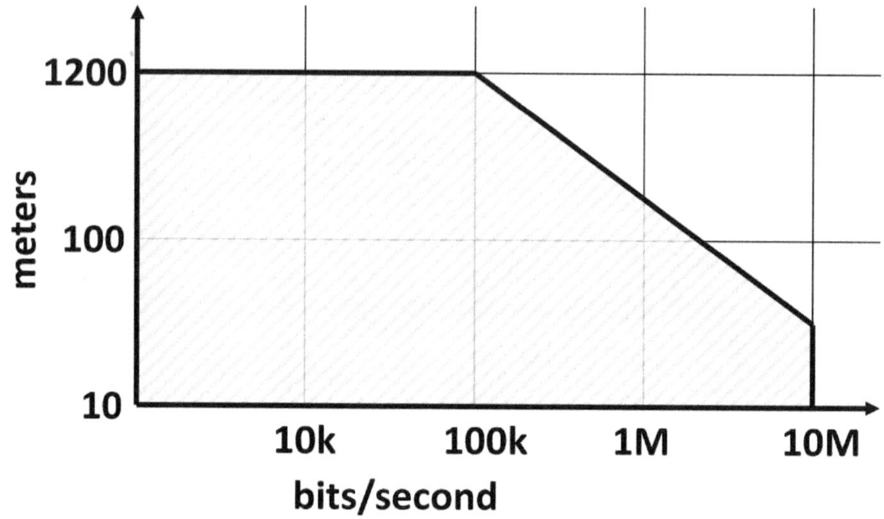

FIGURE 3.5 Transmission rate vs. cable length.

somewhat limits the transmission speed versus cable length parameters for intrinsic safety applications.)

3.3.3 SSI Data Configuration

When SSI is used for rotary encoders, the lower 13 bits are allocated for the number of bits per revolution, and the upper 12 bits are allocated for the number of turns. If less than 13 bits are needed for the bits per revolution, then the appropriate number of lower significant bits in the data stream are set to zeroes. If less than 12 bits are needed for the number of revolutions, then the appropriate number of more significant bits are set to zeroes.

There are at least two options open to designers for standardization of the linear position sensor resolution across a product line:

1. All linear position sensors, no matter the length, can report the digital data with the same resolution. This is practical with SSI because it can handle up to 25 bits (representing a count of over 33 million). So, if the resolution is standardized at 1 micron, then the 25-bit data stream can represent positions of up to more than 33 meters (or more than 330 meters, if the resolution is 10 microns instead).
2. A second option is to always represent the full range as approximately a set number of bits (for example, 16 bits, representing a count of 65,536).

After selecting a resolution format, it is best for a given supplier to standardize on that format, and to keep it the same for all products.

The data can be sent in either binary or Gray code. Since this interface was originally designed for use with encoders, the Gray code format is by far the most popular, and should be the standard. However, some users prefer to use straight binary so that they do not have to convert from Gray to binary in the receiving equipment. So, for that reason, many suppliers offer straight binary format as an option.

3.3.4 SSI Data Sequence

The SSI data versus clock sequence is shown in Figure 3.6.

A clock pulse train is sent by the controller, starting with a high-to-low transition of the clock, at t_0. After the message bits are sent, another pulse train is sent some time later, starting again with t_0.

t_m (monoflop time) = 20 µs ± 5 µs

t_d (delay time) = 540 ns max. for first pulse, 360 ns for subsequent pulses

Msb = most significant bit, Lsb = least significant bit

T = clock period = 100 ns minimum (10 MHz)

The controller, or other device to which the SSI sensor is connected, sends a series of pulse trains on the clock line (the "clock" pair of conductors in the cable). After a pulse train is sent, there is a minimum pause time before another pulse train can be sent. A synchronizing circuit (monoflop) in the sensor recognizes the first falling edge of the clock as starting a new pulse train, representing a new request for the next set of data.

The SSI circuit in the sensor recognizes the synchronization time (or sync time, or pause time) through the use of a mono-stable multivibrator (also called a "one-shot" or "monoflop"). During the pause time (monoflop time), there are no further high-to-low transitions of the clock. When the clock returns to high, it remains high for the duration of the pause time. The minimum pause time is 25 µs. There is no maximum pause time. The controller will send another clock pulse train when it is ready for the next set of data. The monoflop time, t_m, is 20 µs ±5 µs. So, the controller must

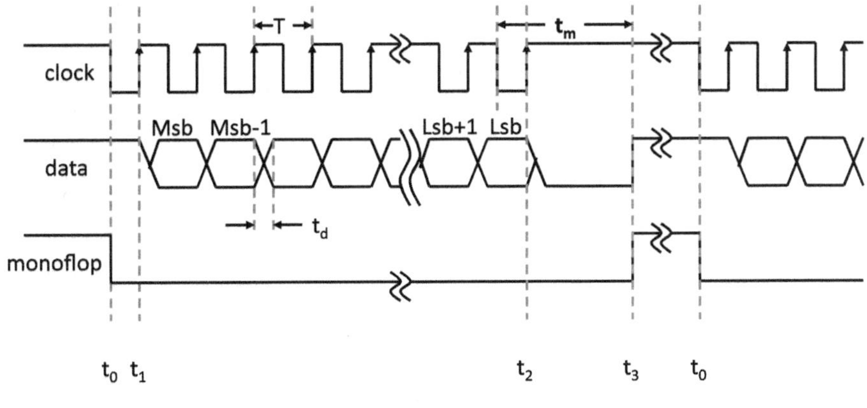

FIGURE 3.6 SSI clock and data timing sequence.

wait at least 25 μs before sending another clock pulse train (after the last low-to-high transition of the previous clock pulse train).

The first falling edge of the clock, t_0, sets the monoflop (set = low logic level), and at the same time, the sensor position parallel data are latched into the shift register. (Each subsequent falling edge of the clock line also sets the monoflop again.) The data in the shift register remain latched until the monoflop resets (the monoflop resets at t_3, after the clock pulse train has ended). The data in the shift register of the sensor are latched so that they do not change while the serial data are being clocked out of its shift register.

The sensor data latching must be completed before the clock line goes high the first time. On the first rising edge of the clock, t_1, the most significant bit (Msb) datum is transmitted. After delay time t_d, the data line transition is stable and the controller can read that bit. At each subsequent low-to-high transition of the clock, the next highest bit is transmitted to the controller. The pulses continuously re-trigger the monoflop so that its output remains low, and prevents storage of new data in the sensor's register.

When the least significant bit is received by the controller, the monoflop is no longer re-triggered. So then, after the monoflop times out, at t_3, the data line returns to high, and new data can be stored into the shift register of the sensor.

3.3.5 Optocoupler

The SSI clock line may be driven by a standard RS 422 driver integrated circuit, but the receiver is optically isolated. A block diagram of a receiver circuit is shown in Figure 3.7.

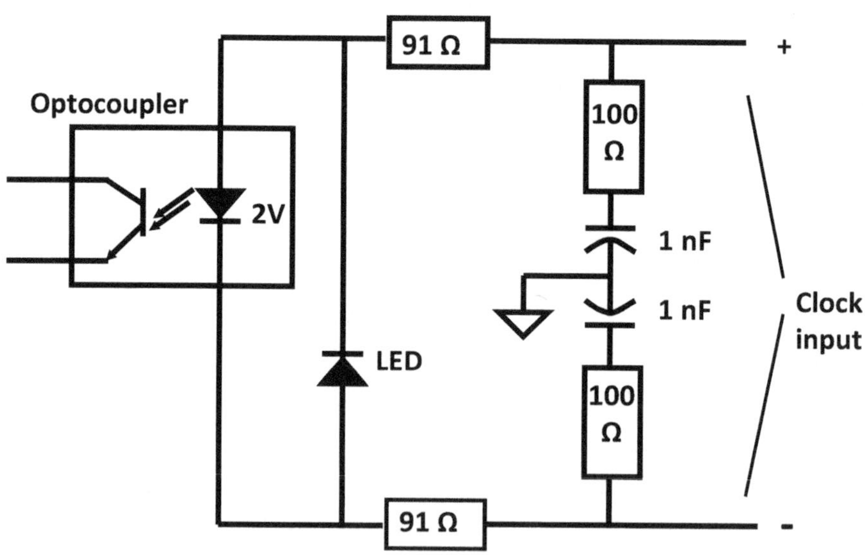

FIGURE 3.7 Optocoupler connection diagram.

The 100 ohms resistors and the 1 nF capacitors serve to terminate the line with a fixed impedance. The two 91 ohms resistors set the LED current within the optocoupler (relative to the actual voltages received). The other LED is included to offer the same impedance to negative-going pulses as the optocoupler LED offers to positive-going pulses.

3.3.6 Other Considerations Regarding SSI

Due to the nature of some types of encoders that use SSI, a few additional features are sometimes incorporated. It is probably best for many sensor manufacturers to avoid including any of these into a linear position sensor product, but some are presented here to provide a better understanding. In an analog linear sensor, such as a string pot (presented in Chapter 5), there is no abnormality (such as a count corruption) that comes from a power failure or turns counting, and there probably is no need for an extremely high transmission rate or a parity bit. But some manufacturers have had to implement additional SSI features in order to make their products work properly. Such additional features are not a part of the commonly used SSI specification, but may include addition of a power failure notification bit, a parity bit, ESSI, and "synchronous" SSI.

Power failure, if added, takes the place of the LSB. It is switched to logic 1 when a power failure has occurred. Then allowance must be made regarding when and how to switch it back to logic 0 after power has returned to normal. The possibility for which it might be used with an analog sensor is if it is important to avoid even one set of bad data to be clocked into the controller, which could happen if a power failure sets the monoflop and corrupts one string of data. Of course, as soon as power returns, the next set of data would be okay again.

Parity bit, if used, takes the place of the LSB.

ESSI: Enhanced SSI provides full-duplex communication, but requires more wires in the cable, and is not very common.

Synchronous SSI sounds strange (synchronous synchronous serial interface), but some sensors are unable to take new data and load it into the shift register in the time of one-half clock period. That means that a set of data is stored at various times when it is ready, until the controller asks for it by clocking out the data. When the data are latched for sending to the controller, the data may therefore have a variable latency (or age). One set of data may have been taken immediately before the controller asked for it, and be relatively fresh (new). The next set of data may have been taken 1 millisecond before the controller asked for it, and is thereby an additional 1 millisecond old. So, because of this variability, the lag time between sampling of the measurand and the time the controller actually receives the data set may change from sample to sample. This can be a problem in a small number of high-speed, servo control applications. So, with "synchronous" SSI, a method is developed to make sure that all data have the same latency. The method depends on the possibilities of data sampling for the particular type of sensor. For example, with a magnetostrictive sensor (see Chapter 12), the amount of time required to receive a signal pulse following an interrogation pulse changes with respect to the location of the position magnet. That is, when the position magnet is farther from the pickup, the signal pulse has more delay time from the application of the interrogation pulse. (In sensors that can perform a complete measurement cycle in less time than one-half SSI clock period, this is not a problem or consideration, since there is no delay.) But when synchronous SSI is desired for a magnetostrictive sensor, it can be implemented in this way: as the sensor recognizes the clock speed and the period from one sync time to the next sync time coming from the controller, the sensor can adjust its own internal sampling time to anticipate the next time the controller will begin a string of clock pulses. Once this has been adjusted, then the next internal sampling of the data within the sensor will become ready to be latched just as the controller will be next asking for it. This will make the latency of all measurements the same. But if the controller sync period is shorter than the time required by the sensor for sampling the data, then there may be some sync periods during which no new data are available. But at least, the data that are transmitted will be fresh. A further correction can remove the problem of new data not being ready for some sync periods. In this version, a microcontroller inside the sensor will recognize that a fresh sample will not be ready in time for the next anticipated sync cycle. So the microcontroller will calculate the expected measurand from the last few measurement cycles, using their position and velocity to approximate the new data so that it can be latched at the required time. (Whew, this is getting complicated! And SSI is one of the more simple digital protocols!)

3.4 CANBUS

3.4.1 Introduction

CANbus was introduced at the SAE congress in Detroit in February of 1986. Robert Bosch, GmbH developed the CANbus communication bus system on behalf of BMW and Mercedes for use in automotive applications. CANbus became an international standard in 1994 (http://vector.com). As a high-speed serial data network, it was designed to replace wiring bundles and provide connection among distributed sensors, controllers, actuators, and indicators, etc. The CANbus specification is covered by several international standards: ISO 11898-1 for the data link layer, ISO 11898-2 for the physical layer of high-speed CAN, and ISO 11898-3 for the physical layer of low-speed, fault-tolerant CAN.

Since it is a bus connection, several CANbus devices can be connected in parallel to the same set of four wires. These comprise two wires (a twisted pair) for power, and two for signal (another twisted pair). There are several modes of operation. In one mode, the user application circuit (for example, the controller) sends out an address to all of the sensors. The sensor having that address acknowledges that it received a message having its own address, and either sends its data at that time or waits for further instruction (depending on how that sensor was programmed to respond). If the controller sends a further instruction, then the sensor responds with the corresponding information. In another mode, any device on the bus can send a request or data to any other device on the bus in a similar manner. In this mode, there could be a time when two or more devices try to use the bus at the same time. Deciding on which sensor gets to respond is called bus contention, and is solved based on a priority level assigned to each device. As each bit of a device address or instruction is consecutively placed on the bus, each device continues to listen until it reads a bit that does not match its own address or that does not match the instruction or data it was attempting to send, in which case it lets go of the bus because it lost the bus contention arbitration. If a device is trying to send a message, but is unable because it lost the bus contention arbitration, or the bus is in use, it will try again after a pause.

Implementation of a CANbus version of sensors already on the market in another version (such as analog output) can improve the suitability of the sensor product for wider deployment in digitally controlled systems in the process industry, factory automation, building automation, and other markets.

Traditional wiring harnesses comprise a bundle of wires and cables that can be several inches thick, so that troubleshooting and repairing a fault may sometimes be very costly. To address this problem, auto manufacturers wanted to develop a more simple and cost-effective means of interconnecting all of the electrical and electronic devices in the vehicle. Subsequent to its introduction at BMW and Mercedes, CANbus gained wider use in automotive applications with additional manufacturers, and versions have been developed that are suitable for many industrial applications, including factory automation. Both CANopen and DeviceNet are versions of CANbus that are commonly called Fieldbus systems, being used in factory automation and other industrial applications. Please note: in this section, the terms bus, network, and net may be used interchangeably.

CAN is an acronym for Controller Area Network. Data are communicated over a balanced (differential) two-wire interface comprising a twisted pair of wires. Often, there is at least one additional pair of wires for power and ground (see the hardware configuration section, later). Since it is a bus connection, several devices can be connected in parallel to the same set of wires. Each device connected to the network constitutes a *node* (see DeviceNet vs. CANopen in this section, for the number of devices, or nodes allowed).

Although CAN is a serial communication standard for digital (smart or intelligent) devices, it is unlike many faster communication standards that provide data rates of up to millions of data bytes in a single frame. CAN has a bit rate with a maximum of 1 mega baud. Baud means bits per second, or bps, of transmission rate. A data byte includes a set of bits to form a word. In ASCII (American Standard Code for Information Interchange), for example (see the end of this section), 8 bits are used for the information in a word, and such a "word" represents one character of the English language. But a complete RS-232 transmission (RS is an acronym for recommended standard) of one byte comprises 11 bits, by including a start bit, stop bit, and parity bit. That is: 8 bits to represent the character, + 3 extra bits = 11 bits for one byte or digital "word". So, 1 mega baud would mean the transmission of about 90 thousand bytes (that is: characters) per second (1 million bits per second/11 bits per byte). Of course, the words transmit, transmission, receive, and so on are not meant in this section to denote a radio or other type of wireless electromagnetic "transmission". The word transmit is used with the same meaning as the word "send", and only means the act of transferring a signal or data from one point to another. CANbus uses a more complex bit structure than only 11 bits per character sent, as explained later in this section.

CANbus is not nearly as fast as some other communication standards, but most industrial applications don't need anything near the speed of the highest digital communication methods. 125 kB (125 kilo baud) is usually sufficient for most industrial applications.

Since CANbus was further specified and developed for many uses, it has international standards that outline several implementations. Two such implementations of CANbus include DeviceNet and CANopen. DeviceNet™ is owned by Rockwell, but originally developed by Allen-Bradley. Its specification is managed by ODVA (the Open DeviceNet Vendor Association, Inc.), with a website at www.odva.org, and follows the international standard IEC 62026-3-3. DeviceNet products use CAN for their data link and physical layers, and CIP (Common Industrial Protocol) for the application layer. Note: the "layers" are further explained later in this section.

The CANopen specification is controlled and overseen by a group called CiA (CAN in Automation, www.can-cia.org). Version 4 of the CANopen specification is standardized as EN 50325-4. (EN is an acronym for European Norm.)

In order to implement CANopen, the exact same hardware circuit may be utilized as with DeviceNet, but modifying the microcontroller firmware to implement the CANopen higher layers. The programming for operating the A/D converter and the communication circuit can remain the same. The information describing the product must also be updated to conform to the CANopen application layer specs. Firmware for the CANopen application layer is available from microcontroller manufacturers, and others. For example, the T89C51CC01 microcontroller is manufactured by Atmel, and Atmel offers the DeviceNet as well as CANopen firmware.

3.4.2 The Basic CANbus

The actual "bus" of a CANbus network is a twisted pair of wires that runs along a length of cable. Devices such as a controller, sensors, actuators, displays, etc. are connected to the bus at various points along the cable length. The CANbus specification as described by ISO-11898 specifies the main requirements of a device to be connected onto a CANbus. ISO is an acronym for the International Organization for Standardization. The CANbus ISO standard implements layers 1 and 2 of the OSI 7 layer Model (this will be described later), that include the data link and physical layers. DeviceNet and CANopen each specify other higher layers, as needed for their respective implementation, in addition to the basic CANbus specification.

The electrical connection of a device to the two data transmission wires of a CANbus as specified by ISO-11898–2 and ISO-11898–3 includes a transmitting module and a receiving module that are capable of sending and receiving messages on the CANbus. The messages are sent as streams of voltage pulses that are timed and spaced to represent the information of the message. Each device, such as a controller, or linear or angular position sensor, that is connected to a CANbus cable must include such transmitter and receiver modules.

The devices connected to the CANbus are called nodes. A node is a portion of the overall system or network. Each node can have one function, or it can have many functions. Depending on the system configuration, different nodes may transmit messages at different times based on the function or functions of each node. For example:

- One node might send a message only on startup or when a there is a system failure.
- Another node might constantly send messages at a fixed rate, such as when it is reporting the position of a hydraulic cylinder in a control loop.
- A different node might send a message or perform a task only when requested to do so by a controller or another node, such as when an over-temperature is detected and then a coolant pump is activated.

3.4.3 CANopen and DeviceNet

As stated, CANbus specifies the basic features needed for electrical connection and placing information onto the bus. DeviceNet and CANopen add a set of specifications of different levels (called layers) to ensure that all devices connected to the net will be able to function together properly, even though they may be supplied by different manufacturers. (CANopen and DeviceNet both add those layers, but with some performance differences between them.)

Many commonly used technical specifications are not open, but are often proprietary. Sometimes the organization that owns the copyright will make the specification available under contract, or sometimes it is not available.

To the contrary, these specifications (CAN, CANopen, DeviceNet) are open standards that are publicly available. The complete specification of an open standard may be easily obtained by anyone wanting to use it. If someone, or a company, has sufficient resources and technical ability, they can properly implement the standard so that all devices that may be designed can be assured to function properly with other devices designed to meet the same standard.

The communication profile (CiA DS 301) and the application layer are listed in the CANopen and DeviceNet specifications. Also included are information for programmable devices (CiA 302), cables and connectors (CiA 303-1), and SI (Le Système International d'Unités, or metric system) units and prefix representations (CiA 303-2).

The application layer and profiles for CAN are not hardware related, but are implemented in software (profiles are explained later in this section).

3.4.4 DeviceNet vs. CANopen

Both DeviceNet and CANopen are called Fieldbus systems, and are commonly used in factory automation. They are both object-oriented (explained later), and function similarly regarding data transmission, configuration, and information for managing the network. Although they can each be used in many similar applications, here are some differences between them:

- CANopen supports up to 127 nodes in a single network, while DeviceNet supports up to 64 nodes.
- A DeviceNet cable can have a length of up to 500 m (baud-rate dependent), and a CANopen cable can run as long as 1000 m (baud-rate dependent).
- DeviceNet has a more strict specification for the physical layout of the network connections, so DeviceNet is somewhat dependent on the physical connections, as well as on the message content. This provides a little higher level of plug-and-play capability, allowing DeviceNet components from different manufacturers to be more easily integrated. Conversely, CANopen is more flexible on the physical layout, and is therefore more dependent on the message content. However, using a standard "device profile" can assure that CANopen products will also be easily integrated.
- Because of this, CANopen can support a higher level of customization toward a specific application. This may enable a somewhat higher level of performance in certain applications. Device profiles are sometimes available for specific applications and might also provide similar benefits.
- DeviceNet has a maximum transmission rate of up to 500 kbps, while CANopen has a maximum transmission rate of up to 1 Mbps. Both are dependent on maximum cable length in the given network.
- Due to higher bit rates possible, CANopen cycle times can be up to two times faster than with DeviceNet.
- DeviceNet is generally more popular in the USA, while CANopen is more popular in Europe.

3.4.5 Object-Oriented

CANopen and DeviceNet are both object-oriented. To explain: in traditional methods of writing software, everything that has to be accomplished is laid out into a set of functions to be performed one after the other. The software for each of the functions is written, aligned with respect to other functions, and interpreted as instructions for the microcontroller. In complex software, it can become difficult to assure that the desired interaction among several related functions will occur properly. Also, the many functions of the software may be difficult to remember when coming back to it at some time in the future, or difficult to understand by someone other than the original programmer.

Conversely, in an "object-oriented" method of writing software, the problems of writing complex code are addressed by using modularity. Each module, or *object*, is designed to accomplish a specific task. An individual object can receive a message, process or act on the information, and may also send an instruction or message to another object. All of the objects together comprise an object-oriented program.

Although such an object in itself is more complex than its underlying *function*, an object-oriented program enables greater flexibility, and is popular in larger-scale software. Due to this modularity, object-oriented programs are simpler to develop, and easier to understand later on, as compared with linear or function-oriented programs.

Objects are classified into several types, including Process Data Object (PDO), Service Data Object (SDO), etc. The objects applicable for a given device on a CANbus are listed in a location in the device that is called the object dictionary. Objects are explained further in a later section of this chapter.

3.4.6 Layers

Every network system has many activities and attributes that must be controlled in order to ensure that devices and software will work together. CANbus follows the "Open Systems Interconnection Reference Model", or OSI Model, for short. The OSI Model has seven "layers". The term layer just refers to a certain area or section of the total specification, meaning that the specification is divided into seven functional areas. Breaking down the specification into these seven layers makes it easier to explain and understand a given part of the total specification. The seven layers encompass the total communication structure of the network. Within each layer, one or more entities implement their particular functionality. Each entity interacts directly only with the layer below it, and provides facilities for use by the layer above it, with each layer being responsible for some discrete aspect of the networking process. See Figure 3.8.

When data is transferred in a network, a specific method must be utilized in order for all devices to receive the information correctly. Such a method is described in the OSI model. The OSI model starts with the physical layer (hardware) as the lowest layer, and proceeding up to the highest layer (application) that will pertain to specific requirements of the user.

Data from one layer can be transferred to an adjacent layer, either above or below the originating layer. Having this capability, each layer can be implemented as an independent component in software. Each message has a message identifier that includes a destination address. Upon receiving a message from an adjacent layer, a given layer checks the address of the intended destination. If that layer's own address is present, then the layer will take the appropriate action regarding the information within the message. If that layer's own address is not present in the message identifier, then the message is passed to the next layer.

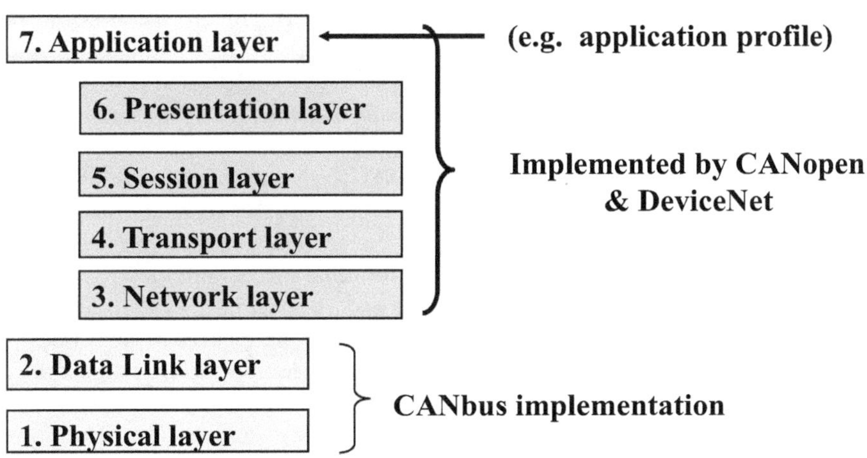

FIGURE 3.8 OSI 7 Layer Reference Model, as used with CANopen and DeviceNet protocols.

Devices on a network are called nodes. In order to transfer a message from one node to another, a given layer at one node transfers the message to the same layer at the other node. For example, a message from layer 5 of node 1 that is to be sent to node 2 would pass down through the lower layers of node 1 and then up through the lower layers of node 2 to arrive at layer 5 of node 2. The lower layers mentioned include various software layers, hardware circuits, as well as the network wires used to connect the nodes.

Layers 1 and 2 are part of the CANbus spec. The upper layers (layers 3 through 7) will be implemented according to CANopen or DeviceNet (for example) specifications.

Microcontrollers used in CANopen and DeviceNet products typically have software available from the microcontroller manufacturer for implementation of CANopen and DeviceNet. This is called a CANopen or DeviceNet "stack". The word "stack" refers to the software for implementing all of the layers, 1 through 7, according to the desired protocol. But most of layers 1 and 2 are in hardware or otherwise already included in CANbus. So, a CANopen or DeviceNet stack will implement all of the communication and network software that is needed for the product (including CANbus, plus either the CANopen or DeviceNet layers). The complete set of software for a sensor would also include whatever is needed for interfacing with the sensing element or transducer, plus information about the sensor that is called for by the CANopen specification and the application profile (more about the application profile later in this section).

A brief description of the seven layers of the OSI Reference Model are presented at this point. The seventh, or highest layer, is usually listed at the top, with other layers in order below it. So, here is a brief description of each of the seven layers of the OSI model:

7. **Application**—the application layer enables an end user to interface with the network. In this level, *applications* can access information and functions that may be available on the network. An *application* is a module of software that can interface with the network. Application layer software can communicate through protocols such as File Transfer Protocol (FTP), Simple Mail Transfer Protocol (SMTP), and Hypertext Transfer Protocol (HTTP).
6. **Presentation**—assures that data originally starting out in different formats will be presented in a readable format. It translates application layer data into an intermediate format, such as ASCII (American Standard Code for Information Interchange), and vice versa (i.e. from ASCII to application layer format). Interface is provided for the layers above and below the presentation layer. It also provides coding that is simplified for communication with the lower layers, data compression, and data encryption.
5. **Session**—the session layer controls the sessions (dialogues) between two applications. It provides for either duplex or half-duplex operation. Duplex means that transmissions can take place in either direction, like when talking on a telephone. But to the contrary, communication can only take place in one direction at a time with half-duplex. A session between two applications is

opened and closed by the session layer. It also controls the session by determining whose turn it is to transmit, and for how long.

4. **Transport**—ensures reliable data transfer, providing error recognition and recovery. It breaks down long messages for transmission as smaller-sized packets, and then rebuilds them to the original message at the receiving end. The transport layer keeps track of the packets and can re-transmit any packets that fail (that is, that are not received error-free). In the transmission control protocol (TCP), which is an example of a layer 4 protocol, the transport layer is the layer that converts messages into TCP segments. Like a post office, it classifies and dispatches the mail that is sent.
3. **Network**—the network layer is used by the transport layer to receive incoming messages from other nodes in the network. The network layer knows the address of other nodes on the network. It adds the correct network address information (provided by the data link layer) to a message to be transmitted, selects the route and quality of service. The network layer also recognizes incoming messages and forwards them to the transport layer. An example of an OSI network layer is the Internet Protocol (IP), which together with the TCP of the transport layer, comprises the well-known TCP/IP.

   Quality of service (QoS)—the QoS of a message can suffer transmission errors and/or delay time. So, an appropriate QoS can be selected ahead of time to optimize the message quality versus the communication speed. Accordingly, selecting a higher QoS for a particular message can reduce the likelihood of errors when that message is received, but this comes at the expense of a slower data transmission rate.
2. **Data Link**—the data link layer packages raw bits of a message from the physical layer into frames, or logically structured *packets* of data for use in higher layers. Likewise, data packets from upper layers are decoded into bits for use in lower layers. The data link layer checks for an acknowledgment from the receiving node that the frame was received, and detects errors. Sometimes, some transmission errors may also be corrected.

   The data link layer has two parts, or sublayers: media access control (MAC) and logical link control (LLC). The LLC helps with synchronization and error checking, while the MAC allows a node to access and transmit information. The data link layer protocol specifies how devices detect and recover from message collisions, and may provide routines to reduce or prevent such collisions.
1. **Physical**—most networks utilize wires, fiber optics, or wireless means for connecting nodes together on a network. The physical layer includes these interconnection media, as well as the hardware circuitry needed to operate and provide the interface. Specifications may include type of wire, connectors, pin layout, timing of data, and so on, as needed to accomplish the hardware interface. With CANbus, data are transmitted differentially over a twisted pair of wires, using an RS-485 type of electrical interface. The physical layer provides hardware bus-contention resolution and flow control.

3.4.7 Message Frames

CAN includes four message types: the data frame, remote frame, error frame, and overload frame.

Data frame—this is the most common type of frame, which is used by one node to transmit information to all other nodes on the bus. The specific bitwise information in a data frame is shown in Figure 3.9.

Remote frame—this is the next most common type of frame. A remote frame, also called a remote transmission request (RTR), is sent as a command from one node to another to request that a message be sent. For example, a controller may need input data from a sensor so that the proper level of control signal can be generated. So the controller might send a remote request for the present pressure or temperature from the appropriate node (sensor). As a command, an RTR is similar to a data frame, but has no data field.

Error frame—the remaining two frame types (error frame and overload frame) are used for error handling. There are many types of errors that are defined. A node sends an error frame when it detects an error.

Overload frame—when a node is busy still processing a message, it generates an overload frame. An overload frame should not appear in a properly functioning network.

3.4.8 The CANbus Data Frame

Any message that is sent along the wires of a CANbus network comprises a series of digital ones and zeroes, called bits. They are produced as differential voltages across a pair of wires. But instead of just calling the bits ones and zeroes, they are instead called dominant and recessive. This terminology reflects the function that when two or more nodes are trying to impress their message onto the network, a node

Field Name	Bits	Description	Field Type
Start of Frame (SOF)	1	indicates the start of a frame	SOF
Identifier (ID)	11	a unique identifier for the message	arbitration
Remote Transmission Request (RTR)	1	must be dominant for a data frame	arbitration
ID Extension bit (IDE)	1	additional 18 bits in the ID field	control
Reserved (r0)	1	reserved, but must be dominant	control
Data Length Code (DLC)	4	number of data bytes (0-8)	control
Data Field	0-64	data, length as shown by DLC	data
Cyclic Redundancy Check (CRC)	15	error checking	CRC
CRC Delimiter	1	must be recessive	CRC
Acknowledge Slot (ACK)	1	any receiver can assert dominant	ACK
Acknowledge Delimiter	1	must be recessive	ACK
End of Frame (EOF)	7	must be recessive	EOF
Interframe Space	≥3	must be recessive	bus idle

FIGURE 3.9 The sequence of fields in a CANbus data frame, in order from top to bottom.

transmitting a dominant bit will be successful, whereas a node trying to transmit a recessive bit at the same time will notice that its bit did not appear on the network. When unsuccessful in transmitting a recessive bit, a node will stop trying to send its message and try again at a later time.

The working of a dominant vs. recessive bit is explained in more detail later in this section (Figure 3.15), but the aforementioned explanation was included here as needed for listing the possible states of bits in a data frame as shown in Figure 3.9.

When a data frame is sent on a CANbus network, it is available to all nodes on the network. The data frame includes from 0 to 64 bits of data (i.e. 0 to 8 bytes of data, each byte consisting of 8 bits. See bits, binary, and so on, later). Each type of information is grouped into its own area, called a field. The fields comprising a frame must always follow the same specific sequence. This sequence is shown in Figure 3.9.

3.4.8.1 Bits, Binary, and So On

The smallest piece of binary data is a *bit.* It can have either of two states: 0 vs. 1, or yes vs. no, or 0 volts vs. 5 volts, or 0 volts/3 volts vs. 3 volts/0 volts (differential), or dominant vs. recessive, etc. This next one is not heard as often: 4 bits comprises a *nibble.* A nibble comprising 4 binary bits represents the same as 0 through 15 in base 10, so including 0, there are 16 possible states. A nibble can represent one digit of a hexadecimal number (HEX, or ##h). Hexadecimal represents the 16 possibilities of 4 binary bits as the numerals 0, 1, 2, 3, 4, 5, 6, 7, 8, 9, A, B, C, D, E, F. An 8-bit *byte* comprises 2 nibbles. As 1 byte or 2 nibbles or 2 hexadecimal digits, FF would represent 255 in base 10 (but it's actually the 256th possible number if you count 0 as the first possible number). A word is often 2 bytes, or 16 bits, or 4 hexadecimal digits, but words can have various lengths, depending on the system being used.

When a CANbus node wants to send data, a data frame as shown in Figure 3.9 is transmitted on the bus. The sequence starts with a start of frame (SOF) bit. This is a dominant state of a length of time equivalent to 1 bit. Prior to that, the bus was in the recessive state, during the Interframe Space from the previous message.

The next 11 bits form the identifier. (But with extended CAN, there is an additional ID field of 18 bits, for a total of 29 bits.) This is a unique binary number to identify the data being sent. The following 1 bit is the remote transmission request (RTR). For a data frame, the RTR must be dominant. (If the message is actually a request for remote transmission, instead of a data frame, then the RTR bit is recessive.) The arbitration field comprises the identifier plus the RTR, and indicates the priority of a message. Message priority is further explained later in this section. The difference between a data frame and a request for data is indicated by the RTR.

The next 6 bits comprise the control field. The Identifier Extension (IDE) is the first bit in the control field. It is always dominant to indicate a standard data frame (if it is recessive, that would indicate an additional 18 bits for the ID field). The following bit is reserved, but must be dominant. The next 4 bits form the data length code (DLC), since the data set can be of a length from 0 to 64 bits, that is equivalent to 0 to 8 bytes, each byte comprising 8 bits. The DLC indicates how many data bytes are present in the message. The 4 bits of the DLC are used to form a binary number that is equivalent to decimal system 0 through 8 (i.e. 0 data bytes to follow would be indicated as binary 0000, 1 would be 0001, 2 = 0010, 3 = 0011, etc. up to 8 = 1000.

Byte	0	1	2	3	4	5	6	7
PDO 1	Position reading, in.				Velocity, in./sec		Hi/Lo limit status	unused
PDO 2	Position reading, mm				Velocity, mm/sec		Hi/Lo limit status	unused

FIGURE 3.10 Two example PDOs that could be sent from a CANbus-enabled string pot, in order from left to right.

With a binary number, each digit, starting from the right, represents a power of 2. The rightmost digit represents 0 or 1, the next is 0 or 2, the next is 0 or 4, and the fourth from the right is 0 or 8, etc. So, to represent the base ten number three, the two rightmost digits would be 1s, e.g. 0011).

The data field follows the control field. Up to 8 bytes can be used in the data field. As an example, the position measurement from a CANbus-enabled string pot would be sent as a type of data frame called a Process Data Object (PDO). PDOs (explained later) transmitted from the sensor could be apportioned according to the following example in Figure 3.10.

As shown in the figure, this example allows 4 bytes for the position data. That would be 32 binary bits, or the equivalent of more than 4 billion increments (that is, 2^{32}). If an increment (the resolution of the position measurement) is made equal to 0.0001 inch (or 0.0025 mm, or 2.5 μm), then the maximum full-scale range (FSR) would be over 35 thousand feet, which would certainly be more than enough range for any string pot. With this resolution and range, all sensors could be accommodated without any changes to scale factor. If a 16 bit A/D is being used in the sensor, for example, the data would only need to be shifted over to where it matched this format. So any receiving node would always see the same units, no matter how long or short is the stroke of the sensor. Alternatively, the resolution could be more coarse, the number of bits allocated could be 16 or 24, and then the number of bytes used would be 2 or 3, respectively, instead of 4.

There are also other bytes available for additional data, if desired, such as those suggested in the figure: velocities and limits. These, of course, are only suggestions and are not requirements. Other measurements or calculated results could be substituted.

The next field after the data field is the Cyclic Redundancy Check (CRC) field. When a message is transmitted, it is important to know whether or not the message was received without errors. The CRC is used to help to confirm this. A standard polynomial function is applied to the contents of the data field before transmission, and the result is a binary number that is transmitted within the 15 bit CRC data field. A node that receives the message also applies the same polynomial function to the data that it received in the message. The receiving node compares the results of its polynomial function to the binary number that it received in the CRC field. If the numbers match, then the data are assumed to have been received without error. If the numbers do not match, then the data included at least one error, so the receiving node discards the data, and asserts a dominant bit in the acknowledge slot (ACK) of the field. The node sending the data sees the dominant acknowledge bit, and thus knows that the message was not properly received by at least one node. The node sending the data may choose to re-send the data frame.

The CRC delimiter is always recessive and comes just before the ACK. The CRC delimiter provides some time, ensuring that a receiving node can complete the CRC function and have enough time to respond in the ACK field. The acknowledge field comprises 2 bits: the slot bit and the delimiter bit. They are both recessive, so they can be overwritten if another node is trying to transmit a dominant bit.

Next is the end of frame (EOF), comprising 7 recessive bits. This is followed by any bus idle time, or Interframe Space.

The bus is determined to be free when no dominant bits are sent for at least 3 bits following the EOF. When the bus is free, the next dominant bit will be recognized as starting the next message frame.

3.4.9 Profiles

It can be fairly complex to implement all of the information required to conform to the CANopen or DeviceNet standards. Even after that, there is no inherent guarantee that the resulting device would be useful in a given network system unless the other devices on the bus understand how to talk to it and what to ask for, etc. Such a problem is largely avoided by using one of the standardized application profiles. They specify the parameters that are usually important for a given application. The encoder profile is one example, and specifies how information regarding a linear or rotary encoder (or other similar type of position sensor) can be encoded so the data format is standardized.

CiA members supply standardized profiles to help designers in developing systems that will use CAN devices. Standards for DeviceNet are controlled by the ODVA, and are available on their website. The next few paragraphs pertain to CANopen, for simplicity, but the same procedures can be applied to DeviceNet using the respective DeviceNet information.

Microcontrollers, protocol stacks, and tools are available from many semiconductor manufacturers for implementing the firmware and software of CANopen devices. Functionality and inter-operability of a CANopen device is described in the standard application profiles. Functions that are specific to a given manufacturer may also be included, but a user will need to have access to the specific programming requirements.

Table 3.1 lists some of the application profiles that are standardized with CiA.

The CANopen application profile for encoders that was mentioned earlier (CiA 406) is suitable for use with position sensors other than encoders, such as string pots, for example. Application profiles suitable for a given type of sensor may be found by members on the www.can-cia.org website.

A list of CANopen application profile specifications that have been released may be found here: www.can-cia.org/index.php?id=systemdesign-profiles. CiA membership is required for access to some application profiles, but many are available without charge.

A sensor designed according to the mentioned CiA 406 encoder application profile might measure the position, and transmit the position data to all nodes on the network. It could do this at a pre-defined periodic rate, called the cyclic mode.

TABLE 3.1
A Listing of Some CiA Application Profiles

CiA Number	Device Type
401	I/O modules
402	electric drives, servo controller, stepper motor, etc.
404	transducer and closed-loop controllers
405	programmable logic controllers (PLC)
406	rotating and linear encoders
408	proportional valves and hydraulic transmissions
410	inclinometers
412	X-ray collimators
413	truck gateways
414	yarn feeding units
415	road construction machinery
416	door control
417	lift control systems
418	battery modules
419	battery chargers
420	extruder downstream devices
422	municipal vehicles
425	medical diagnostic add-on modules

In the *cyclic mode*, the sensor transmits a measured position value repeatedly on its own. In this mode, there would be a set time between consecutive transmissions, that can be from 1 to 65,536 ms. (That's a number of ms equal to 2^{16}, or the number of possibilities with 16 bits of binary data.)

Alternatively, in *sync* (synchronization, or synchronized) *mode*, the sensor would transmit the current measurement value upon receiving a sync message from a control or host node. In sync mode, if there are several nodes responding to the sync message, then they respond in the order of the CAN identifiers of their messages.

In the *polled mode*, the sensor would remain quiet until it receives an RTR PDO, which would be sent by another node when it requires information from the sensor. When polled, the position sensor would read the current position, calculate any parameters that have been set, and return the measured value using the same CAN identifier.

The sensor could also receive configuration information via Service Data Objects (SDO), if so desired and configured in the CANopen microcontroller. The CANopen encoder profile allows the encoder parameters to be programmed via the CANopen network. For example, the encoder could be instructed to count down, instead of the normal counting up, for a given direction of position change. Or the resolution could be changed. These are possibilities, but may not be necessary for most position sensor products. And in fact, they may sometimes only serve to complicate the product more than necessary.

The encoder application profile includes a software limit switch (min. and max.), which allows the programming of two setpoints.

3.4.10 Connecting CANbus Devices

During the automation of a process having several sensed parameters and/or control functions, the bus connection nature of CANbus significantly simplifies the installation and wiring of the associated devices. Thereby, high reliability and relatively high-speed communication can be maintained. So, through identification, command, and response message information, all devices on the net can communicate with each other. Each message from a device has a pre-programmed priority, or importance level.

CANbus, including DeviceNet and CANopen, provides a good protocol for communication of absolute sensor position data in industrial applications for several reasons:

1. The differential voltage signaling (RS 422/485) are relatively immune to corruption by electrical noise that may be present in the environment.
2. It is relatively simple to implement devices onto the network, requiring minimally only two interconnection wires on a common bus configuration. The two wires are the signal wires, called CANH and CANL (CAN high and CAN low), although the bus is often implemented with four wires, configured as two sets of twisted pairs. When four wires are used, there is a twisted pair for signal and another twisted pair for power. Typically, the two power wires would be +24 volts DC and power common. Sometimes, additional ground or shield wires, etc. are also implemented. In any case, all devices on the net connect to the same two signal wires in the same way.
3. Data transmission among messages is arbitrated according to preset priorities. So, more critical messages are set to have higher priorities. Such higher priority messages therefore have greater control over bus transmissions.
4. Data transmission can be as fast as 1 MHz (or 1 MBPS, 1 mega bits per second), but the speed depends on the total length of the bus.

On the minus side, the hardware and software are somewhat complicated, but microcontrollers have been developed to implement most of the hardware, and the software is typically available from the microcontroller manufacturer.

CANopen and DeviceNet in their complete definitions are quite complex, and require a great deal of effort to implement all or most of the possible functions. One mode of operation is the leader/follower configuration for the nodes on the network. In this case, there is usually one leader and one to many followers. This is the traditional implementation of what could be called a full implementation. Basically, the leader is in charge and tells the followers what to do. However, many implementations do not require all the network management features, etc. that the leader/follower architecture provides, so in many cases, CANbus networks are implemented with standalone followers that do not require leader functionality. This simplifies the implementation. Such is the case with many position sensor implementations, and they therefore could be implemented as follower devices, since they may have no need to be the controlling device.

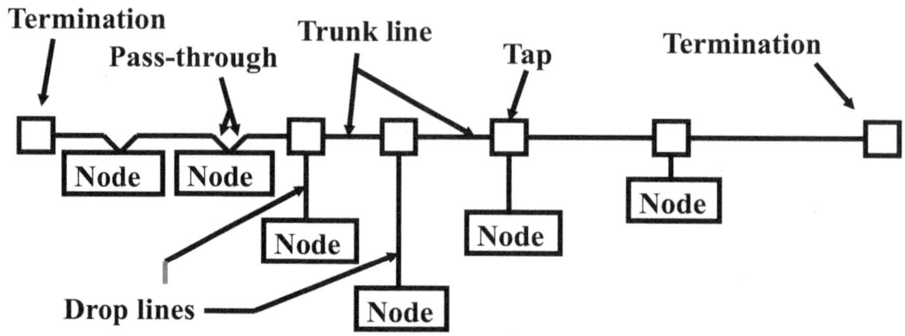

FIGURE 3.11 Connecting devices on CANbus (each device is called a node).

TABLE 3.2
CANopen Transmission Rate vs. Cable Length

Bus Length	Bit Rate (Baud Rate, etc.)	Bit Time
<40 m	1,000 kbit/s (1 MB, 1 MBaud, 1 mega baud, 1 MHz, 1 megahertz)	1 μs
40 to 300 m	500 kbit/s (500 kB, 500 kBaud, 500 kilo baud, 500 kHz)	2 μs
300 to 600 m	100 kbit/s (100 kB, 100 kBaud, 100 kHz)	10 μs
600 to 1000 m	50 kbit/s (50 kB, 50 kBaud, 50 kHz)	20 μs

The peer-to-peer communication of CANbus means that each device on the bus can connect to any other device on the bus. The devices are interconnected on a common bus, with the connection of each device being called a node. See Figure 3.11.

The bus is a cable with at least two wires (or more, for additional uses). Often, four wires are arranged as one twisted pair for power and another twisted pair for signal. In a system of many devices, as shown in Figure 3.11, the main length of the bus is called the trunk line, or trunk. Lines that are branched off are drop lines. There is a limit on the overall length of the bus lines (see Table 3.2), which depends on the bit rate. The overall length is the sum of the trunk length plus all of the drop line lengths. There is also a limitation on the length of a drop line (to avoid reflections). For example, the length of a drop line must be limited to 0.3 m at a bit rate of 1 MHz.

In addition to the mechanical configuration of a tap and a drop line, a CANbus device may be fitted with two connectors and used in a pass-through configuration. To avoid reflections from the ends of the line, termination resistors are connected across the ends of the data lines. E.g. if the cable has a characteristic impedance of 120 ohms, then a 120 ohm resistor is connected across the data pair of wires at any "stub-ends". If a device is located at the end of a line, the impedance of the device can be set to the proper amount so that the device impedance will terminate the line.

3.4.11 Hardware Configuration

Since digital data will be used in CANopen communications, the sensor position data must be in digital form. The position information can be digitized in an A/D

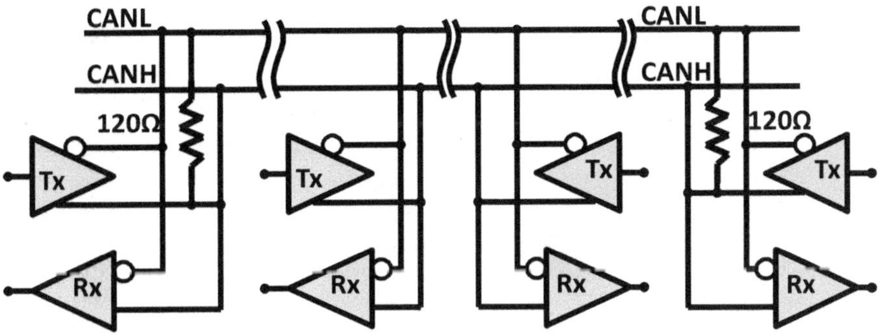

FIGURE 3.12 CANbus hardware circuit functions.

converter, and then sent to the microcontroller. This is the way that many position sensors operate even when they provide an analog output (by adding a DAC). So, this part of the circuitry can remain the same when adding CANbus capability. Then the additional digital communication protocol hardware and software is implemented instead of the DAC and its associated circuitry and software.

Alternatively, since the data part of a CAN message that would represent the position data will possibly hold 32 bits, a higher resolution A/D converter can be used, if desired. The hardware connection for CANbus (DeviceNet, CANopen) is shown in Figure 3.12.

The hardware circuit functions shown are usually implemented by designing with an RS 422/485 interface circuit. Figure 3.12 shows four nodes connected to a CANbus network. Tx and Rx stand for Transmit and Receive, respectively. Each node has a transmit circuit (Tx) and a receive circuit (Rx). Each node has the ability to send, or transmit, a message (through using the Tx circuit), as well as to receive a message (through using the Rx circuit). CANH and CANL stand for CAN high and CAN low, respectively. Those are the names of the two data transmission wires in the cable. A 120 ohm termination resistor is connected across the two data transmission wires at each end of the network. Drop lines must be kept short, with their lengths dictated by the baud rate, in order to avoid additional points of reflection. The termination resistors are only used at the two ends of the bus, and help to minimize reflections of the electrical signals from the ends of the cable. Without the termination resistors, there is a great impedance mismatch between the end of the wire and the adjacent air. This may cause reflections to occur, and thereby increase errors in the messages on the network. Such an error could be caused if a reflection (or "echo") of a previous message bit arrives at a node at the same time as a new, valid bit. The interaction of the two signals can corrupt the signal and thus cause an error. The error detection and correction capabilities of the system can correct a small number of errors. But under certain conditions, a very large number of errors can be generated by a continuous stream of reflections. If the errors can't be corrected, the affected nodes may go off-line. Reflections are avoided by installing the termination resistors.

CANbus position sensors are sometimes available with two connectors installed, as shown in Figure 3.13.

Pin#	Connection
1	CAN low
2	CAN high
3	N.C.
4	N.C.
5	+24VDC
6	Ground

FIGURE 3.13 CANbus position sensor having two connectors, the two connectors wired identically.

The two connectors have identical wiring. This configuration can facilitate easy wiring in applications where several sensors can be connected by simply plugging cables from one device to the next. One cable brings the wiring into a device through the first connector, and the second connector is utilized to bring the wiring into the next device, and so on. Often, one connector is male and the other female, as shown in the figure.

When using a CANbus network, there is a maximum cable length that can be used. There is also a maximum transmission rate, that is given in bits per second (BPS or baud) or in transmission frequency (such as 500 kHz). If frequency is given, and since it is the clock rate (the timing signal) for data transmission, the given frequency is equivalent to a baud rate of the same number.

Although data rates and cable lengths must be kept within these maximums, the use of a cable length that is more than 40 m will limit the maximum transmission rate to less than 1 MHz. The cable length is the sum of the lengths of all cables in the network. For CANopen, the maximum data transmission rate vs. cable length is shown in Table 3.2. So, according to the table, a transmission rate of 1 MB (mega baud) can be supported with a cable length of up to 40 meters in length. In order to figure out how long it takes a message to make a one-way trip to another node (calculating clock time, but ignoring propagation time), multiply the bit time of Table 3.2 by the number of bits comprising the message, according to Figure 3.9. For example, with a transmission rate of 500 kB (2 μs bit time), and 111 bits for the frame (with 64 bits of data, and 3 bits interframe), the total time would be 222 μs, or about 4,500 messages per second. To transfer requested data from one node to another, at least two such messages are required, with the second one being the response from the node transmitting the requested data.

3.4.12 Bus Contention

When two or more nodes try to send a message at the same time, only one node will be allowed to successfully transmit its message; this is called bus contention. (Two or more nodes are "contending" for permission to use the network.) The unsuccessful nodes will wait and try again later, so none of the messages are lost.

As mentioned earlier, CANbus devices send their messages as voltage pulses over a pair of wires. All nodes on the network monitor those pulses. While a node is trying to transmit a message on the net, it also monitors each pulse that appears on the net. This is done in order to confirm that the message it is trying to transmit is actually the one that is appearing on the net. If another node is sending a message at the same time with a higher priority, then the higher priority message will be successful. The first node, that was sending the lower priority message, will notice that it is unsuccessful as soon as the messages encounter a bit in which they are different (see bus arbitration, later). At that point, transmission of the lower priority message will stop, and that node will instead, listen to the message. When the bus later becomes idle, the unsuccessful node will try again. If a given message was not overtaken by a higher priority message, then the given message was successfully sent. The method by which a higher priority message prevails is called bus arbitration.

3.4.13 Bus Arbitration

The transmitted data message does not contain information indicating the sending or receiving nodes, but instead, each transmitted message has a unique identifier, as listed in Figure 3.9. No other messages on the network can be assigned the same identifier.

The numerical value of the message identifier establishes the priority level of the message. The lower the numerical value, the higher is the level of priority. The priority level is not determined by the type of device or the node, but by each message itself. When a device is providing information that is of high priority, that device has been programmed to send such a message with an identifier indicating high priority.

When more than one node tries to send a message at the same time, *bus arbitration* takes place, by application of a *bitwise arbitration* method (see bitwise arbitration, later. CANbus uses *non-destructive bitwise arbitration*, in which each successive bit is checked immediately as it appears on the network, instead of waiting for a complete message to be sent before evaluating it. Thus, when two or more nodes are trying to send their messages at the same time, the highest priority message is not damaged or delayed (but the unsuccessful, lower priority messages are somewhat delayed).

CANbus is a *multicast* type of system, so that all nodes have access to all messages, enabling all nodes to read the identifier of each message. When a multicast message appears on the network, every node checks the identifier to see if the message is intended for them. When receiving a message intended for it, a node will process the message; messages not intended for a given node will be ignored by that node.

3.4.14 Message Priority

In a control system that reads information from various sensors, some of the sensed parameters may be changing quickly, while other parameters may change more slowly. A measured position might be one of the fast-changing parameters. Slower-changing parameters might include the condition of an air filter, or the ambient temperature of the surrounding environment. A rapidly changing parameter will often be required to be reported more frequently than a parameter that changes slowly. A message reporting a rapidly changing parameter would usually be given a higher priority, especially if a specific response is required.

Non-destructive bitwise arbitration is a hardware-centered method, explained in a following paragraph, to provide message collision resolution (although with CANbus, no actual collision occurs). The method used by CANbus to determine message priority, and to optimize use of the available capacity of the bus, is called Carrier Sense, Multiple Access with Collision Detect (CSMA/CD).

3.4.15 CSMA/CD

CSMA/CD is a protocol for supervising network access in a local area network (LAN), such as Ethernet. Before a node can send a message, it first monitors the bus for *traffic* (messages being sent). This is the *carrier sense.* If the network is *idle* (no messages being sent for a specified period of time), all nodes have an equal opportunity to start sending a message (this is the *multiple access*). In Ethernet, when two nodes try to transmit at the same time, a collision is detected, both messages are discarded, and each node waits for a random amount of time before trying again (the random amounts of time increase with each successive collision).

But in CANbus, the bitwise arbitration allows the higher priority message to succeed, and only the lower-priority message needs to be sent again.

3.4.16 Non-Destructive Bitwise Arbitration

As mentioned earlier, CANbus uses CSMA/CD, and also incorporates non-destructive bitwise arbitration. With non-destructive arbitration, the message that succeeds in the arbitration is not damaged.

Bitwise arbitration means that each bit is evaluated as it is received by each node. So, if a node that is transmitting notices that a bit on the network does not match its intended message, then that node has lost arbitration and stops transmitting immediately (and then listens to the message). Figure 3.14 shows an example of bitwise arbitration.

Figure 3.14 depicts the voltages impressed onto the network pair of wires by two nodes, A and B, that are both attempting to send a message at the same time. The start of frame (SOF) begins with bit time 1, during which both A and B are asserting a dominant (low) signal. In bits 2 through 9, the two messages are identical so far, so both A and B continue to transmit their respective messages. In bit 10, it is the first bit in which the messages differ. Message A has a dominant bit, and this will appear on the bus. Node B attempted to send a recessive bit in bit 10, but was overwritten by the dominant bit from node A. Node B therefore lost the arbitration, stops sending its message, and listens to the message appearing on the bus. Node B will attempt to send its message again later.

As an aid to understanding the dominant vs. recessive idea, refer to the simple sketch of Figure 3.15.

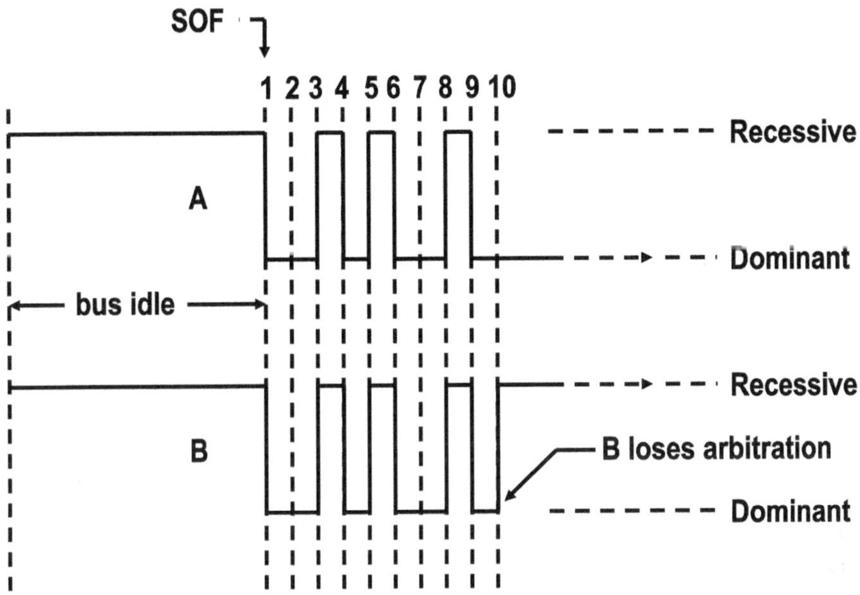

FIGURE 3.14 Non-destructive bitwise arbitration. SOF is start of frame. Message B has lower priority, and loses arbitration when it changes to recessive (high) at bit 10. The numbers 1 through 10 represent the starting points of successive bit times.

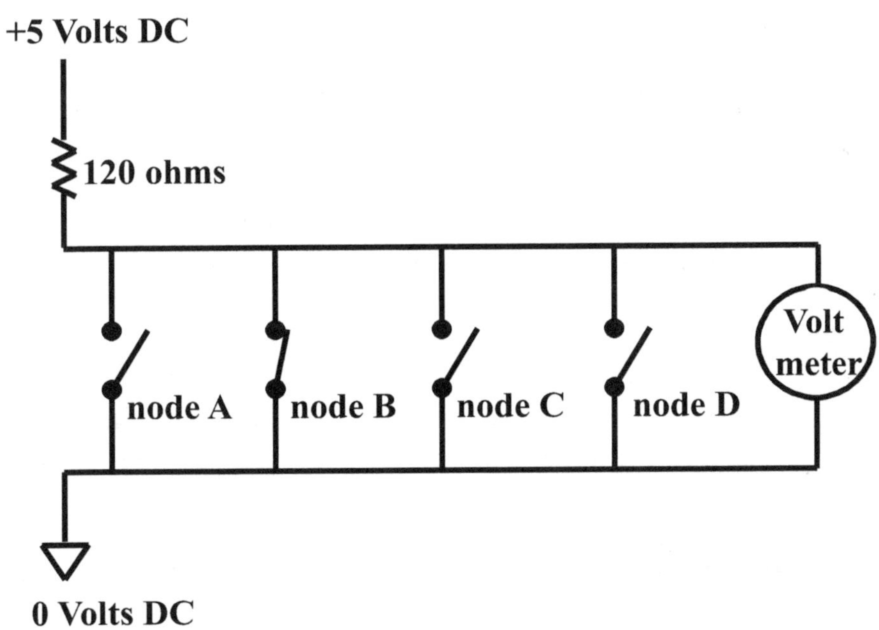

FIGURE 3.15 Dominant vs. recessive bit; closed switch at node B is dominant.

Figure 3.15 is not an accurate representation of the dominant and recessive states in a CANbus network, since CANbus uses a differential voltage signal, but Figure 3.15 is intended to present the general idea of how a dominant bit can overcome a recessive bit.

In the figure, the voltmeter reads the voltage across the "bus". Zero volts is the dominant state, and +5 volts is the recessive state. Any node can transmit the recessive state at a given time by opening its switch, or transmit the dominant state by closing its switch. A message would comprise a stream of consecutive opening and/or closing of the node's switch. If all nodes are transmitting the recessive state (all switches open) then the recessive state (+5 volts) will appear on the bus (as read by the voltmeter). But if any one or more node(s) is transmitting the dominant state (its switch is closed), then the dominant state (0 volts) will appear on the bus. In the figure, node B is dominant and the other nodes are recessive. That means that the message of node B is succeeding at the time shown, and the other nodes have lost in this hardware-based arbitration.

3.4.17 Bit Encoding

Non return to zero (NRZ) with *bit stuffing* is an encoding method used by CANbus in order to provide protection from external disturbance as well as to produce messages with a minimum number of voltage transitions on the bus. Bit stuffing is the addition of non-information bits into data. The location of the stuffed bits is also encoded so that they can be removed at the receiving end.

CANbus uses a method of *bit encoding* that is called non-return-to-zero (NRZ), as opposed to other types such as Manchester, or pulse width modulation. Manchester return-to-zero coding (also called phase coding) is used in Ethernet. In order to maintain synchronization, it avoids long time periods without any transitions of the signal, by ensuring that each bit of data requires at least one voltage level transition. See Figure 3.16.

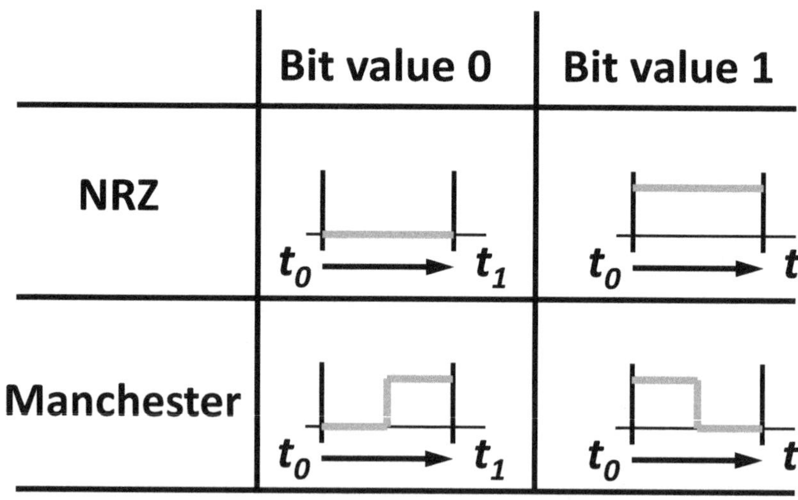

FIGURE 3.16 NRZ compared with Manchester bit representation.

NRZ is the sending of binary code having only two possibilities (in the case of CANbus, the dominant and the recessive states), and requires fewer (voltage) transitions per bit and less time to send a given message as compared with return-to-zero. With NRZ, only one bit time (or time slot) is required to send each bit of a message. However, if a great number of consecutive bits are the same, synchronization could be lost. So, to address this possible problem, CANbus uses bit stuffing.

As can be seen in figure 3.16, NRZ has a constant voltage during the complete bit time. Whereas Manchester bit coding uses a "going high" voltage transition to indicate a logic 0, and a "going low" voltage transition to indicate a logic 1.

Bit stuffing is accomplished in CANbus by adding a bit of the opposite type (a complementary bit) after five consecutive bits of the same type (dominant or recessive) appear in a message. The receiver also expects this, so that it can unstuff the appropriate bits after they are received.

3.4.18 CANopen Objects

CANopen is fairly common as an industrial sensor communication protocol, and so further details are presented here. But many sensors utilize the DeviceNet protocol, so interested readers can find further DeviceNet information on the Open Device Vendor Association webpage (www.ODVA.org).

It was noted earlier in this section that CANopen is an object-oriented protocol, objects being independent programs, each having a distinct purpose and performing a particular function. In a CANopen follower-configured product, up to seven types of communication objects are likely to be incorporated: service data object (SDO), Process Data Object (PDO), sync object, emergency object, network management object, time stamp object, and nodeguard object.

According to the CANopen protocol, a standardized device description must be listed in an Object Dictionary (OD) for each device. The OD lists all of the commands that can be sent to a device. The object dictionary has a communication profile and a device profile. Setting of the identifier, configuring PDOs, and other communication parameters are described in the communication profile. Application objects, error codes, and default PDO mappings are specified in a device profile.

Everything necessary to control the operation of a position sensor, including full-scale input range, output range, resolution, frequency response, and so on, is included in an ASCII text file called an electronic data sheet (EDS).

The OD of a device is based on a table that has the same structure for all types of devices. Important data, parameters, and functions of a device can be accessed using a logical addressing system (index, sub-index) over the CANbus. Access to the OD is provided by means of the SDO protocol.

The OD can be read through a hexadecimal 16-bit index and having an 8-bit sub-index (or sub-entries). Sub-entries are normally used to combine values of the same type, such as with an array, or to access connected values, such as with a data record. An object in the OD therefore has a 24-bit address, comprising the 16-bit index and 8-bit sub-index.

The OD is partitioned into standardized areas, each having 12 bits of entries (4,096). From 1000 to 1FFF is the communication area, covered by CiA in DS-301, and also by EN 50325-4. Device objects that are specific to the manufacturer are in 2000 to 5FFF, and the area from 6000 to 9FFF is standardized in the device profiles.

There is a value associated with every OD object, that can be read or written with the aid of an SDO transfer. The read function can be used to configure and control a device, as an object dictionary entry can directly represent a property or a function of the device (e.g. start-up of a drive). With read access to an object dictionary entry, the device returns the value of the entry. The data type and meaning of the value must be known when asking for this information, so an electronic data sheet (EDS) is supplied for the device. An EDS describes each object dictionary entry with an address (main/sub-index), parameter name, data type, access type and the default value. The EDS is an ASCII text file that is supplied by the manufacturer of the device, and can be read by CANopen configuration tools.

Here is a brief description of the seven types of communication objects previously mentioned:

Process Data Object (PDO)—A PDO may include from one to eight bytes of data (8 bytes = 64 bits). For a position sensor, this data might include the position, velocity, acceleration, status of any limits that have been set, and so on. PDO messages from a given sensor would include an identifier that is unique to that device (sensor), but can be received by any device(s) on the network. Sending of a PDO may be initiated in many ways, such as when a limit is exceeded, using a timer to send messages periodically, in response to a remote transmission request (RTR), and others. PDOs can be transmit or receive (TPDO, or RPDO, respectively). In the standard configuration, there are four each of TPDOs and RPDOs available. But if specifically configured, there are up to a total of 512 PDOs available. PDOs can be sent synchronously, after a sync message, or asynchronously after an internally generated trigger, or after receiving a request to send a PDO.

Service Data Object (SDO)—The SDO transport protocol is specialized to allow transmission of an SDO of any size. The main usage of an SDO is in device configuration. If the data being transmitted by a sensor are not according to a standard application profile (CiA 406, for example), then an SDO message can be transmitted with the appropriate configuration information.

Sync Object—When there is a leader (control) node and one or more follower nodes, the control node can transmit a sync object periodically. A sync object has high priority and contains no data bytes. The sync object causes the followers to become synchronized, by causing them to respond with their pre-determined PDOs in the order of their message priorities.

Network Management Object—In a leader-follower configuration of nodes, the leader can send five different types of messages to the follower in order to control their mode of operation. These are described next, under network management.

Time Stamp Object—A time stamp object includes the time of day, and is sent to devices (such as position sensors) to bring them all to the same time setting. The length of the data field is six bytes, and the identifier is already defined as 256. It is usually sent by a leader node to all follower nodes.

Emergency Object—When an error is detected within a device, it may transmit an emergency object. An emergency object is only transmitted one time for a given error occurrence. There may exist one or more nodes that are programmed to respond to the message of an emergency object, or there may be no nodes programmed to respond to it. When any node has been programmed to respond to an emergency message it finds on the bus, the action to be taken is application-specific, and is not defined in the CANopen specification.

Nodeguard Object—A nodeguard object may be transmitted by a follower node in order to determine if the leader is active, or a leader may request a follower to transmit a nodeguard object through an RTR message. This is explained further under nodeguarding.

3.4.19 Network Management

The simplest way to implement a position sensor CANopen product is to configure it as a follower node. A follower node is basically a node that does not have the capability to tell other nodes what to do. Every follower node contains a state machine of four possible states: initialization, pre-operational, operational, and prepared. There are five network management (NMT) messages which allow a CANopen leader to change the state of one or more followers simultaneously. These NMT commands are start remote node, stop remote node, enter pre-operational state, reset node, and reset communication.

Initialization—With power on, a CANopen follower first performs its own initialization sequence (depending on what is required for that particular type of device), and if completed without problem, then enters automatically into the pre-operational state.

Pre-operational—When entering the pre-operational state, the node performs its internal I/O functions, looks for communication via an SDO, implements nodeguarding (see later), and will respond to nodeguarding messages.

Operational—In the operational state, the PDO and SDO functions will be active, and if needed, will send emergency messages, etc. Of course, this is the normal state in a functioning system.

Prepared—In the prepared state, the node is disabled. There is no SDO, PDO, or emergency communication. It continues to perform its internal I/O functions.

Nodeguarding—When there is a leader node and one or more follower nodes, nodeguarding enables mutual monitoring of the operational status of the leader or follower nodes. On a cyclical basis, the leader node can request the followers to transmit the nodeguard object through means of an RTR. If the follower is operating normally, it will send the nodeguard object. Likewise, if a follower only needs to transmit its data message infrequently, it may want to verify occasionally that the leader node is operational. In that case, the follower node may transmit the nodeguard object on a cyclic basis. If the leader is operating properly, it will respond. If the leader node doesn't reply to such a nodeguard message, the follower node can stop sending messages in order to reduce the message load on the bus.

TABLE 3.3
CANopen Identity Object

Index Number	Sub-Index Number	Description
1018 h	0	Number of Entries
	1	Vendor ID
	2	Product Code
	3	Revision Number
	4	Serial Number

Identity Object—The identity object of a CANopen device lists information pertinent to identification of the device, including a vendor ID, product code (type of product), revision number, and serial number. See the example of an identity object in Table 3.3.

The index number and sub-index numbers are used to find the information in the object dictionary.

Sub-index 0 lists the number of entries that are included in this file.

Sub-index 1 lists the vendor identification, which must be implemented by any manufacturer of a CANopen device. This identifier is unique to that manufacturer, and is assigned by CAN in Automation (CiA), upon request. It is an 8 byte number (i.e. 8 hexadecimal digits), such as 000001A4.

Sub-index 2 lists the product code that is assigned by the manufacturer. It is optional, but must be included if the device is to support LSS (Layer Setting Services).

Note on LSS: CANopen divides the 11-bit identifier (of Figure 3.9) into a 4-bit function and 7-bit node ID. The LSS protocol is used to assign the node ID and baud rate of the target CANopen device (a follower), such as when the device has no mechanical means (rotary switches) for setting these parameters. This requires that a CANopen device in the network acts as a leader, and the remaining devices are followers.

Sub-index 3 lists the revision number of the product, and is assigned by the manufacturer. It is optional, but must be included if the device will support LSS. If implemented, it should have two parts. The first part (the lower portion of the rev number) changes when the revision was implemented to fix bugs. The second part (the higher portion) of the revision number changes when the revision includes changes to the function of the device.

Sub-index 4 lists the serial number that is assigned to the device by the manufacturer. It is optional, but must be included if the device is to support LSS. If used, the serial number must be unique within that particular product code for that manufacturer. The serial number may include a revision number.

3.4.20 Minimal Functionality Devices

In order to accommodate the lesser requirements of a simple follower type of device, the CANopen application layer profile (CiA 301) also specifies the minimal functionality that a CANopen device must provide. The configuration effort for simple networks is reduced by using mandatory default identifiers. These identifiers are available directly after initialization, but may be modified if needed. The network supports up to 127 nodes (zero is used for broadcast messages). Initialization of the device is simplified so that the device goes directly into its pre-operational state. A message from the leader switches the device to operational status. A set of DIP (dual in-line package) switches is used to form the CAN 11-bit identifier: the 4 MSBs indicate the device function, and the 7 LSBs indicate the node identification.

3.4.21 Error Detection

Five types of errors are defined by CANbus: Form, Stuff, CRC, Acknowledge, and Bit.

Form Error—Also called format error, is sent when a form violation is detected. It comprises the sending of six consecutive recessive bits, starting immediately after the bit having the error is detected. A form violation occurs when a dominant bit is found where there is supposed to be a recessive bit in the end of frame, CRC delimiter, acknowledge delimiter, or Interframe Space.

Stuff Error—As mentioned earlier, if a message would contain more than five consecutive bits of the same type (dominant or recessive), the CANbus protocol calls for an opposite bit to be automatically stuffed. So, if six consecutive bits of the same type are found after the start of frame and before the CRC delimiter, a stuff error has occurred. In this case, a stuff error is sent, and the message is repeated.

CRC Error—Cyclic Redundancy Check is an error-checking method for detecting accidental changes in data. A block of data has a check value attached that is based on a calculation. All CANbus nodes on a network will use a standard algorithm to calculate CRC on a message received. A transmitting node uses the same algorithm and includes the calculated value in the CRC field. If a node finds that the calculated CRC value does not match the transmitted CRC field, the node reports a CRC error by asserting a dominant bit in the CRC delimiter slot.

Acknowledge Error—A transmitting node sends a recessive bit in the acknowledge slot. If the message is received by another node, that node asserts a dominant bit in the acknowledge slot. If the acknowledge slot remains recessive, it means that the message has not been correctly received by any node on the network.

Bit Error—A bit error can only be detected by a transmitting node. Every receiving node reads back each transmitted bit. If a bit is detected that differs from the transmitted bit, outside of the identifier field, then a bit error has occurred. In such a case, an error frame is sent, and the message is sent again after a wait time.

3.5 PROFIBUS

PROFIBUS is an open digital communication protocol that is widely used in factory and process automation. It is a bus system based on *profiles* (also called *application profiles*), and hence the name PROFIBUS. The profiles are specifications from manufacturers and users, defining performance features and other properties of sensors, devices, and systems. PROFIBUS has mainly a leader/follower type of operation, but also has a token-passing function in order to support multiple leaders. The complete PROFIBUS standard is contained in the international standard IEC 61158 or the EN 50170. The PROFIBUS User Organization (PNO) is a non-commercial organization created by PROFIBUS manufacturers to publish documents and provide other support and education to members. Another group, PFOFIBUS International (PI), is a large association of fieldbus users. Besides providing educational information, PI also helps with setting standards and developing new technology.

PROFIBUS began in 1987 in Germany (http://us.profinet.com/technology/PROFIBUS) in cooperation with the German government, to establish a serial data communication fieldbus, and standardize the interface to be used with field devices. The PROFIBUS FMS (Fieldbus Message Specification) protocol was defined first, and specifies the most demanding communication requirements. Later, the PROFIBUS DP (the DP stands for Decentralized Peripherals) was defined for a more simple and faster configuration. By 1995, the system gained popularity.

PROFIBUS DP is suitable for implementing sensors, actuators, and other devices in applications having a central controller in factory automation systems.

It includes many diagnostic functions that can be helpful in factory automation.

PROFIBUS PA (the PA stands for Process Automation) is suited for use in hazardous areas that may have flammable gases, dusts, and/or fibers present (see Section 2.18). The physical layer of PROFIBUS PA allows limiting of current to a low level as required for intrinsic safety equipment. The lower current level limits the number of sensors that can be on a bus line, and limits the communication rate to 31.2 kbits/s. The physical layer is controlled under the international standard IEC 61158-2.

PROFIBUS FMS is the third type of PROFIBUS, and is mainly for data communication between controllers, and so will not be further described here. The FMS stands for Fieldbus Message Specification.

PROFIBUS provides services at layers 1 (physical), 2 (data link), and 7 (application) of the OSI model; layers 3 through 6 are not defined. However, some of the functions of layers 3, 4, and 5 are included in layer 2.

The PROFIBUS physical layer is commonly implemented using a standard RS-485 type of hardware connection. This requires four wires: a twisted pair for data, plus two wires for power. Other wired options include RS485-IS (Intrinsically Safe), MBP (Manchester coding, Bus Powered), MBP-IS, and MBP-LP (Low Power). The MBP connections are two wire, having power and data transfer using only one pair of wires. The MBP data transmission rate is much slower than the RS485 rate.

Another option for PROFIBUS data transfer is by use of a fiber-optic cable instead of wires. Data transmission by fiber optics may be selected when the system will operate in an environment of high electromagnetic field intensity (which could otherwise cause electrical interference with the data), high voltage potentials (fiber-optic cables provide electrical isolation), long transmission distances (over 10 km), or high data transfer rate.

Since up to 32 devices can be installed per node, a station address is assigned to each device. This can be programmed as a *hard address* using switches within the device, or as a *soft address*, while assigning parameters during system startup.

All devices execute a startup routine during which time they try to join the network. If the device is a follower, it has a timer. If a leader does not communicate with the new follower before its timer times out, then the follower changes to a safe status. A leader must go through another startup sequence before that follower can try again to join the network.

The data link layer utilizes a Fieldbus data link (FDL), with a leader-follower implementation, as well as token passing. In a leader-follower arrangement, a controlling device (leader) requests information or sends commands to sensors and actuators (followers). With token passing, a leader must be the holder of the token (an identifying signal) in order to act as a leader. The token can be passed from one device to another as needed.

3.6 HART

The HART protocol was originally developed by Rosemount in 1985, and is an acronym for "Highway Addressable Remote Transducer". The HART Users Group was formed in 1990. Ownership of the HART technology was transferred to the HART Communication Foundation (www.hartcomm.org), a not-for-profit organization, in 1993. HART was developed as an open protocol to improve the commonly used two-wire transmitter system by adding bidirectional digital communication capability. HART devices can operate on a 4 to 20 mA current loop, as well as communicate digital data, on a single pair of wires. (For information about the standard 4 to 20 mA current loop, and how is it configured, see Section 3.1.) The digital words are impressed on the 4 to 20 mA analog lines as low-level sinusoidal oscillations at a much higher frequency than the analog signal. So the higher-frequency oscillations do not affect the indication of meters or other analog devices that may be reading the 4 to 20 mA signal. The current loop signals are allocated the frequency band between 0 and 25 Hz, but most current loop signals have a frequency response at around 10 Hz. The oscillations are FSK (Frequency Shift Keying) signals according to the Bell 202 telephone standard. A logical zero is indicated as a 2,200 Hz oscillation, while a logical one is 1,200 Hz.

Hart digital communication is half-duplex, so transmit and receive functions cannot happen at the same time, but must take turns. Within the sensor, the received FSK signals are filtered and then converted to digital data.

A HART sensor can operate in a standard current loop (one sensor per loop), but can also be set to operate in a multi-drop mode. In the multi-drop mode, the loop current is *parked* at a constant level (such as 4 mA), and up to 15 devices can be connected in parallel. Only digital communication is used in the multi-drop configuration. (But up to 256 devices can be installed on a loop in systems implementing a *long form* address, described later.)

HART devices are widely implemented in process control, and other areas where digital signaling rate of several times per second is acceptable, such as in power generation facilities, pulp and paper mills, and oil pipelines. HART devices include both sensors and actuators or positioners.

A HART device can be a leader or a field device. The leader device can be a PC with HART modulator and software, a controller, or other device that asks field devices to report their data. Requesting a field device to report its data is also called polling the device. A field device can be a follower or a device operating in burst mode. A follower responds to messages from the leader, also called poll/response. A field device operating in burst mode reports its data periodically without the need to be polled.

HART message frames may be STX, ACK, or BACK. STX (start of transaction) is a message from a leader to a follower. ACK (acknowledge) is a message sent from a follower in response to an STX message. A BACK (burst acknowledge) message is one transmitted by a field device in burst mode, without need for an STX.

A HART data transaction comprises a leader command and a follower response. Communication access is governed by token passing among the devices on a channel (a *channel* is a pair of wires). In order to transmit a message, a device must be the holder of the token. The transmitted message includes the information for the next passing of the token. If no communication takes place within the period of a preset timer, the token expires and control of the channel is again open. The protocol supports various device parameters, including the measured value of a process variable, device identification, device status, diagnostics, and calibration settings.

Cable lengths of up to 3,000 meters can be used when a cable having a single twisted pair of wires is used with a single sensor. Maximum cable length is less when more than one pair of wires is in the same cable, and when using a multi-drop configuration. Intrinsically safe sensors can be implemented with the proper installation of the rated safety barriers, but maximum cable length may be limited in order to stay within the maximum capacitance requirement.

The HART protocol is generally unsuitable for motion control, and is not widely used for industrial applications, because it is relatively slow when compared to other digital communication techniques such as CANbus. But HART is commonly used in the process control industry because process parameters generally change more slowly than in motion control and other industrial applications.

With HART, the primary process variable (PV) can be communicated through the loop current and also digitally, as well as communicating additional secondary, tertiary, and quaternary process variables (SV, TV, QV, respectively) digitally, and adding other capabilities as needed. Such capabilities of the digital communication may include, for example, to query the field device of its type (e.g. pressure transmitter), its measurement range (e.g. 0 to 50 PSIG), and the actual signal amplitude presently being reported (e.g. 5.203 mA) via the current loop. In addition, many HART devices can also be calibrated, error-checked, re-ranged, and so on by sending appropriate commands to the device. In this operating mode (that is, not multi-drop), the normal 4 to 20 mA analog signal is available, and provides the fastest signal response time, which is usually around 10 Hz.

3.6.1 HART Data

A HART data transaction comprises a leader command and then a follower response (also called *poll/response*). Access to the network in a multi-drop configuration is governed by token passing among the devices on a channel (a channel is a pair of wires). The transmitted message includes information to indicate which device will receive the token next. If there is no message sent within a preset time period, the token expires, and control of the channel is open to the next device that wants to use it.

In the multi-drop configuration, measured information from a sensor or control signal to an actuator is accomplished only through the digital data. That's because every device on the network is parked at a constant 4 mA.

3.6.2 Process Variables

Multivariable field devices can provide more than one digital process variable to the control system. The additional process variables can include additional measurements, or some variables may be calculated. For example, a differential pressure transmitter mounted to an oil tank can also be fitted with a sensor to measure temperature, and it can then be possible to calculate the liquid level and/or volume (if supplied with strapping tables for the particular tank, and a specific gravity vs. temperature table for the measured liquid). For example, some pressure transmitters can map the primary process variable (PV) and the current loop to indicate either the pressure or the level. Support for device variables in multi-variable field devices was added to the HART specification with revision 7.0.

3.6.3 HART Network Connections

Several field devices connected together in a multi-drop configuration are shown in Figure 3.17. The twisted pairs of wire are just shown as single lines, for clarity. In this case, the power supply for the network is supplied by the leader device.

Multi-dropping requires that the loop current of all of the devices on the loop be parked, usually at 4 mA. The leader can communicate with and configure each device in its loop. When needed, a handheld device can be temporarily connected at any place along the network wiring, but easy access is usually only available at a node where a device is already connected. When a handheld device is connected on the network in addition to the usual (primary) leader, it is called the secondary leader.

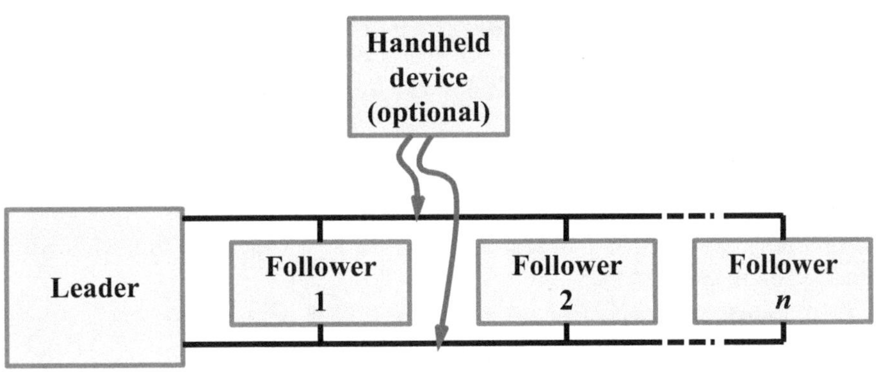

FIGURE 3.17 HART leader and field device (followers) connections in a multi-drop configuration.

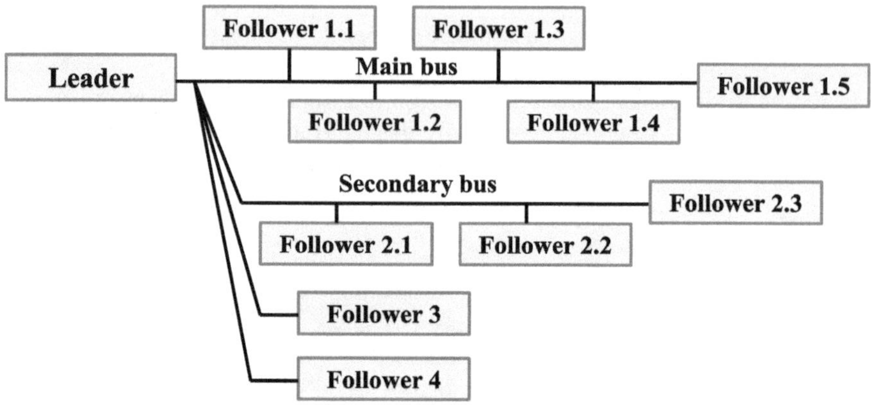

FIGURE 3.18 Main and secondary buses, plus additional devices.

A network having a large number of field devices can comprise one or more main buses, as well as additional secondary buses. See Figure 3.18.

The main bus and secondary bus in the figure are each a twisted pair of wires, with the pairs wired in parallel at the leader device or at another convenient location. Follower devices 3 and 4 each have their own twisted pair, also wired in parallel with the buses and the leader. With long cable distances, the total cable capacitance and resistance RC time constant must be calculated to ensure that it is less than 65 μs. This is to make sure that the HART signaling frequencies can be transmitted and received with an acceptable level of attenuation. As an example, 500 ohms total network resistance × 0.13 μF total cable capacitance = 65 μs.

3.6.4 Seven-Layer Model

In order to function properly, a network system requires a description of how all of the devices on the network will be able to communicate in specified way. HART follows the open system interconnection (OSI) seven-layer model. The transport of messages among devices on the network is described by the OSI model. This model is utilized by many network protocols, but will be described here as it pertains to the HART protocol.

The OSI model is divided into seven functional areas, or layers. Within each layer, a specified set of functions is completed. Each layer can communicate directly with the layer above and below it. The OSI seven layers were shown in Figure 3.8, and generally explained in that section. Here, the layers included in HART will be described more specifically.

Many protocols do not implement all seven of the OSI layers. HART protocol implements layers 1, 2, 3, 4, and 7. So layers 5 and 6 are not implemented in HART. It is often stated that HART implements only layers 1, 2, and 7. In that assessment, the functions included here as layers 3 and 4 are instead considered to be incorporated into layers 2 and 7, respectively.

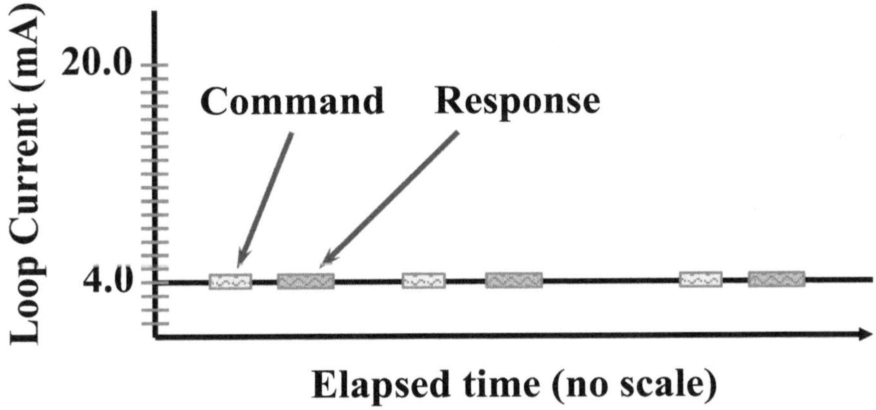

FIGURE 3.19 FSK signals, loop current parked,

The physical layer is the lowest layer (1), and includes specification of the cables containing the twisted pair, the FSK format, frequencies, voltages, currents, wire impedance, and so on. The highest layer (7) contains the user applications.

The transmission of FSK signals while the loop current is parked is illustrated in Figure 3.19.

When there is only one device connected on a loop, the device can be set so that the loop current varies with the measurand by setting its polling address to zero. But when devices are networked (multi-drop), the loop current setting of each device is usually a constant 4 mA (it is parked). As can be seen in Figure 3.19, the FSK oscillations are superimposed upon the 4 mA loop current.

Data rate for a HART device is 1,200 bits per second (bps). The FSK oscillations have an amplitude of about ±0.5 mA if sent by a field device, or about ±0.1 V (that is, 0.1 Vpp) if sent by a leader (the control device). The FSK levels are listed as a voltage for the control device because it usually implements a 250 ohms load resistor, and ± 0.5 mA across 250 ohms of resistance causes a voltage drop across the resistance of ± 0.1 volt.

So, using FSK, the HART physical layer converts a digital message transmission of ones and zeroes into a carrier frequency modulated between 1,200 and 2,200 Hz. Likewise, the oscillations of a received FSK signal are demodulated into a digital message for the device to read.

Starting with HART revision level 6 in 2001, phase-shift keying (PSK) is also possible in addition to FSK. PSK can support a transmission rate of 9,600 bits per second (bps), and results in a maximum transmission rate of about five times faster than with FSK. All versions of HART devices must support FSK. Support of PSK is optional, and has not been widely used so far. Revision 7 of the HART specification added WirelessHART®, using the 2.4 GHz ISM band (Industrial, Scientific, and Medical band).

The data link layer (2) of a HART system describes an asynchronous half-duplex protocol by which a leader sends a command to a field device, and by which the field device responds as necessary. So each complete communication includes a command

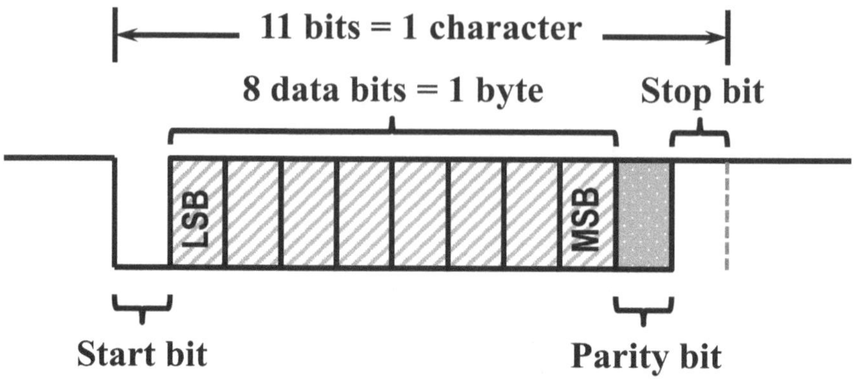

FIGURE 3.20 Total number of bits to transmit one character.

and a response (also called poll/response). The digital ones and zeroes (in binary, called "bits"), of a message are formed into ASCII characters (American Standard Code for Information Interchange). The characters are the same as are marked on the keys of a keyboard (numbers, letters, punctuation, and so on). A set of 8 binary bits is called a byte. (FYI: a set of 4 bits can be called a "nibble" or a hexadecimal character.) An 8-bit byte can also be considered as two 4-bit hexadecimal (HEX) characters. A HEX character has 16 possibilities: 0, 1, 2, 3, 4, 5, 6, 7, 8, 9, a, b, c, d, e, f. A set of 8 bits (a byte) is used to represent each of the 256 possible characters. In order to properly send the data bits without error, additional bits, called start, parity, and stop bits, are sent with each character. See Figure 3.20.

As shown in Figure 3.20, the total number of bits that must be transmitted to send one ASCII character (or one byte of data) includes the 8 data bits, plus one start bit, one parity bit, and one stop bit. So, the communication of one character comprises the time of sending 11 bits. To support synchronization, the start bit is always low. The stop bit is always high. Data bits, of course, vary as needed to indicate the data. The parity bit in HART is odd. That means that the total number of one bits (not zero bits) in the message is odd (the parity bit is adjusted as needed to make sure that the total is odd). So if an even number of one bits is detected, there has been a parity error, indicating that byte of the message has been corrupted.

When a HART message is sent, each character is organized with start/stop/parity bits as was shown in Figure 3.20, and these are combined to form a message as shown in Figure 3.21, forming 1,200/2,200 Hz signals that were illustrated in Figure 3.19.

Rather than comprising normal characters according to Figure 3.20, a string of consecutive logic ones having a length of 5 to 20 bytes (but usually 5 bytes) forms the *preamble* shown in Figure 3.21. It provides enough time for all of the HART devices to get ready to detect the next logic zero as the beginning of the start byte (or start character). This allows all of the HART devices to become synchronized to receive a message. As with all normal characters of any messages, the first bit of the start character is a logic zero (see Figure 3.20). The type of message is indicated by the *start* character. The message could be from leader to follower, follower to leader, or a burst message. Also indicated are short or long form address, and the number of bytes in the expansion field.

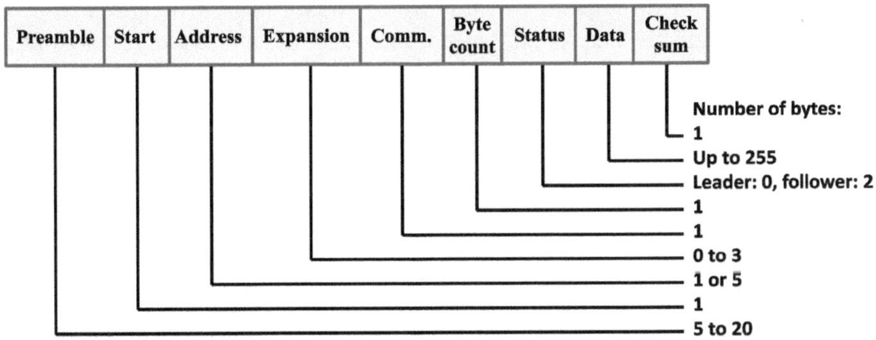

FIGURE 3.21 The structure of one complete HART message.

Both the leader and field device addresses are included in the *address* field. The first bit is 1 for primary master or 0 for secondary master. The second bit is 1 for burst mode or 0 for poll/response mode. In burst mode, a field device transmits it message repeatedly at a programmed time interval. The address can be short or long form. The field device address is 1 byte in the short form, including a 4- or 6-bit address in addition to the leader and burst bits. There are 2 bits that are zeroes in earlier versions through revision 5 (allowing a 4 bit polling address), but is 6 bits in later versions, allowing the polling address to increase the number of devices addressable on the loop from 15 to 63. If the polling address is zero, then the current is not parked, and the loop current is indicative of the process variable.

In the long address, there are 38 bits to indicate a unique identifier for that device, in addition to the leader bit and the burst bit. So for a long form, 1 bit + 1 bit + 38 bits = 40 bits, which is the 5 bytes (8 bits each) length for the long address field. HART revision 5 or later requires the device be long form address capable.

The *expansion* field, introduced in HART revision 6, can contain up to 3 bytes of additional information, but the nature of this information is not yet defined. The number of bytes of the expansion field is indicated by bits 5 and 6 of the start delimiter.

The *command* byte, contains the command for the particular message, instructing a device what to do (such as report the pressure or temperature, and so on). 0 to 30 are universal HART commands. 32 to 126 are common practice commands. 128 to 253 are device-specific commands. The list of possible commands for a device will be shown in its instruction manual. Only the universal command 0 is allowed to use polling addresses.

The *byte count* indicates the number of bytes that will be in the following status and data bytes.

The *status* field is only included in messages from a field device, not in a message from a leader. The status field (0 bytes if leader, 2 bytes if field device) indicates the field device status. It comprises information about errors in the outgoing message, as well as the status of a received command, and status of the device. It is also known as the *response code*.

The *data* field can include up to 255 bytes of information, but most are within 25 bytes, which was a limit in older revisions. A data field may not be present, depending on the command.

The *checksum* field indicates the result of a longitudinal parity check from the start character up to the checksum, but not including the checksum (a longitudinal parity check is also called an Exclusive OR check). The longitudinal parity check is used for error checking, in addition to the parity bit.

The network layer (3) of a HART system manages the sessions among devices that are sending/receiving messages, as well as providing message transport, routing, and security. The *internet protocol* (IP) is a widely used example of the network layer. Combined with the transport layer *transmission control protocol* (TCP), the well-known TCP/IP system is formed.

The transport layer (4) of a HART system ensures the successful completion of a message between two devices on the network. In the transport layer example mentioned earlier, TCP, messages are converted into TCP segments. Often compared with a post office, it classifies and dispatches mail (messages).

The application layer (7) of a HART system provides a definition of the commands and responses, as well as describing the types of data and reporting the device status. Common protocols used to implement the OSI application layer include Simple Mail Transfer Protocol (SMTP), Hypertext Transfer Protocol (HTTP), and File Transfer Protocol (FTP).

The set of HART pubic commands include the following groups:

- *Universal commands* must be recognized and supported by all HART devices. This includes commands such as to read the manufacturer and device type, the primary variable the type of units, and so on.
- *Common practice commands* are supported by many, but not all, HART devices. They include such functions as set zero, set span, select which of up to four variables to read, and others.
- *Device specific commands* relate specifically to the particular device, and may include, for example, start, stop, setpoints, and travel limits.
- *Device family commands* include some standard functions for various device types, for generic use instead of using device-specific commands.
- *Non-public commands* are those intended only for factory use, and not for use in the field.
- *Wireless commands* are for HART devices that are capable of operating with WirelessHART, and all such devices must implement all of the WirelessHART commands.

The current versions of various HART protocol documentation, downloads, and available training are listed at: www.fieldcommgroup.org/technologies/hart/documents-and-downloads-hart. A list of HART specifications is shown at: www.fieldcommgroup.org/hart-specifications. An order form is available at the bottom of the webpage. A complete list of the HART protocol specifications comprise a set of over 20 documents, at a cost of $975 to non-members. They are available at no charge to FieldComm group members.

Additional documents and products are available for purchase from the HART Communication Foundation to support the development of a HART compatible device; including an applications guide, calibration guide, test kit, and so on, as listed on the order form mentioned.

3.6.5 Device Description Language (DDL)

A HART DD must be written in compliance with the specifications of the HART Device Description Language (DDL). Many HART devices have functionality that is greater than that supported by the universal and common practice command set. The additional functions may include specific device configuration, diagnostics, and features specific to that model of device. The individual device description (DD) is an electronic data file that specifies how to access these additional functions for the specific model of HART device. The DD provides means to access all of the data and functions of a device, and may also provide graphic display features and menus for handheld and other devices. Most leader devices have the capability to read the DD for all of their field devices.

A library of all of the registered devices and their DD is maintained by the HART Communication Foundation, and updated quarterly. Access to the library information is available to all members of the HART Communication Foundation (HCF). A list showing the latest release of HART registered devices is available at www.fielDCommgroup.org/registered-products.

3.6.6 Long Form Address versus Short Form Address

As mentioned earlier, the HART address field can be short form or long form. In versions before revision 5, only the short form address was available, and comprises 8 binary bits, the least significant 4 of which originally formed the polling address to select the device. With a 4-bit polling address, up to 15 devices can be selected on a network (address 1 through 15). If the polling address is zero, the device is not networked, and the loop current is indicative of the process variable. If the polling address is from 1 to 15, then the loop current is parked, usually at 4 mA, so the device can be networked with up to 14 other devices on that pair of wires.

All newer devices (after revision 5) use a 5 bytes *long form address*, having a 38-bit unique identifier plus the leader and burst bits. In field devices having a long form address, command 0 can be sent to their short form address after first power-up, and the device will respond with its long form address, so that the long form address can be used by the leader in the future. The long form address can also be acquired through use of command 11.

The components of the short form and long form address frames are illustrated in Figure 3.22.

Both the short and long form address frames start with a leader indicator bit (1 for primary leader, 0 for secondary leader), and then a burst mode bit (1 = burst mode, 0 = poll/response mode). A leader device sets the leader bit as appropriate, and follower devices must include this bit unchanged in their response. In a message from

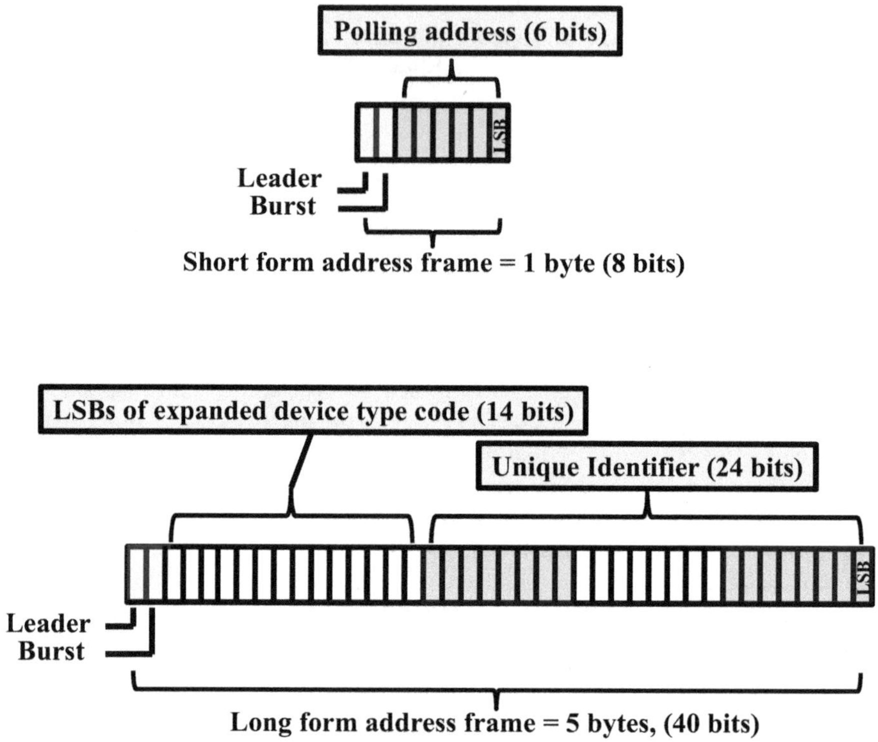

FIGURE 3.22 Components of the HART short form and long form address frames.

a field device, the burst mode bit may be set as 0 or 1, depending on its mode. In a message from a leader device, the burst mode bit is always 0.

In the long form address, the manufacturer of the device must ensure that the unique identifier number is unique for a device of that particular device type. The device type is assigned by HCF for each new field device. The complete form of the manufacturer's ID is 16 bits, and is reported in an extension of the command 0 response.

HART protocol also implements a broadcast address, in addition to addresses based on the unique identifier of a device. A broadcast address is a long form address having 38 zeros and is accepted by all field devices as being addressed to them. There is usually no response required to a broadcast message.

3.6.7 Communication Speed

The speed for wired HART FSK communication is normally 1.2 kbps (but can be up to 9.6 kbps if using PSK, after HART revision 6). The several fields comprising a message were shown in Figure 3.21. A typical message from a leader and response from a follower, each with 4 bytes of data, might include the following:

Leader: 5 byte preamble, 1 byte start, 5 byte address, 0 bytes expansion, 1 byte command, 1 byte count, 0 byte status, 4 bytes of data, and 1 byte checksum = *18 bytes.*

Follower: 5 byte preamble, 1 byte start, 5 byte address, 0 bytes expansion, 1 byte command, 1 byte count, 2 byte status, 4 bytes of data, and 1 byte checksum = *20 bytes.*

So, the example command plus the response totals 38 bytes. Since each byte comprises 11 bits (as explained earlier and shown in Figure 3.20: 1 start, 8 data, 1 parity, 1 stop = 11 bits), then transmitting 38 bytes means transmitting 418 bits. At a transmission rate of 1.2 kbs, that equals 348 ms per complete communication of command/response. There is always some small amount of time between messages, so it will round out to about 380 ms per complete command/response message set. Table 3.4 lists some typical amounts of time to complete a set of command/response messages to all followers on a network.

The update times of Table 3.4 pertain to digital messaging using FSK signals on the loop. But in a case where there is only one device on the loop, and having its polling address set to zero, the loop current follows the process variable. In this case, the response time is faster, usually around 100 ms or less. This is fast enough to accommodate most process control applications.

When networked, according to the longer times listed in the table, it may be difficult to operate an industrial control system with HART devices when there are several devices in the network. But a single HART device operating digitally is also fast enough for many process control applications (0.38 seconds communication response time).

3.6.8 Installing Leader and Field Devices in a Wired System

The primary HART leader is indicated by a 1 in the leader bit in the long or short form address, and a secondary leader (usually a handheld device) is indicated by a 0 in the leader bit. The primary leader is always installed. In a wired system, a secondary leader can be applied across the loop whenever and wherever required without interrupting the network, by clipping its two leads across the loop wires.

TABLE 3.4
Time to Receive Updates from All Devices on a Network

Number of Followers	Time to Get Updates from All Followers (Seconds)
1	0.38
4	1.52
8	3.04
15	5.7

3.6.8.1 Wiring

The loop wiring may be a twisted pair of unshielded copper wires (of at least 32 AWG or 0.2 mm diameter) when the total network wiring length is less than 100 meters. A shielded twisted pair of at least 24 AWG or 0.51 mm diameter is suitable for wiring lengths of up to 1,500 meters. For wiring lengths of up to the limit of 3,000 meters, a shielded cable with a twisted pair of at least 20 AWG or 0.51 mm diameter should be used. Cable capacitance must also be below the limit for the wire length, as shown in the table on the HART Communication Foundation site at: http://en.hartcomm.org/hcp/tech/using/usinghart_wirelength.html.

3.6.8.2 Primary or Secondary Leader

A leader device may be one of the many HART devices available on the market for that function, or could be a PC, laptop, or smartphone having a HART modem and appropriate software. An example is the wired USB to HART modem for a PC (could be a laptop, rack-mount PC, notebook, pad, or other type) shown at: www.procomsol.com/online_store/hm_usb_iso.

Before a field device (follower) is to be installed in the field, it is best to first check it by using a handheld in the control room or lab. Verify that all of the manufacturer's information is correct for the application, and communication signals and loop current are all working properly. If reconfiguration or calibration is needed, it may be quicker to accomplish same while still in the control room or lab. As each device is installed in the field, proper communication with the leader device should be verified.

3.6.8.3 Calibration

HART calibrators provide a simulated process variable, and allow one to check for proper response from the device under test. After installation, the field device may be routinely checked for calibration accuracy as part of a preventive maintenance program.

3.6.8.4 Troubleshooting

When troubleshooting a newly completed HART installation that is not performing properly, there are several simple things to check first.

1. Make sure that all addresses are properly assigned.
2. If using a PC, check that the proper com port is selected (1, 2, . . . 8).
3. Make sure the power supply is capable of supplying the power needed, by all of the devices added up, at the same time.
4. Ensure that cable resistance and capacitance are appropriate for the cable length.

3.6.9 WirelessHART

As noted earlier, support for WirelessHART was first implemented with HART revision 7.0. This mesh network communication protocol is often used in process automation applications, is low power, and operates on the Industrial, Scientific, and Medical (ISM) radio band at 2.4 GHz. Transmission is possible over a distance of up to 200 meters between devices, and multiple communication paths may be present. If multiple paths are present, allowance is possible for correction of communication errors (that is, if one wireless path produces an error, another path may be selected that is not producing an error at that time).

Instead of passing a token as with the FSK wired version, WirelessHart uses a time division multiple access (TDMA) system.

Each WirelessHART network includes Field Device(s), Gateway(s), and a Network Manager.

A WirelessHART Field Device (such as a temperature, flow, humidity, or position sensor) can have built-in WirelessHART functionality, or can be a standard HART device with an adapter added for WirelessHART.

A WirelessHART Gateway communicates with WirelessHART Field Devices within the wireless network mesh to receive and store data, automatically managing the network to ensure that the field devices have a reliable path to send data. The gateway is often DIN-rail mounted, and has the capacity to interface with multiple WirelessHART Field Devices, such as with 25 or 100 devices. The gateway can also be used to adapt to legacy host systems such as Modbus RTU.

The Network Manager is a main component of a WirelessHART system, being mainly responsible for the performance of the system. An example of a WirelessHART network is shown in Figure 3.23.

A WirelessHART network includes field devices, and may include a handheld device. The field devices may have WirelessHART built in, or may use an adapter.

For more detailed information on WirelessHart, see www.fieldcommgroup.org/technologies/hart.

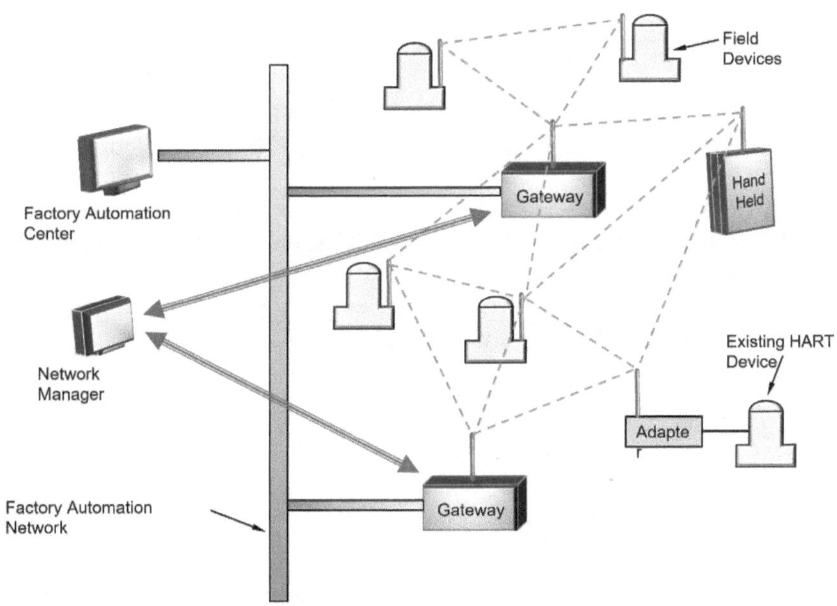

FIGURE 3.23 WirelessHART Network with Field Devices, Gateways, and Network Manager.

3.7 INDUSTRIAL ETHERNET

Ethernet is ubiquitous for various LAN (Local Area Network) and Internet applications, and includes routers, cables, network cards, and so on. Ethernet speeds include 10, 100, and 1,000 Mbit/s, and can also utilize fiber-optic cables.

3.7.1 EtherNet/IP

EtherNet/IP is an industrial implementation (Ethernet/Industrial Protocol) of the Ethernet protocol. It has become popular in both factory automation and process control, and is managed by ODVA (ODVA was referenced in the CANbus and DeviceNet sections in this chapter). The Industrial Ethernet family also includes other protocols, such as EtherCat and Modbus, but EtherNet/IP is presented here.

EtherNet/IP development began in the 1990s within ControlNet International, Ltd., then joined with ODVA in 2000, and was released in 2000. In 2009, it was under the control of ODVA.

Using the Internet Protocol suite (TCP/IP), as well as IEEE 802.3 (Ethernet), a complete specification of Ethernet/IP can be downloaded from the ODVA website (www.odva.org, but may require membership). Since Ethernet itself has been in widespread use for personal computers and the Internet, it is very well known, and will not be presented here in detail.

3.7.2 OSI Model

CIP is the Common Industrial Protocol for industrial automation applications, and is supported by ODVA. EtherNet/IP is the use of CIP with Ethernet. When implementing EtherNet/IP, TCP/IP and IEEE 802.3 define the transport, network, data link, and physical layers of the OSI model (refer to Figure 3.8). EtherNet/IP defines the session, presentation, and application layers, as shown in Figure 3.24.

So, while Ethernet addresses physical networking, EtherNet/IP defines industrial communication.

CIP provides EtherNet/IP with Device Profiles to assure that all product applications will be implemented in a uniform way. CIP supports a set of factory automation applications, such as for motion control, safety, synchronization among devices, configuration, and other information. EtherNet/IP provides improvement to real-time communication so that it may better support factory automation and control applications.

3.7.3 Connections

Position sensors having an EtherNet/IP interface will typically have three electrical connectors.

Figure 3.25 shows an example of the connections that may be found for the pins of the connectors. Ports 1 and 2 have the same connections, so that one cable can be connected to Port 1 from one device on the bus, and then a second cable can go out from Port 2 to another device on the bus. Position sensors are often powered from

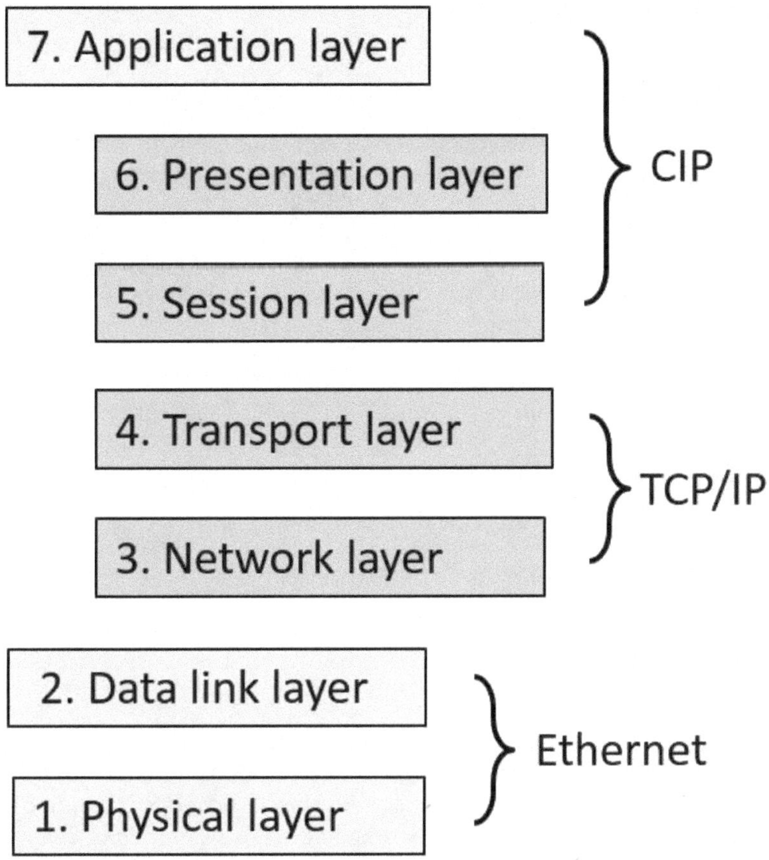

FIGURE 3.24 OSI model for EtherNet/IP.

a standard 24 VDC power source. A given device may be able to work with a range of voltages, such as 10 to 30 VDC. Factory automation position sensors will usually utilize a rugged, sealed type of connector, such as an RJ-45 type.

3.8 MODBUS

3.8.1 Introduction

Modbus is one of the earliest serial communication protocols for industrial use. It is openly published, royalty free, and is still commonly used in process control and other factory applications. The Modbus protocol was developed by Modicon, and published in 1979. Modicon is now Schneider Electric. The Modbus protocol has been managed by a trade association of users and suppliers called the Modbus Organization (https://modbus.org) since transfer of the protocol to it in 2004 from Schneider Electric.

Port 1:

Pin#	Function
1	Tx +
2	Rx +
3	Tx -
4	Rx –

Power:

Pin#	Function
1	V+ (10 to 30VDC)
2	no connection
3	0V (DC Ground)
4	no connection

Port 2:

Pin#	Function
1	Tx +
2	Rx +
3	Tx -
4	Rx –

FIGURE 3.25 EtherNet/IP connections.

Modbus was designed to operate Programmable Logic Controllers (PLCs). A PLC is a ruggedized industrial computer that was developed to replace relay logic systems. Before the PLC, sets of hard-wired relays and switches were used to control industrial equipment, motors, valves, lights, and so on. Relay functions are represented as coils and the associated contact(s). There are also switches (such as hand-operated switches, position-sensing switches, and so on), timers, AND/OR logic functions, accumulators, and many other functional blocks. A simple relay logic circuit is shown in Figure 3.26.

In the figure, +24 VDC power is applied along the left rail. The common rail is along the right. Toggle switch is a hand-operated switch (shown as normally open) that will be used to cause activation of Lamp #1. The coil of a relay is named R_1. The contact labeled R_1 is the (normally open) contact that is part of relay R_1. When power is applied across the coil of relay R_1 (by closing the toggle switch), contact R_1 becomes closed, thus activating the lamp. Of course, in this simple circuit, the toggle switch could directly be used to activate the lamp, but many circuits that are much more complex can be implemented with relay logic. Such hard-wired relay logic, although no longer commonly used, usually comprised a panel with racks of relays and other devices having point-to-point wiring among the devices. "Re-programming" required revising the hard-wiring.

In a PLC, relays and other functions are implemented in software rather than hard-wiring, and a more convenient physical arrangement allows adding various input and output modules as needed; see Figure 3.27.

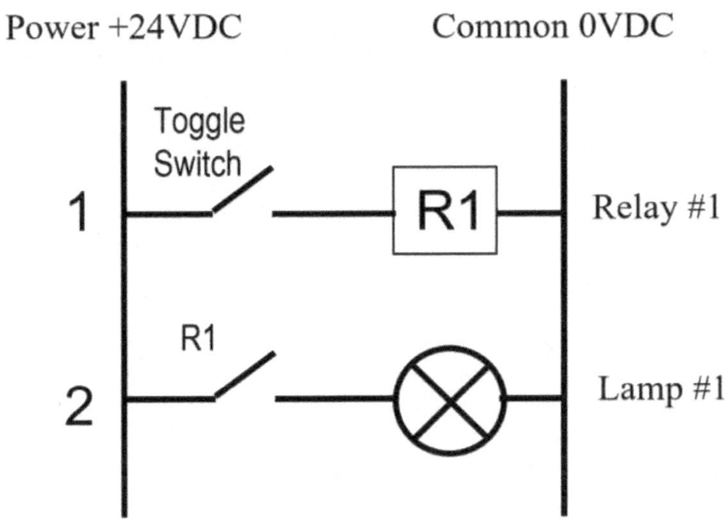

FIGURE 3.26 Relay logic.

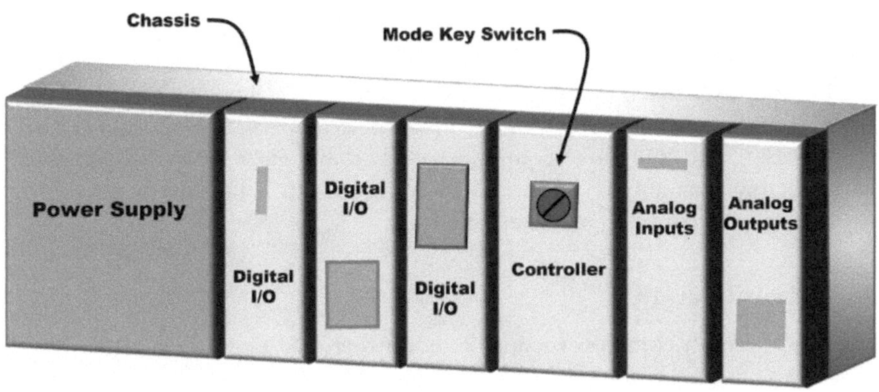

FIGURE 3.27 PLC with modular construction.

The controller module is shown with a mode key switch. A mode key switch is often implemented having three positions: Run, Remote, and Program. The PLC of Figure 3.26 has a modular construction, supporting customization by adding the appropriate modules as needed for the particular installation. But all-in-one configurations are also popular, comprising all of the functions typically required for industrial applications, and making installation easier.

Modbus communicates serial data by any of several means, including a serial port (RS-232, RS-485, or RS-422), Ethernet, or the Internet Protocol Suite (such as TCP/IP).

3.8.2 Modbus Versions

Modbus RTU is a very common implementation of Modbus, whereby a controller can address up to 247 devices. One cable can be connected with devices of varying types, such as temperature sensors, humidity, pressure or flow sensors, and other input or output devices. A Remote Terminal Unit (RTU) provides the functions of supervisory control and data acquisition (SCADA). Ladder logic is often used as a replacement for discrete relay logic. In ladder logic, a discrete (on/off) output is called a coil (as in a relay coil), and a single bit input is called a contact (as in a switch contact).

There are many versions of Modbus, some of which are the following:

- Modbus RTU—is the most common version of Modbus, with little data overhead, using compact binary data in serial communication. It follows a command/data format, and includes CRC. Data framing is accomplished by idle periods between data sets.
- Modbus ASCII—uses serial communication, but incorporates ASCII characters instead of straight binary data. A longitudinal redundancy check is incorporated. Data framing is accomplished by a leading colon (:) and a set of data are followed by a carriage return and line feed (CR/LF).
- Modbus TCP/IP or Modbus TCP—communication is via a TCP/IP network. Checksum is only included in a lower layer.
- Modbus over TCP/IP, or Modbus over TCP (or Modbus RTU/IP)—similar to Modbus TCP, but includes a checksum within the data set.
- Modbus Plus, or Modbus+—an extended version that is related to Modbus, but uses token passing, and is proprietary to Schneider Electric. Enables communication over a twisted pair at up to 1 MB/s. Uses transformer isolation, and therefore requires edge-triggering rather than voltage level detection. Can only be implemented by partners of Schneider Electric, as it requires the use of a proprietary chip set.

3.8.3 Communication

Since it is the most common version, some further information regarding Modbus RTU is presented.

Each device on a cable is assigned a unique address, 1 through 247. Communication of data follows this sequence:

1. The leader (controller) initiates a request, along with a device address.
2. The follower device recognizes its own address, performs the requested action, and initiates its response.
3. The leader receives the response.

Data are transmitted serially, at baud rates from 1200 to 115,200 bits per second.

3.8.4 Object Types

Object types such as coil, discrete input, input register, and holding register have corresponding capabilities and number of bits.

3.8.5 Commands

Modbus commands can tell a device to change the value in one of its coil or holding registers, read data from a discrete input or from a coil, or command a device to return values held in its coil or holding registers. A sample of some Modbus RTU commands is shown in Table 3.6.

3.8.6 Frames

An Application Data Unit (ADU) contains a device address plus a Protocol Data Unit (PDU), plus a CRC error check. The PDU contains a function code plus the data. An idle time during which no data are sent, for a duration equivalent to at least 28 bits, indicates that the address data will be coming next, followed by the remaining information as shown in Table 3.7.

TABLE 3.5
Modbus RTU Object Types

Object Type	Access Type	Size (# of bits)
Coil	Read/write	1
Discrete input	Read only	1
Input register	Read only	16
Holding register	Read/write	16

TABLE 3.6
Some commands of Modbus RTU.

Command	Number of Bits	Description
01	1	Read coils
02	1	Read contacts
03	16	Read holding registers
04	16	Read input registers
05	1	Write a single coil
06	16	Write a single register

TABLE 3.7
Modbus RTU Message Frame

	Number of Bits	Description
Start	≥ 28	No data for at least 28 bit times (mark)
Address	8	Address of the device (1 through 247)
Function	8	Function code as in Table 3.6
Data	$n \times 8$	Data length depends on type of message
CRC	16	Cyclic Redundancy Check
End	(≥ 28)	Idle time between frames ≥ 28 bit times

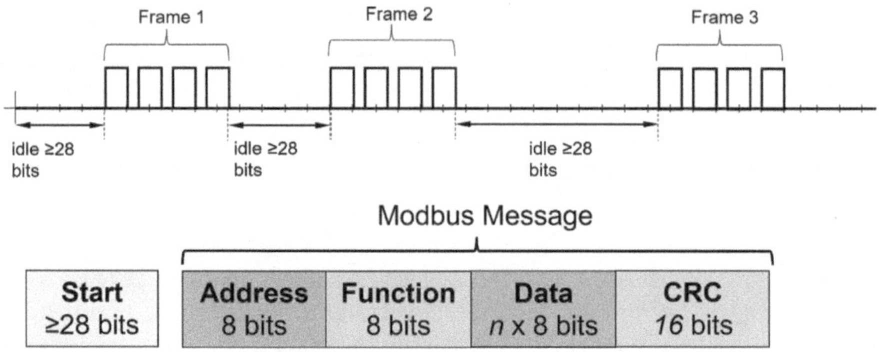

FIGURE 3.28 Modbus RTU message frame timing.

A Modbus message is sent in a frame that begins after a minimum idle time equivalent to that of 28 bits. Figure 3.28 shows three message frames, separated by idle times. The figure also shows a block diagram of the contents of one message frame.

3.9 QUESTIONS FOR REVIEW

1. **If a CANbus bit does not appear on the bus, that device must have been sending a:**

   a. Recessive bit
   b. Burst
   c. Node
   d. Null frame
   e. Query

2. **A HART long form address frame contains:**

   a. Only numbers
   b. Calibration data
   c. 5 bytes
   d. A physical address
   e. A time stamp

3. **The output of an operational amplifier may be protected against ESD by using a:**

   a. Higher voltage
   b. Reverse polarity
   c. Zener diode
   d. Voltage follower
   e. Higher gain

4. A position sensor that receives power and sends a signal on one pair of wires is:

a. Incremental
b. Fast
c. Modular
d. Loop-powered
e. Paired

5. SSI data can be sent using:

a. Voltage or current
b. Decimal numbers
c. ASCII
d. MOVs
e. Binary or Gray code

6. This is commonly used to convert a 4 to 20 mA signal to 1 to 5 volts:

a. Op amp
b. VFC
c. Application layer
d. 250 ohms resistor
e. RS-485

7. In the HART communication protocol, the parity bit is always:

a. Odd
b. Logic zero
c. The reciprocal
d. A data bit
e. An even number

8. Specifications of features and other properties of PROFIBUS sensors are called:

a. PROFIspecs
b. Profiles
c. Manchesters
d. NRZ
e. Descriptors

4 Resistive/Potentiometric Sensing

4.1 RESISTIVE POSITION SENSORS

Resistive linear and angular position sensors are very popular, relatively inexpensive, and are also the most easily understood type of position sensor. They are normally three-wire devices, and are also called *potentiometric position sensors*, *linear potentiometers*, *potentiometers*, or *pots*. So here, the terms resistive position sensor and potentiometric position sensor are considered to be interchangeable. A linear or angular potentiometer can be called a position *transducer*, since it controls an electrical output directly from the physical variable input. But, since the sensing element also provides a usable electrical signal with no additional signal conditioning, it is also a position *sensor*.

The basic concept is the same as that used in non-digital versions of the volume and tone controls on many radios and other electronic devices having an audio output. A voltage is applied across a resistive element, and an electrically conductive wiper slides along the resistive element, the wiper making electrical contact with the resistive element. This allows a voltage potential to be read from the wiper, with respect to one end of the resistive element. As the wiper voltage varies, it indicates the position of the wiper along the resistive element.

In an audio circuit, the voltage across the resistive element of a potentiometer is usually AC (alternating current), having amplitude and frequency variations indicative of the audio signal. But in a resistive position sensor, the voltage applied across the resistive element is normally DC. The wiper voltage is approximately determined by the linear or angular distance of the wiper from one end of the resistive element, divided by the total length or arc of the element, and multiplied by the total voltage across the ends of the element. In electronics engineering terms, this kind of function is called a voltage divider, and is shown in Figure 4.1.

Figure 4.1(a) is a two resistor voltage divider circuit. The output voltage at the connection between the upper and lower resistors is defined by the formula:

$$V_{out} = V_T \, R_a / \left(R_a + R_b\right) \tag{4.1}$$

Figure 4.1(b) is the symbol for a potentiometer, where the arrow that happens to be pointing to the middle of the resistor is called the wiper. The wiper can be moved up or down, to change the voltage divider setting. The same voltage divider formula also applies for the potentiometer output voltage, with the resistance below the wiper connection being called R_a and that above the wiper called R_b. With a potentiometer, the value (Ra + Rb) remains constant as the wiper is moved.

DOI: 10.1201/9781003368991-4

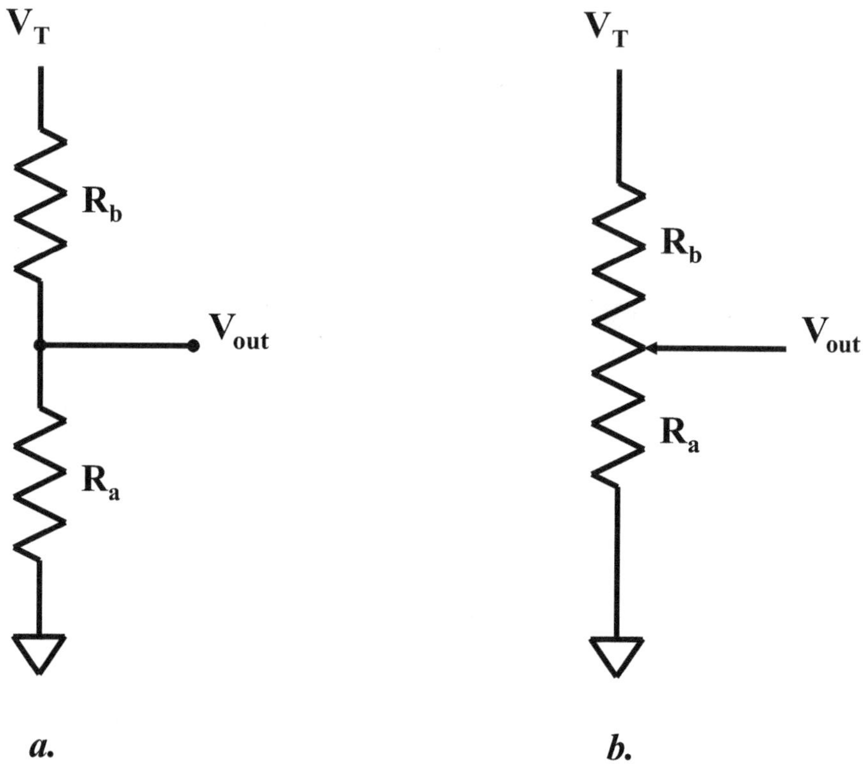

FIGURE 4.1 A two resistor voltage divider circuit, (a), and a potentiometer circuit, (b).

4.2 RESISTANCE

Electric current is often compared to the flow of water in a pipe. A volume of water flowing in a pipe per unit time, such as in liters of water per second, or gallons of water per second, is analogous to electrical current flowing in a wire, such as in coulombs per second, or amperes. 1 ampere = 1 coulomb per second, where 1 coulomb is equivalent to the charge of 6.24×10^{18} electrons.

A drop in water pressure across a length of pipe (e.g. in Pascals or pounds per square inch) is analogous to a voltage drop across a resistance. A higher pressure difference across a length of pipe can force a higher water flow rate through the pipe if other parameters remain unchanged. For example, if the water pressure at the supply end of a pipe is 50 PSIG (pounds per square inch, gauge pressure) and the other end of the pipe is open to the atmosphere, then the pressure difference across that length of pipe is 50 PSID (pounds per square inch, differential). The pipe diameter, roughness, and straightness control how easily the water can flow through a length of the pipe. A longer pipe, a smaller pipe diameter, greater degree of inner wall roughness, and curves or obstructions in the pipe increase a resistance to water flow through the pipe that is analogous to electrical resistance to current flow through a wire. A pipe with a higher resistance to water flow requires a higher pressure difference (across the ends of the section of pipe in question) to achieve a given flow rate. Similarly, a higher voltage potential difference across a given resistance of a length of wire is required to obtain a higher current flow through the wire. In electric circuits, the voltage is equal to the product of current and resistance:

$$E = I\,R \tag{4.2}$$

For example, 10.0 V = 0.10 A × 100 Ω.

This is a re-arrangement of Ohm's Law, traditionally written as $I = E/R$, where E is the voltage in volts (V), I is the current in amperes (A), and R is the resistance in ohms (Ω).

Hint: in electrical circuits, *voltage* is always measured *across*, and *current* is always measured *through*. That is, one would measure the voltage (or voltage drop) across a resistor, or one would measure the current through a resistor. One would never try to measure the voltage through a resistor or any component. Likewise, one would never try to measure the current across a resistor or any component (because neither of these would make any sense).

If several resistances are connected together in *series*, the total resistance, R_T, is equal to the sum of the individual resistances:

$$R_T = R_1 + R_2 + R_3 \tag{4.3}$$

When several resistances are connected in *parallel*, the total resistance is equal to the reciprocal of the sum of the reciprocals of the individual resistances:

$$R_T = \frac{1}{1/R_1 + 1/R_2 + 1/R_3} \tag{4.4}$$

This may seem difficult to calculate, but is easy with an algebraic calculator that has the reciprocal function (that is: *1/x*). If three parallel resistors are 10, 100, and 1,000 ohms, for example, take the following steps on a calculator to find their total parallel resistance (the symbol ® is used here to indicate pressing the reciprocal button (*1/x*). Sometimes a second function key must be pressed to select the ® key function):

1. enter 10, ®, +
2. enter 100, ®, +
3. enter 1,000, ®, +
4. =, ®

And then the answer is shown. This adds the reciprocals together, then takes the reciprocal of that. Very easy. (The same procedure works for adding series capacitances.)

It is common for some people to state that an electric current follows the path of least resistance. This is incorrect. When this statement is heard, the speaker should be educated so the statement will not be repeated again (one should try to refrain from slapping the speaker, as they are only parroting something heard from another who was never corrected). When more than one path is available, an electric current will divide among all of the available paths in amounts inversely proportional to the resistance of each path. So, with a given voltage applied across a set of several parallel resistance paths, a higher amount of current will flow in a lower resistance path, while a lower amount of current will flow in a higher resistance path.

4.3 HISTORY OF RESISTORS AND RESISTIVE POSITION SENSORS

Ohm's Law is named for George Simon Ohm, a German physicist, who first published this relationship in 1827. An electrical parameter called resistance was found to be the controlling factor in determining the amount of current that would flow in a circuit having a given voltage potential difference. Later, as an electrical component, early resistors were formed of coils of wire, followed by versions using a carbon composition core having wire leads attached and coated with an insulating layer ("carbon comp" resistors). Subsequent refinements in precision and manufacturing process included resistive elements in the form of carbon film, metal film, and cermet types. Many fixed resistors are used in the circuit design of sensors. The most common fixed resistors today are surface mounted "chip" resistors of metal film construction. Chip resistors are common in sizes such as 0204, 0402, 0603, 0805, and 1206. These are based upon Imperial dimensions, with length and width in inches. For example, an 0603 chip resistor has dimensions of 0.060 inches in length, 0.030 inches in width (and 0.018 inches in height). The same parts also have a metric equivalent name, such that an 0603 Imperial size resistor has the same dimensions as a metric size 1608. (See www.resistorguide.com/resistor-sizes-and-packages/#) The package size generally determines the maximum power dissipation of a fixed resistor, with 0402 having a power rating of 0.062 watts and 0603 at 0.100 watts, for example. Resistors used in the design of sensor circuits are usually thick film, but some may be thin film. This information on fixed resistors is included to help in the general design of sensor circuits.

A *thick film* resistor is generally manufactured by screening a ruthenium oxide paste onto a ceramic substrate and then firing it to a glass-like condition. This results in thick film resistors being nearly insensitive to moisture. General-purpose thick film resistors used in sensor circuits typically have a resistance tolerance at 25°C of ±1%, and a temperature sensitivity of ±100 ppm/°C. This temperature sensitivity means that for every 1°C of temperature change in the same direction, the resistance of a given resistor could change by 0.01%, or by 1% for a 100°C change.

A *thin film* resistor is usually manufactured by sputtering a nickel-chromium alloy onto a ceramic substrate. The film thickness is only about 1,000 angstroms thick, and so the film is less than one-thousandth the thickness of that of a thick film resistor. The film is then laser-trimmed to provide the desired resistance. This allows thin film resistors to achieve a much lower resistance tolerance level. It is common for some thin film resistors used in critical circuits in sensors to have a tolerance at 25°C of only ±0.1%, or even ±0.05%. And these accurate resistors may have a temperature coefficient as low as ±25, or even ±10 ppm/°C. Thin film resistors may be somewhat more sensitive to moisture than thick film resistors. The author typically specifies

100 ppm, 1% resistors for general use, and 0.1%, 25 ppm resistors for the more critical analog parts of a given circuit.

Wirewound fixed resistors are sometimes used when a higher power rating is required than can be practical with a film resistor, typically those that must dissipate more than 0.5 watts. But such resistors are rarely used in modern sensor designs because sensor design engineers strive to develop sensors that operate at a low-power level. (See www.resistorguide.com for more information on fixed resistors.)

Wirewound variable resistors and resistive potentiometers were developed as electrical components, and rotary versions were used in early electrical and electronic circuits. They were also constructed with a variety of resistive element types, as described here in the resistive element Section 4.5. It was a small step to start using a rotary potentiometer as a rotary position sensor. Linear and rotary potentiometers were also adapted for use as sensors to sense parameters other than position. Figure 4.2 shows a pressure sensor, designed by the author, which uses a wirewound resistive element.

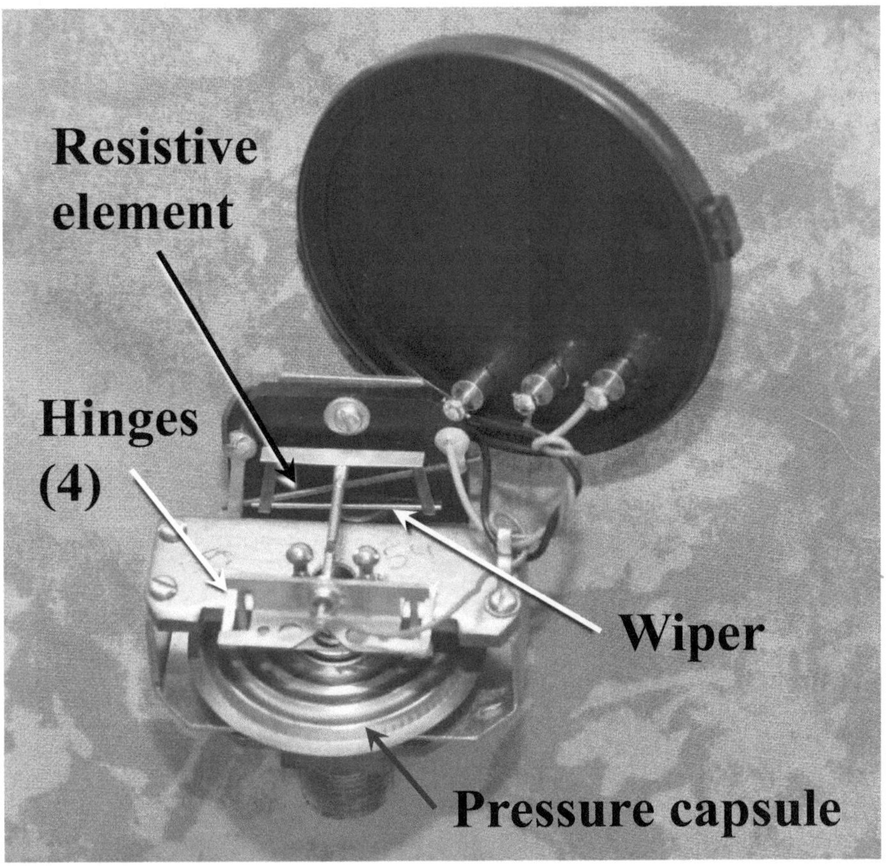

FIGURE 4.2 A pressure sensor that utilizes a wirewound resistive element and wiper to form a potentiometer. The black disc is the housing cover, incorporating the three terminals for electrical connection. (The main housing is not shown.)

In Figure 4.2, a pressure capsule acts as a first transducer, converting an applied pressure input into a linear motion output, such that increasing pressure causes the capsule to expand and thus move the wiper of a potentiometer (moving up, in the figure, with increasing pressure). The potentiometer acts as a second transducer, converting a linear motion input into a variable voltage output (voltage output measured at the wiper, when a voltage differential is connected across the resistance element). Together, the two transducers form a sensor.

The wiper moves along a wirewound resistive element. Three terminals provide electrical connection to the wiper and to each of the two ends of the resistive element. If a voltage of 5 VDC is applied across the resistive element, for example, an output voltage of approximately 0 to 5 V can be measured at the wiper, corresponding to an input pressure of zero to full scale.

Linear position measurements can be made by using a rotary potentiometer with a toothed gear (pinion gear) mounted onto the potentiometer shaft. The pinion gear would engage the gear teeth on a movable rack, so the position of the movable rack would be indicated by the potentiometer rotation. With refinement of the potentiometer for use as a linear position sensor, a linear resistive element was utilized with a contact wiper tracking along its length. Further improvements were added to increase lifetime and performance (some of these are described in Section 4.4).

4.4 POSITION SENSOR DESIGN

Figure 4.3 shows the basic construction of a resistive linear position sensor having a rod for actuation. It comprises a resistive element, a parallel conductor, and wiper fingers as major components. The parallel conductor is positioned parallel to the resistive element. Also needed are a mechanical arrangement to keep the major components slidably positioned with respect to one another, a rod, and a housing, with bearing, seal, and wipe incorporated as needed to facilitate the performance and longevity of the major components.

In addition to the special considerations pertinent to designing each major component, as explained in the next few paragraphs, it is also important to execute the general design of a sensor in such a way to eliminate any major sources of wear and inaccuracy. The main culprit in causing these problems in a potentiometric position

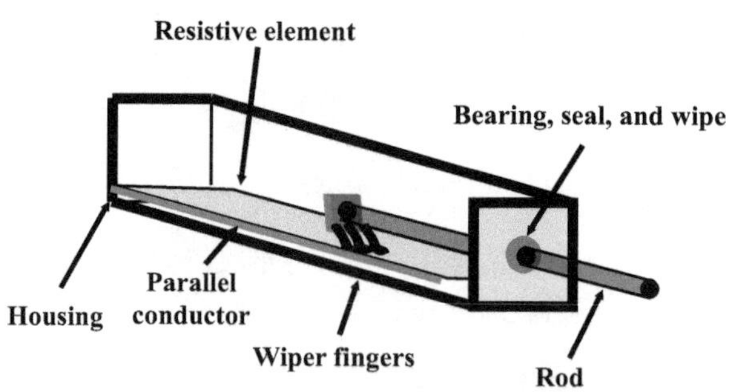

FIGURE 4.3 Rod-type resistive linear position sensor construction.

sensor is the admittance of dirt to the sliding parts and to the resistive element. Dirt is excluded in rod-actuated designs such as that depicted in Figure 4.3 by adding one or more seals and wipes that clean the rod (the rod also can be called the shaft) as it is retracted into the bearing, and seal the rod-to-bearing sliding joint. A wipe is a tough fiber or elastomer annular element that rubs on the rod to remove debris that may cling to the rod as the rod is retracted into the housing. A seal is a tight-fitting elastomer ring positioned around the rod between the wipe and a bearing. The seal removes any smaller contaminants that may have gotten past the wipe, and prevents the entry of liquids into the housing. The bearing can be a ball or roller bearing, but is usually a relatively tough and somewhat hard plastic bearing material that resists wear, and is compatible with the material and surface finish of the rod. The bearing is shown simply in Figure 4.3, but must be designed to accept side loads on the rod, and to ensure that the wiper tracks over the resistive element throughout the stroke.

A miniature version of a resistive linear position sensor having rod-eye ends is shown in Figure 4.4.

The rod is shown extended about half-way in the figure. This particular sensor measures 88 mm from eye to eye when the rod is fully retracted, and 123 mm when fully extended, for a measurement range of about 35 mm. Some models are available with or without a spring-return feature.

Some resistive linear position sensors are of the rodless type, as shown in Figure 4.5, a small *car* (usually plastic) rides within a channel along the top of the sensor housing. In this type, it is a little more difficult to keep the dirt out. A flexible strip is positioned lengthwise along the top of the housing, forming a flexible seal, such that the top of the housing would otherwise be open if the flexible strip was not in place. The housing is often made from an aluminum extrusion. A wiper assembly is attached to the underside of a car that moves smoothly within its guiding channel. As the car moves, the flexible strip is picked up ahead of the car and replaced against the housing behind the car. Unfortunately, there always remains a small gap in the sealing area near the front and back of the car through which it is still possible for small particles of dirt to enter.

As the wiper slides along and makes electrical contact with the resistive element, it also contacts a parallel conductor track (or plated wire) that runs parallel to the

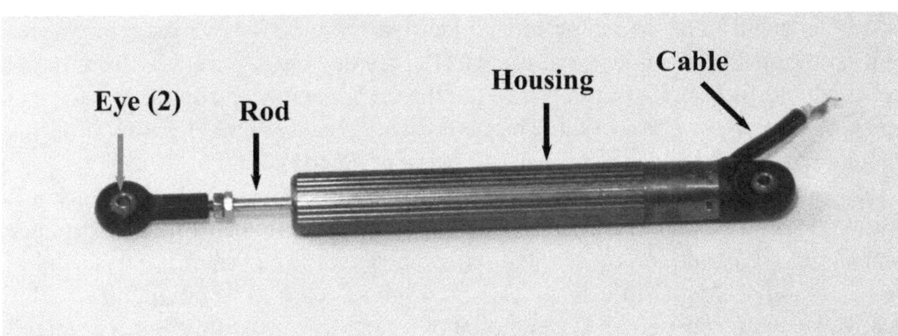

FIGURE 4.4 A miniature version of a resistive linear position sensor.

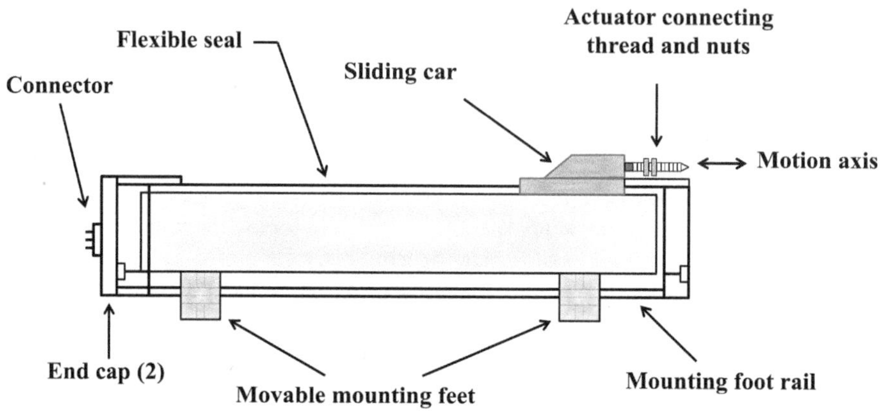

FIGURE 4.5 Rodless-type resistive linear position sensor.

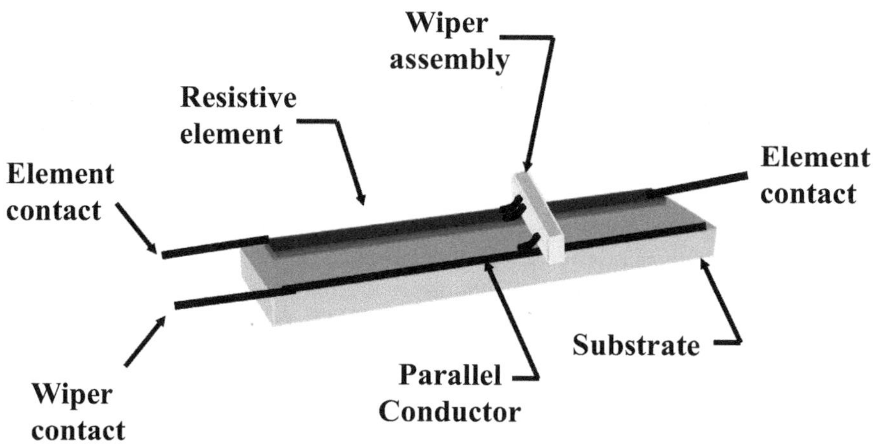

FIGURE 4.6 Connections to the wiper, and to the left and right ends of the resistive element.

resistive element. The wiper provides electrical connection between the selected point on the resistive element and the parallel conductor. A wire lead from the parallel conductor is brought out to one of the three connector pins on the outside of the housing as the wiper connection. The other two connector pins (for a total of three) are connected to the ends of the resistive element. See Figure 4.6.

Besides resolution, nonlinearity, and hysteresis (see Chapter 2), there are a few additional electrical characteristics that must be controlled in order to produce a reliable and well-functioning resistive position sensor. The wiper is fabricated of a suitable metal alloy and in a shape that are both selected to be compatible with the resistive element, causing the least amount of wear while providing low contact resistance. This may be a hardened spring-copper alloy, and plated with another alloy in the contact area. When the wiper is positioned all of the way at one end of its normal travel, there may be a part of the resistive element remaining between that point and the connection point on the resistive element for the connector pin. This non-usable portion of the resistive element, plus the resistance of the contact, form an *end resistance*. There is an end resistance at each end of the resistive element. This means that if the resistance is measured between the wiper pin and one other pin while the wiper is at that end of its travel, the resistance will not be zero ohms as one might expect. But there will be a minimum resistance that is usually less than 1% of the nominal resistance of the resistive element. This end resistance slightly limits the range of output voltage from the wiper, so that the lowest output voltage is not quite as low as zero volts, and the full-scale output is not quite as high as the supply voltage. When designing a resistive linear position sensor, the end resistance should be minimized but, if the wiper is allowed to go too far, then the opposite problem occurs: excess end travel. In this case, there may be a conductive area adjacent to the resistive element past where the lead connects. If the wiper is allowed to move past the point where the connector lead contacts the resistive element, there will be a portion of wiper movement for which there is no change in output voltage at the wiper. This is described as the mechanical stroke being longer than the electrical stroke. The difference is called the overtravel. It can be tolerable in some applications, where the mechanical zero can be adjusted. But, since there is essentially no change in output in the overtravel areas, it could be a source of error when there is no mechanical adjustment available.

Contact resistance is the resistance between the wiper contact and the resistive element and parallel conductor. If loading of the potentiometer output (the wiper) produces a measurable current in the wiper circuit, then the voltage drop across the contact resistance will cause an offset in the output voltage. This offset can be adjusted out in some systems, but it is desirable to design the load circuit so that the offset is negligible for the application. Minimal loading is also important to retain the linear performance of the sensor. The load resistance should therefore be in a range of more than 50 times the resistance of the sensor resistive element.

4.5 THE RESISTIVE ELEMENT

In early resistive position sensors, the resistive element was either wirewound or constructed of a substrate onto which a carbon surface was formed. A wirewound element has the limitation of providing a relatively coarse resolution, because the minimum increment of resistance selection that is possible is one turn of the resistance wire, as the wiper moves from one turn to the next adjacent turn. Some wirewound position sensors suffer as well from a difficulty of maintaining good wiper contact with the resistance wire while the wiper remains stationary for long periods, due to slight oxidation in the contact area. Occasional movement of the wiper is sometimes required in order to clean the resistance wire surface and maintain a good contact with the wiper.

Types of position sensor resistive element construction that are available now include wirewound, carbon film, conductive plastic, metal film, cermet, and hybrid.

Wirewound elements are manufactured by winding a wire around a nonconductive mandrel. The wire usually has an electrically insulating coating, which is removed along one edge after winding. In the completed sensor, the wiper rides against the area of the winding where the insulation has been removed. The mandrel can be made of ceramic, plastic, or of metal that has an insulating coating. Ceramic is the most stable mandrel material, plastic is the cheapest, while metal conducts heat and is useful for higher power dissipation. The temperature coefficient of resistance of a wirewound element is relatively low, at about 50 to 100 ppm/°C.

The disadvantages of a wirewound resistive element are the coarse resolution, noise due to breaking and making contact between adjacent windings, and the tendency to lose contact between the wiper and the resistive element when the wiper is not moved for long periods of time (days). The resolution is coarse because the minimum step from one setting of the wiper to the next is equal to the resistance of one turn of wire. If a wirewound element has 500 turns, then the finest resolution possible is 0.2% (1/500). Noise is higher than with other element types because the wirewound element is inductive and the make-and-break action when the wiper moves from turn to turn causes voltage pulses on the output due to the rapid change in wiper current, d_i/d_t. When the wiper is not moved for long periods of time, an oxide layer may form on the surface of the wire resistance element. This increases contact resistance, and can cause an open circuit between the wiper and resistance element. Moving the wiper will clear the oxide and return the pot to the normal contact resistance.

Carbon film resistive elements are manufactured by mixing a paste of carbon and clay, and then screening the paste onto a nonconductive (usually ceramic) substrate material. This is followed by firing the assembly to harden the carbon/clay mixture into a ceramic composite. Carbon film has the advantage of yielding a finer resolution than with a wirewound element because the wiper can follow a continuous path (not passing from one wire turn to the next, as with a wirewound element). Noise with carbon film is also less than with a wirewound pot, because the carbon film surface is relatively smooth, and the element is non-inductive. One disadvantage is that a carbon film resistance has a temperature coefficient of resistance (about 200 ppm/°C) which is larger than wirewound, metal film, or cermet, but this is not so important when used in a voltage divider circuit unless there is asymmetrical end resistance or external added resistance for trimming that has a different temperature coefficient. Although electrical noise is much lower than with a wirewound element, there is still a substantial amount of noise that is generated as the wiper(s) scratches along the somewhat rough surface. The noise is a rapid change in resistance produced by large differences in contact resistance.

Conductive plastic elements are very smooth, allowing for very low noise generation from moving of the wiper contact. They are made by adhesive mounting a conductive plastic film onto a rigid nonconductive substrate. The conductive plastic film is a composite mixture of carbon or another conductive powder with plastic. A wide range of resistance values is possible, but conductive plastic elements have the largest thermal coefficient of resistance (about 300 to 500 ppm/°C).

Metal film resistive elements are produced by depositing a thin film of metal alloy onto the surface of a ceramic substrate, usually by sputtering. This makes a very durable assembly. The resistance can be adjusted by cutting away some of the metal film. This construction also makes possible the correction of the change in resistance per mm, for a more linear performance of the position sensor (by making partial lateral scratches near the film edges as needed to adjust resistance in certain spots).

The performance is similar to that of a carbon film pot, but is more rugged and has a lower temperature coefficient of resistance (typically about 50 to 100 ppm/°C).

Cermet is a mixture of conductive metal particles with clay. Varying amounts of silver, chromium, and lead oxide, for example, are mixed with the clay to provide a desired range of resistance. The mixture is screened onto a ceramic substrate and fired. This is nearly as rugged as metal film, somewhat adjustable, and has a relatively low-temperature coefficient of resistance (about 50 to 100 ppm/°C).

Since a cermet element is very rugged, with a low-temperature coefficient, and whereas a plastic film has the smoothest surface, it is possible to adhere a conductive plastic film to the surface of a cermet element to get the best properties of each. This is called a hybrid or multilayer element. One disadvantage of this construction is that the conductive plastic surface may not last as long as a cermet surface.

Another type of hybrid resistance element construction comprises the application of a conductive plastic layer over a wirewound element. This provides the high resolution and low wiper movement noise of a conductive plastic element with the low thermal coefficient and low nonlinearity of a wirewound element.

Resistivity of a material is the electrical resistance of the material per unit area or volume. Area resistance is used for measuring thin films, and is called surface resistivity. Since it is common to measure the resistance of thin films, or surface resistivity, a simplified unit is used. It is indicated in ohms per square (Ω/sq.). When measuring ohms per square, it doesn't matter if the linear unit is inches or mm, since both the length and width are affected (thus making the linear unit scale irrelevant). For example, if evaluating the resistance of a rectangular film of width w and length l, the number of "squares" is equal to The total resistance from end to end is then:

$$R = \rho \frac{l}{w} \tag{4.5}$$

where R is the total resistance and ρ is the resistivity in Ω/sq.

Volume resistivity is used for three-dimensional materials, such as wire, and is called bulk resistivity. Since it is a common form for which resistance is measured, there is also a simplified unit of resistivity for wire. Having a uniform cross-sectional area along its length, the resistivity of a given wire material and diameter can be indicated in the units of ohm meters (Ω m), or ohm centimeters. The resistance of a wire is then:

$$R = \rho \frac{l}{A} \tag{4.6}$$

where ρ is resistivity in Ω m, for example, and A is the cross-sectional area.

4.6 THE WIPER

The wiper of a resistive linear position sensor has two areas of electrical contact: one set of wipers contacts the resistive element, while the other set contacts the parallel conductor. Multiple fingers are formed in each set of wipers in order to improve the contact reliability, and also to reduce the contact resistance. If one wiper finger is not making contact with the surface of the resistive element at a particular instant, for example, then at least one of the other fingers is likely to be making contact at that time. This way, nearly continuous contact can be ensured. The degree of difficulty in maintaining contact is partly due to nonuniformity of the surface of the resistive element, the element having high and low spots, as well as areas of better and poorer surface conductivity. The other important part of maintaining contact is due to the ability of the wiper surface to maintain a clean and conductive surface in the contact area. This is accomplished by selecting the best alloy for plating onto the wiper contact area, and is selected to be electrochemically compatible with the resistive element material. The electronegativity of the wiper plating alloy must be very close to the electronegativity of the resistive element material in order to avoid a battery effect. (Electronegativity is a measure of the affinity of an atom in a molecule to attract electrons.)

There is also a trade-off in the wiper design between higher contact pressure providing better electrical contact, versus lower contact pressure providing longer cycle life (due to less wear on the surface of the resistive element at lower wiper contact pressure).

The base material of the wiper is a spring material of which the spring rate and physical dimensions are chosen to find a balance between contact pressure and other performance factors. Higher contact pressure improves (reduces) contact resistance, but worsens (increases) hysteresis by causing flexing of the wiper assembly. Higher contact pressure also reduces lifetime of the resistive element by rubbing off some of the resistive coating and forming particles that interfere with electrical contact.

The chemical composition of the plating on the wiper must be chosen to maintain low surface resistivity in combination with the chosen composition of resistive element surface, especially when the wiper is not moved over long periods of time. Suitable plating materials include gold, palladium, or silver, and can be laid down as multi-layers, alloys, or mixtures.

4.7 LINEAR AND ROTARY MECHANICS

As mentioned earlier, it is typical for the potentiometer housing to be made of an aluminum alloy extrusion. This has several advantages in cost, producibility, and design flexibility. The tooling cost for an aluminum extrusion is very low when compared with the tooling cost for a metal die-casting or a plastic injection molding. A metal housing also has an EMC advantage over a plastic housing because an electrically conductive housing provides electromagnetic shielding.

For a linear resistive position sensor, an extrusion can easily be designed to include a slot into which the resistive element can be mounted. Channels can also be included that provide holes at the ends of the extrusion suitable to accept screws to attach end plates for the housing. Metal end plates complete the shielding scheme. Channels can also be implemented into the extrusion for the purpose of adding the mounting feet, such that each foot can be slid into place and fixed in that position by tightening a screw.

In a rod-type position sensor, a bushing is needed at the rod entry end, and a guide is needed on the inside tip of the rod to keep the rod parallel with the housing. The tip guide rides against the inside wall of the housing to maintain alignment, and can also accommodate the wiper assembly. To stop the undesired entry of particulate debris, and to protect the bushing from wear due to hard particles, one or more each of elastomeric and/or plastic wipes and seals are mounted outside of the bushing.

For a rotary resistive position sensor, a metal or plastic cup may be used, with a front plate attached to complete the housing. The cup would enclose the resistive element and wiper assembly, and provide mounting of the three electrical terminals. The front plate contains the bearings to hold the actuation shaft, as well as any wipes and seals that may be implemented to exclude foreign materials from entering the housing.

4.8 SIGNAL CONDITIONING

A resistive position sensor is normally manufactured without any internal signal conditioning circuitry. If one lead of the resistive element is connected to zero volts, and the other to a regulated supply voltage, then the wiper output is a voltage that is variable between those two points and in proportion to the position. There is a need for a proper input circuit (that is, the receiving circuit) in the device that reads the pot output. This input circuit is a load on the pot output. If the load resistance is too low, it causes a lower signal level (if the pot is not at zero or full scale) and increases nonlinearity of the position indication from the pot. Loss of signal level can be corrected with a gain adjustment in the receiving circuit, but the nonlinearity is more difficult to accommodate. In order to eliminate this problem, the load resistance has to be much larger than that of the resistive element. For example, a common value of resistance for a pot is 10k ohms. In this case, the input impedance of the receiving circuit should be at least 500k ohms, but would be better to be 1 M ohms or more. This is relatively easy to obtain by using the non-inverting input of an operational amplifier, or by using an instrumentation amplifier. A problem with a high impedance input, however, is that electrical noise is easier to pick up. Electrical noise is generated from motors, switches, and solenoids (among other things) in industrial equipment. Electronic filter circuits are typically added to the input signal conditioning circuit in order to limit the effect of the noise. Using a shielded cable or a twisted pair of wires to carry the signal from the pot into the input of an instrumentation amplifier is another way to limit admitting electrical noise into the circuit.

It is also possible to change the output of a potentiometer to a lower impedance, thus reducing the likelihood of picking up noise. This can be done by adding an operational amplifier (op amp) circuit, within the housing of the pot, as a buffer. The conductors bringing the wiper signal to the amplifier are very short, and are shielded by the potentiometer housing. So there is very little noise induced at this point. Then the low output impedance of the op amp circuit (on the order of 1 ohm or less) becomes the sensor output impedance. The op amp circuit can be a voltage follower, with the output tied to the inverting input, and the power supply rails connected to the potentiometer resistive element. Such a circuit used by the author to condition the output of a potentiometer is shown in Figure 4.7.

One consideration with a potentiometric sensor wired as in Figure 4.7 is that the pot then needs to be identified as to which lead is wired to positive, and which to

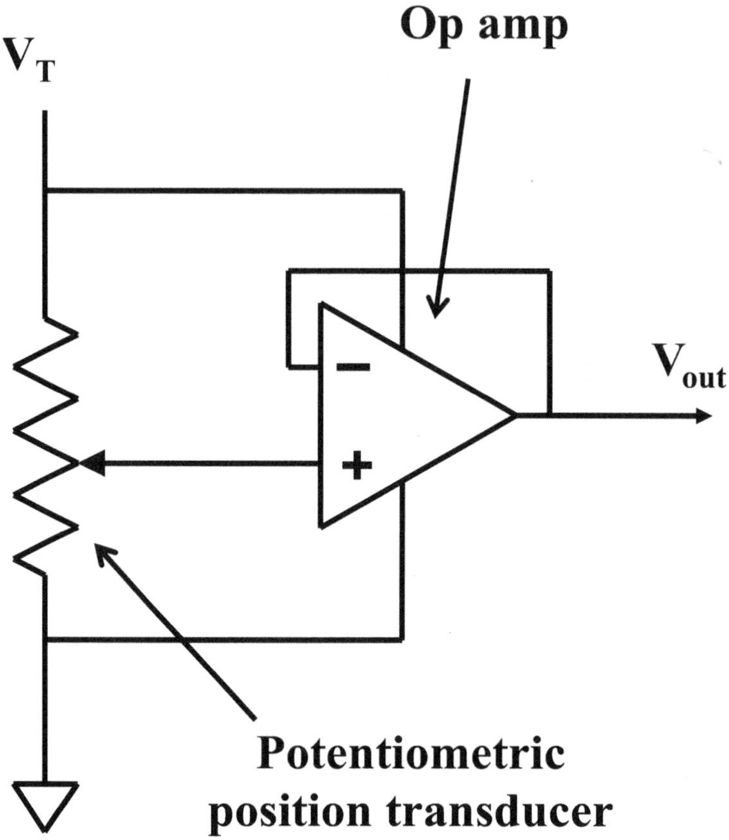

FIGURE 4.7 A resistive position sensor circuit with low output impedance.

negative or common, so that a given direction of motion would provide the desired voltage change (increasing or decreasing voltage) at the output terminal. A possible solution would be to power the op amp using a diode bridge to steer the power supply voltage. That way, the pot could be wired with either end as positive, and the op amp would still receive the correct power supply voltages. The slight disadvantage of the diode bridge would be that the op amp output could not swing to the rails, being limited to about 0.6 volts shy of reaching the supply voltages (if using an op amp with rail-to-rail output).

4.9 ADVANTAGES/DISADVANTAGES

Resistive position sensors are a popular choice because of their relatively low cost, simple wiring, and because they are easy to understand. Wirewound pots are still common, but sensors with cermet elements are often better for industrial use where finer resolution and high cycle lifetime are needed, especially in closed loop control systems.

It should be noted, however, that there is always the problem of long-term wear when using a contact-type of sensor. So, for high reliability applications where periodic replacement of the sensor is not desired or is difficult, a non-contact-type sensor should be installed.

4.10 TYPICAL PERFORMANCE PARAMETERS

4.10.1 Nonlinearity

The nonlinearity of a potentiometric position sensor is mainly controlled by the uniformity to which the resistive element is manufactured. In a wirewound element, the wire must have uniform resistivity, but just as important, is the accuracy in uniformly winding the wire onto the mandrel. The wire must be carefully layer wound with each turn adjacent to the previous turn, and with the same pitch. The turns must be tight to the mandrel to prevent movement when the wiper moves over them.

In carbon film or cermet types, the resistive paste must be homogenous and applied to the substrate in a uniform geometry. Uniformity of the conductive mixtures is mainly determined by the uniformity of the conductive particles and their dispersion throughout the paste. Uniformity of application is determined by the thickness as well as the width and straightness to which the paste is applied to the substrate.

It is a little more difficult to produce a plastic film element having low nonlinearity, and these are therefore used in lower cost/lower performance applications. Cermet elements, however, can be adjusted after firing. This is done by first carefully measuring the linear response, and calculating the needed changes to reduce the nonlinearity. Then scratches are cut along the edge of the element, in an area that is not used by the wiper, to slightly increase the resistance in those areas, as needed. A manual operation or a mechanized one can be used to apply the scratches mechanically with a sharp-tipped stylus. More accurate results can be obtained by making the scratches through laser etching in an automated process.

In order to provide the lowest nonlinearity error, as required in higher accuracy products, the nonlinearity can be adjusted by laser etching to reduce the nonlinearity error to the level of about +/– 0.05% of full range.

Overall resistance is not as important as uniformity, since the potentiometer is typically used in a voltage divider circuit. The resistive load, however, can cause additional nonlinearity if it is too low, as mentioned earlier. An example is considered where a resistive element is connected across a power supply, with terminals at zero volts and V+, and a load resistor is connected from the wiper to zero volts. The load resistor has essentially no effect on the wiper voltage when the wiper is at either extreme position (ignoring contact resistance). However, when the wiper is at other positions along the stroke, the load resistor is in parallel with the lower part of the element and changes the voltage divider ratio. The result is an increase in nonlinearity. As also stated earlier, errors are reduced by making the resistance of the load resistor much higher (at least 50 times) than that of the resistive element, or a circuit such as was shown in Figure 4.7 can be utilized so that the wiper is looking into a very high equivalent impedance.

4.10.2 Hysteresis

The hysteresis of a potentiometer comes from the mechanics of the wiper to element contact, and from the wiper tracking mechanics. Normally, there are three or more fingers on a given wiper assembly (these fingers will collectively be called "the wiper"). The wiper rubs against the resistive element with a carefully controlled (by design) amount of contact force. The force is applied by spring tension of the flexed wiper. The wiper is made from a spring-type of metal and is then plated with an alloy to enhance conductivity to the surface of the resistive element, to reduce oxidation, and to reduce wear. The alloys typically used include platinum, palladium, gold, silver, and rhodium, among others.

Given the contact force, contact area, and surface friction, there is a resulting amount of force required to move the wiper across the element. This force acts against the flexural strength of the wiper to cause a movement of the wiper and its mount. As the measurand moves upscale, the wiper is flexed slightly in the downscale direction. Likewise, as the measurand moves downscale, the wiper is flexed slightly in the upscale direction. This was shown in Figure 2.10. The difference between the upscale and downscale flexing is the major source of hysteresis error in a potentiometric position sensor. The hysteresis just described can also be called backlash (and includes gear backlash when the sensor incorporates a set of gears). The hysteresis/backlash characteristic has a different shape from the sensor hysteresis curve that was shown in Figure 2.8, because the amount of hysteresis/backlash of a potentiometer is mostly independent of the amount of change in the measurand. The curve shape for a potentiometric sensor hysteresis/backlash is shown in Figure 4.8.

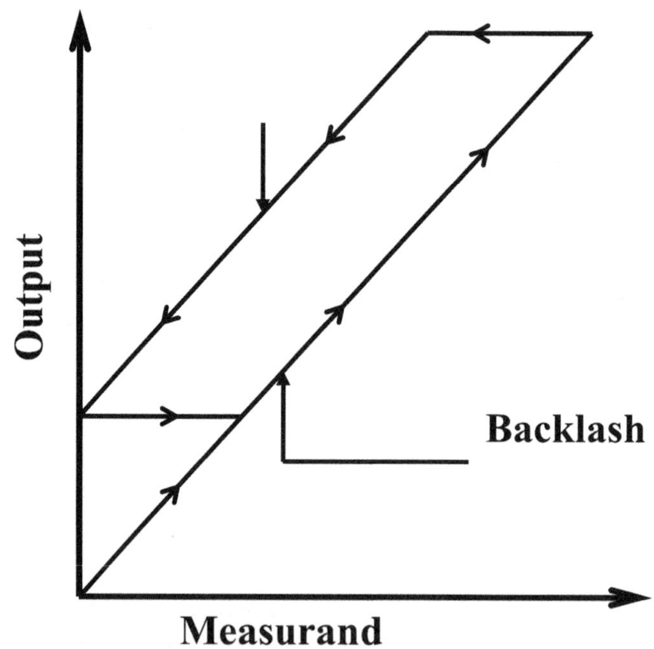

FIGURE 4.8 The shape of a sensor output vs. measurand characteristic including backlash, which is normally considered to be a part of hysteresis.

4.10.3 Wear/Lifetime

Since a potentiometric position sensor is a contact-type device, a major consideration in its use is how long it will last before wearing out. When a sensor is new, the surface element is clean and smooth, the wiper is evenly coated with the selected alloy, and the tracking mechanics are tight. As the potentiometer wears over many cycles of use, all of these initial conditions change. The main problems to be encountered due to wear when a resistive position sensor has undergone a number of cycles nearing its rated lifetime are noise and dead spots. As debris from sensor component wear accumulates on the surface of the resistive element, the wipers make and break contact with the resistive element surface as they move over the debris. This increases the resistance variation and uncertainty of the output for a given position, and can be a source of electrical noise as the wiper is moving. If all of the wipers that are either on the element or the parallel conductor rest on debris particles, then an open circuit can result. When this problem occurs repeatedly at one location, or an open circuit results from the resistive coating being worn off at one location, it is called a dead spot.

4.10.4 Dead Zones

There may be electrical dead zones on each end of travel where a further change in the input measurand does not cause a further change in the output signal. This is possible when the design allows the wiper to ride up onto the conductor lands, which are used for connecting the electrical element to the potentiometer end pins.

Mechanical dead zones may also exist in which further travel of the wiper is mechanically prevented before the full expected range of the input measurand is reached. In this case, it is not mechanically possible to move the actuator rod to the extreme end of the stroke. This may be due to misalignment of the sensor when mounting it.

4.11 SPECIFICATIONS AND APPLICATION

A typical set of specifications for a linear potentiometric position sensor is shown in Table 4.1.

A resistive rotary/angular position sensor would have similar specifications, but the units relating to measuring range would be in degrees of rotation instead of mm or inches of linear travel. When matching a type of potentiometer to a real-world application, it is necessary to specify the required full-scale range, nonlinearity, hysteresis, and total resistance, but consideration must also be allowed for the type of resistive element, expected number of lifetime cycles, and mounting ability.

A wirewound element can provide good long-term reliability, but may not be suitable for use in servo control loops or where very fine resolution is needed. In a servo control loop, there can be a constant dithering between two adjacent turns on the element when the actual control point being sought lies between the two points of contact available (that is, from one turn of the wire winding to the next turn). In this case, a higher resolution pot with a more nearly continuous output could be used.

Some advantages of the resistive position sensor over other types of position sensors include the ease of understanding the sensor performance, ease of application

TABLE 4.1
Specification of a Typical Resistive Linear Position Sensor

Mechanical	
Total mechanical travel	150 to 1200 mm
Starting forces	0.45 kg
Total weight	0.36 to 2 kg
Vibration	20 g rms/0.75 mm 5–20 Hz
Shock	50 g, 11 ms half sine wave
Backlash	0.025 mm
Life	1 billion dither operations
Electrical	
Theoretical electrical travel	150 to 1200 mm
Independent nonlinearity	0.1%
Total resistance	5,000 ohms
Resistance tolerance	20%
Operating temperature	–65 to 105 C
Resolution	nearly infinite
Insulation resistance	1,000 M ohms @ 500 VDC
Dielectric strength	1,000 V rms
Recommended wiper current	<1 micro amp
Electrical connection	Binder series 681
Maximum applied voltage	30 VDC

to a specific use, the relatively low cost, no electronic circuit needed to operate the sensor, and they have a high level output signal voltage directly from the transducer, so that the transducer element itself can be considered as a complete sensor.

Some disadvantages include the limited frequency response due to the mechanics, high operating force due to friction, high dynamic force due to the mass of the moving parts, and the limited lifetime due to wear.

4.12 MANUFACTURERS

Some manufacturers of resistive linear and angular position sensors include the following:

ASM	www.asm-sensor.com
Bourns	www.bourns.com
Data Instruments	www.datainstruments.com
Dynamation Transducers Corp.	www.dynamationtransducers.com
Fiama	www.fiama.it/en
Gefran	www.gefran.com
Honeywell Sensing & Control	http://sensing.honeywell.com
Megatron	www.megatron.de/en/home.html
Midori	www.midoriamerica.com
Novotechnik	www.novotechnik.com

Opkon	www.opkon.com.tr/eng/default.asp
Penny & Giles	pennyandgiles.com
Revolution Sensor Company (design)	www.rev.bz

4.13 QUESTIONS FOR REVIEW

1. **If the load resistance is too low, the performance of a potentiometer will become:**

   a. Sluggish
   b. Unstable
   c. More smooth
   d. Nonlinear
   e. Scratchy

2. **A rotary potentiometer can measure linear position by using a pinion gear and a:**

   a. Rack
   b. Interferometer
   c. Oval wheel
   d. Differentiator circuit
   e. Bushing

3. **A voltage potential that is indicative of a potentiometer setting is available at its:**

   a. Ends
   b. Wiper
   c. Housing contact
   d. Pointer
   e. Control rod

4. **A non-usable portion at the end of a resistive element, plus contact resistance form:**

   a. Hysteresis
   b. The null zone
   c. Voltage division
   d. A current limit
   e. End resistance

5. **A potentiometer can be made insensitive to load resistance by adding a:**

   a. Filter capacitor
   b. Diode

c. Voltage follower
d. Silicon substrate
e. Whisker

6. Wirewound elements are manufactured by winding wire around a nonconductive:

a. Mandrel
b. Capacitance
c. Wiper
d. Crystal entity
e. PCB

7. Potentiometer wiper finger(s) rub against a resistive element, as well as:

a. The control knob
b. The housing seal
c. The spring
d. A parallel conductor
e. An insulator

8. Surface resistivity is measured in ohms per:

a. Unit length
b. Square
c. Volt
d. Wire diameter
e. Cm

5 Cable Extension Transducers

5.1 CABLE EXTENSION TRANSDUCER HISTORY

This chapter is separate from the resistive/potentiometric sensor chapter, due to the special mechanical arrangement used, and the performance and application of this type of sensor being substantially different from the resistive position sensors of Chapter 4. A cable extension transducer (CET) can also be called by a variety of other names, some of which include "cable displacement transducer", "string pot", "draw-wire sensor", "yo-yo pot", and "string encoder". A CET is both a sensor and a transducer, because it provides a ready-to-use output voltage signal when connected with a power supply, but will be called a transducer here, since that is a part of its most widely accepted name: cable extension *transducer*.

Cable extension transducers were first developed in the 1960s for use in the aircraft manufacturing and aerospace industries. The earliest applications were for measuring the position of flight control surfaces and other aircraft mechanisms. After that, additional applications were developed, including factory automation, automotive testing, medical devices, and industrial machinery. And later, further applications were developed for hydraulic cylinder positioning, and robotics.

There are many earlier patents describing devices for accurately and uniformly winding of cables, tapes, or strings onto a spool or drum while under spring tension. Some of these ideas were then adapted for use as a sensing device by adding a rotary potentiometer to measure and indicate the rotation of the spool or drum. A constant DC voltage was applied across the resistance element of the potentiometer. So, when the cable or string is pulled out by a given distance, that distance would be indicated by a voltage at the wiper terminal of the potentiometer. Most ruggedized versions use a stainless steel cable for the cable extension. A typical CET is a contact-type sensor, having a movable wiper that rubs against a resistance element. It provides an absolute measurement, and is mostly implemented by powering from a DC voltage supply and having a simple analog voltage output from its wiper (although other outputs are available, including digital and bus protocols).

5.2 CABLE EXTENSION TRANSDUCER CONSTRUCTION

A CET typically employs a multi-turn, rotary-style potentiometer, coupled with a drum and cable. They can have full-scale ranges from 0 to 38 mm up to 0 to 60 meters.

Less expensive CETs are manufactured with plastic housings. More expensive and rugged versions may have a metal housing. Since the housing is a functional part of the mechanism, having features to hold the drum and wire guiding components, a rugged housing can be important.

DOI: 10.1201/9781003368991-5

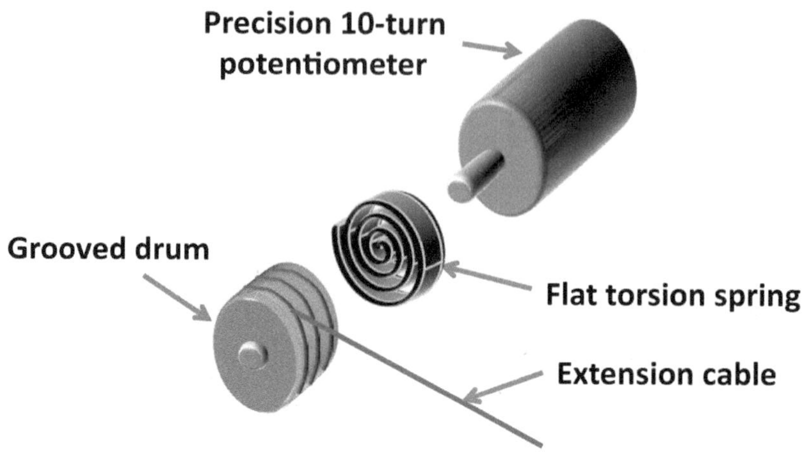

FIGURE 5.1 Construction of a cable extension transducer.

FIGURE 5.2 A cable extension transducer having a plastic housing, and with a 317 mm full stroke range. Cable fully retracted at *a*. Cable partly extended at *b*.

The CET of Figure 5.2 has a plastic housing. On this model, the cable can be pulled out to 317 mm, but is shown fully retracted on the left, at (a). On the right, at (b), the cable is shown partly extended by means of a toothpick to hold the cable in that position against the force of the internal torsion spring. The torsion spring

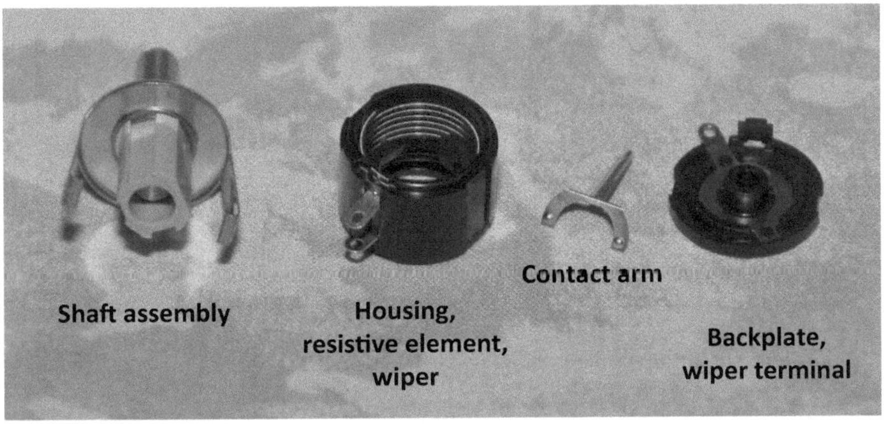

FIGURE 5.3 Internal components of a ten-turn wirewound potentiometer.

retracts the cable back into the housing as needed to hold a tension on the cable. The toothpick can be seen behind the thin stainless steel cable of the CET. In this transducer, a standard ten-turn potentiometer can be seen protruding at the side of the transducer housing. A potentiometer like this one usually has a wirewound resistive element inside. This example has a resistance of 10k ohms, which is a typical value. The three terminals that can be seen on this potentiometer are the terminals that will be used to connect the transducer (the wiper, plus the two ends of the resistive element).

The ten-turn potentiometer commonly used in a CET has an extended-length wirewound resistive element that is coiled as a helix inside the potentiometer housing. See Figure 5.3.

The shaft assembly as shown includes the top plate of the housing, and also includes a plastic insulator attached to the shaft (plastic insulator can be identified in the figure by its three channels and a flat). One of the insulator channels captures the elongated portion of the contact arm. The contact arm makes electrical connection between the stationary terminal/ring of the back plate and the rotating wiper that moves along the helical-coiled resistive element. A closer view of the resistance element and wiper are shown in Figure 5.4.

Since a typical ten-turn potentiometer uses ten turns of a wirewound resistive element, a finer resolution is provided than with a single-turn wirewound potentiometer. The resistance element comprises a thin resistance wire that is wound onto a flexible mandrel or core cable. See Figure 5.4, showing the resistance element and contact being made with it by the wiper. Finer resolution may also be achieved by using a hybrid resistive element, such as a conductive plastic coating over a wirewound element.

In a ten-turn wirewound potentiometer, the resistance wire is usually wound onto a cable core so that the completed element can be coiled into a helix for mounting within the potentiometer housing. The resistance wire is usually fabricated of a copper-nickel, nickel-chromium, or copper-manganese alloy. Since the cable core is metal and electrically conductive, a thin insulating film is applied to the cable core before the resistance wire is wound onto it. A close-up view of a section of a wirewound resistance element of such a potentiometer is shown in Figure 5.5.

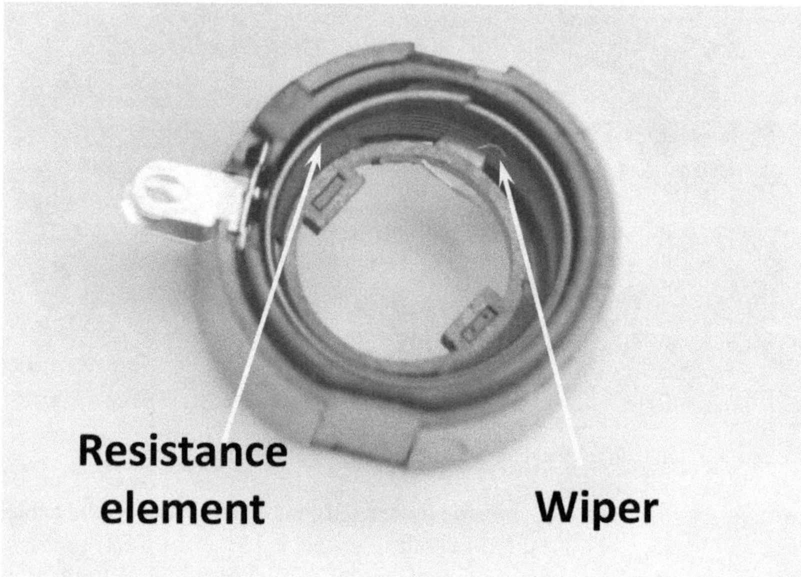

FIGURE 5.4 Resistance element and wiper of a ten-turn potentiometer.

FIGURE 5.5 Close-up view of resistive element within a ten-turn wirewound potentiometer reveals that it comprises a thin resistance wire that is wound onto a flexible, insulated cable core.

In the figure, a contact strip can be seen at the top center of the portion of the resistance element shown. This is where electrical contact is made at each end of the resistance element, and so there is a similar contact at the opposite end of the resistance element (not shown). These contacts are welded to the resistance wire. The resistance element continues to the left, a little past the spot where the contact is made, so that the element can maintain a uniform shape throughout the range of wiper motion. The mechanics of a wiper-guiding assembly will prevent the wiper from moving past the welded contact point.

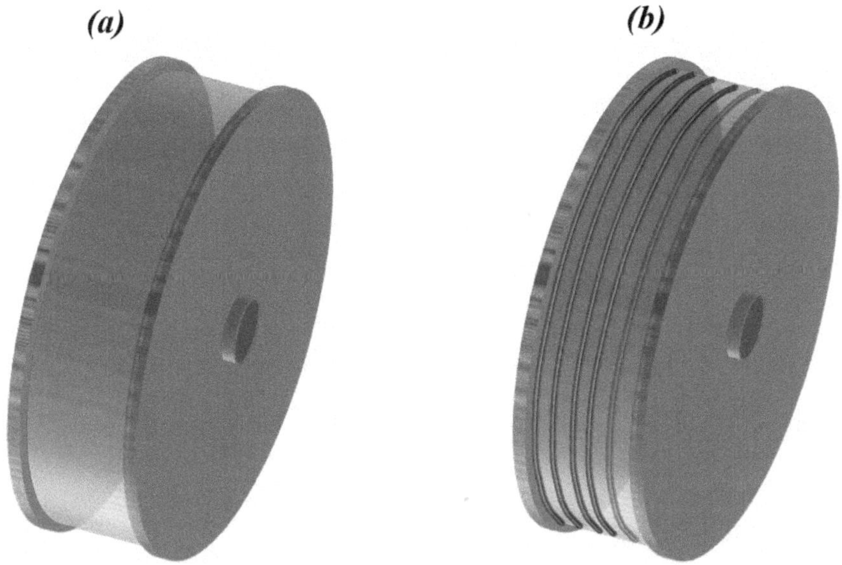

FIGURE 5.6 Non-grooved drum at (a) and grooved drum at (b).

As the extension cable is wound onto the drum (shown in Figure 5.1), it is important that constant tension be maintained on the extension wire, and that the cable be uniformly positioned onto the drum. This will prevent the cable from becoming tangled inside the CET. If the extension cable becomes tangled inside the CET, then the CET will need to be repaired or replaced, depending on the construction of that particular model. Many lower cost CETs are designed to be replaced rather than repaired.

In order to ensure uniform winding of the extension cable onto the drum, many drums are designed to include a spiral grooved winding surface so that each cable turn will be wound into its respective groove. Figure 5.6 depicts a drum without grooves, a., and one with grooves, b., cut into the drum surface.

5.3 SIGNAL CONDITIONING

A CET is normally used without any signal conditioning, so a schematic would be the same as that shown in Figure 4.1(b). CETs generally utilize a ten-turn potentiometer having a total (end to end) resistance of 1k, 2k, 5k, 10k, 20k, or 50k ohms, with 1k, 5k, and 10k ohms being the most popular. It is important to make sure that the circuit reading the wiper voltage has a much larger input impedance than the resistance of the pot (at least 50 times, but preferably 100 times). In cases where there is a somewhat low input impedance in the circuit connected with the potentiometer, the wiper voltage can be buffered as was shown in Figure 4.7.

Some manufacturers offer the availability of many output types, including:

voltage divider (standard)
voltage divider, buffered output voltage

analog voltage (0 to 5 or 0 to 10 VDC)
3-wire current (usually 4 to 20 mA)
2-wire current loop (usually 4 to 20 mA)
incremental (square wave-in-quadrature)
RS-232
RS-422/RS-485
SSI
CANbus
DeviceNet

And some manufacturers also offer a combination of position and velocity. For a velocity measurement, the first derivative of the position signal is derived from an analog differentiator circuit (as was shown in Figure 1.11), or can be calculated digitally in one of the versions implementing a microcontroller (including any of the digital output versions).

5.4 APPLICATION

When installing a CET, the cable must be attached to a member of which the position is to be measured. Various cable end terminations are available to facilitate this, as shown in Figure 5.7.

Using the appropriate extension cable end termination will make installation easier, and may extend the life of the extension cable. If, for example, the measured member moves in a way such that the end of the cable is repeatedly flexed at one point, then the cable may eventually break. But using a swiveling attachment may prevent that problem. Using one of the various snap connectors, it can be easy to detach and re-attach the extension cable when needed. The button, shown at Figure 5.7(g), can also be a magnet for easy attachment to a ferromagnetic surface.

In some of the more demanding industrial applications, a highly rugged CET model with a metal housing may be required, as shown in Figure 5.8 (such as the Celesco model PT9101).

The ruggedized version includes a very heavy duty housing and sturdy mounting feet. The housing material is usually stainless steel or powder-coated aluminum. The extension cable is stainless steel or thermoplastic. A thermoplastic cable may be desired for chemical compatibility, or to provide electrical insulation in case there may be a high voltage present at or near the target of which the position is to be measured. In this large type of housing, measurement lengths can be up to a little more than 60 m.

In some applications, it may be necessary to evaluate any deflection of the cable due to gravity, wind, shock and vibration, and so on, noting the associated error that may be so induced.

5.4.1 Gravity

Since the cable has weight, there will be some amount of sag when the cable is extended horizontally over a long distance. The amount of sag is determined by the

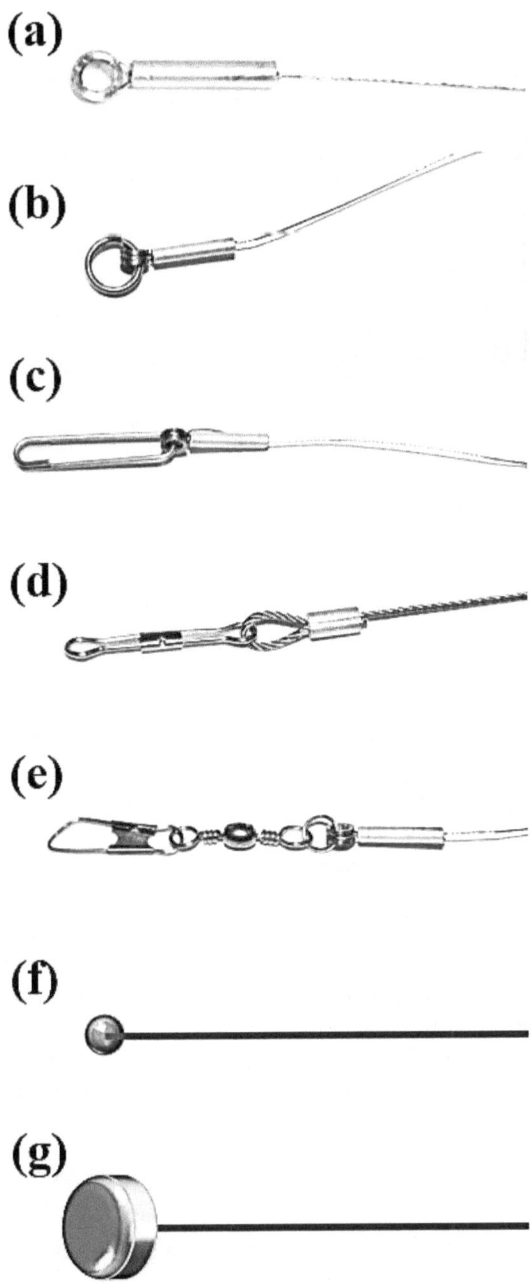

FIGURE 5.7 Various cable end terminations: (a) eyelet, (b) split ring, (c) scissor snap, (d) slide snap, (e) snap swivel, (f) bead, (g) button.

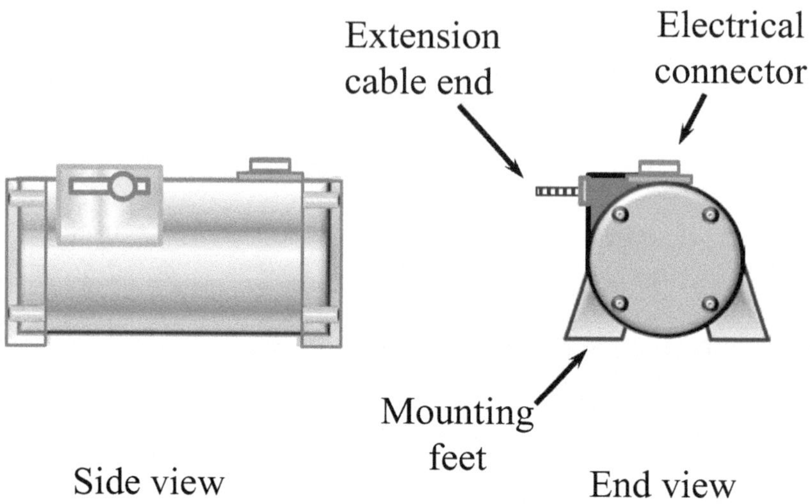

FIGURE 5.8 A heavy duty, ruggedized, industrial CET model.

weight of the cable being extended, the spring force pulling on the cable, and the angle from horizontal over which the cable is extended. With the cable pulled out vertically, there is no error induced by sag. The maximum error occurs when the cable is pulled out horizontally. With sag in the cable, there will be a greater amount of cable pulled out than the straight-line distance being measured, so a small amount of error will result due to this difference. Information on calculating the amount of error to be expected from sag is available from the manufacturer of the particular product.

5.4.2 Wind

Just as cable sag can induce an error, the same type of error can be induced by wind blowing the cable to the side, for example. But since wind error can be variable, there is not a simple conversion factor to apply. In general, long measurement range CETs may not be suitable for high wind conditions if small wind-induced variations would cause undesirable error. Wind may also induce vibration of the extension cable at its fundamental resonant frequency for that extension, and/or harmonics, as shown in Figure 5.9.

5.4.3 Shock and Vibration

Shock and vibration can cause the extension cable to vibrate at the fundamental resonant frequency and/or harmonics of the length of cable that is extended. See Figure 5.9.

The considerations regarding shock and vibration effects on CET error are the same as with wind, as explained earlier. When the entire length of extended cable vibrates as one, it is vibrating at its fundamental frequency, and the extended length

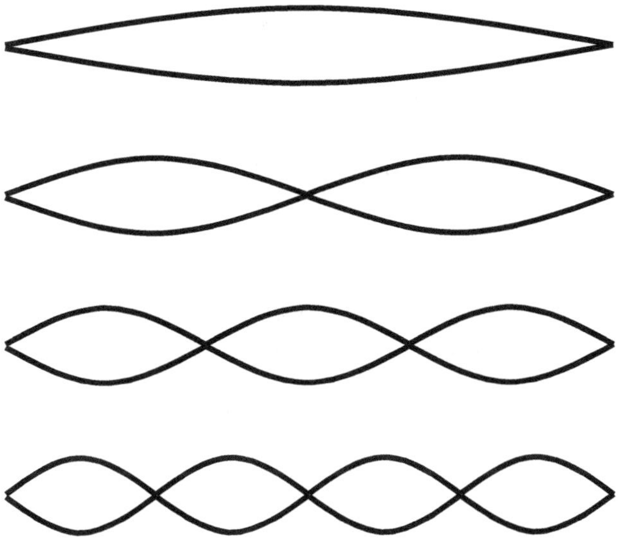

FIGURE 5.9 The extension cable may resonate when subject to wind, shock, or vibration.

of the cable is the wavelength. As shown at the top of Figure 5.9, the fundamental is the lowest frequency for that particular cable type at that extended length and tension. The fundamental frequency is also called the first harmonic. The fundamental frequency is dependent on the mass of the extended cable, the extended length, and the cable tension. At the same time as it is vibrating at the first harmonic, the cable may also vibrate at integer numbers of additional harmonics. Shown are the second through fourth harmonics, but the number can also go higher.

5.4.4 Stretch and T/C

The cable of some CETs is pre-stretched to reduce any additional amount of stretch after installation. Although a greater cable tension is beneficial to reducing sag, some stretching of the cable after installation can result when the tension is high. If the amount of stretch is known, and is well within the elastic limit of the cable, then adjustment can be possible in the calibration of the system. Likewise regarding the thermal coefficient of expansion (*t/c*) of the cable. If the temperature is known, it can be possible to adjust the system calibration based on a measured temperature.

5.4.5 Adding a Pulley

In applications where it may be difficult to mount a CET in-line with the direction desired for cable extension, it is a simple matter to guide the extension cable as needed, through the utilization of one or more pulleys. This is shown using one pulley in Figure 5.10.

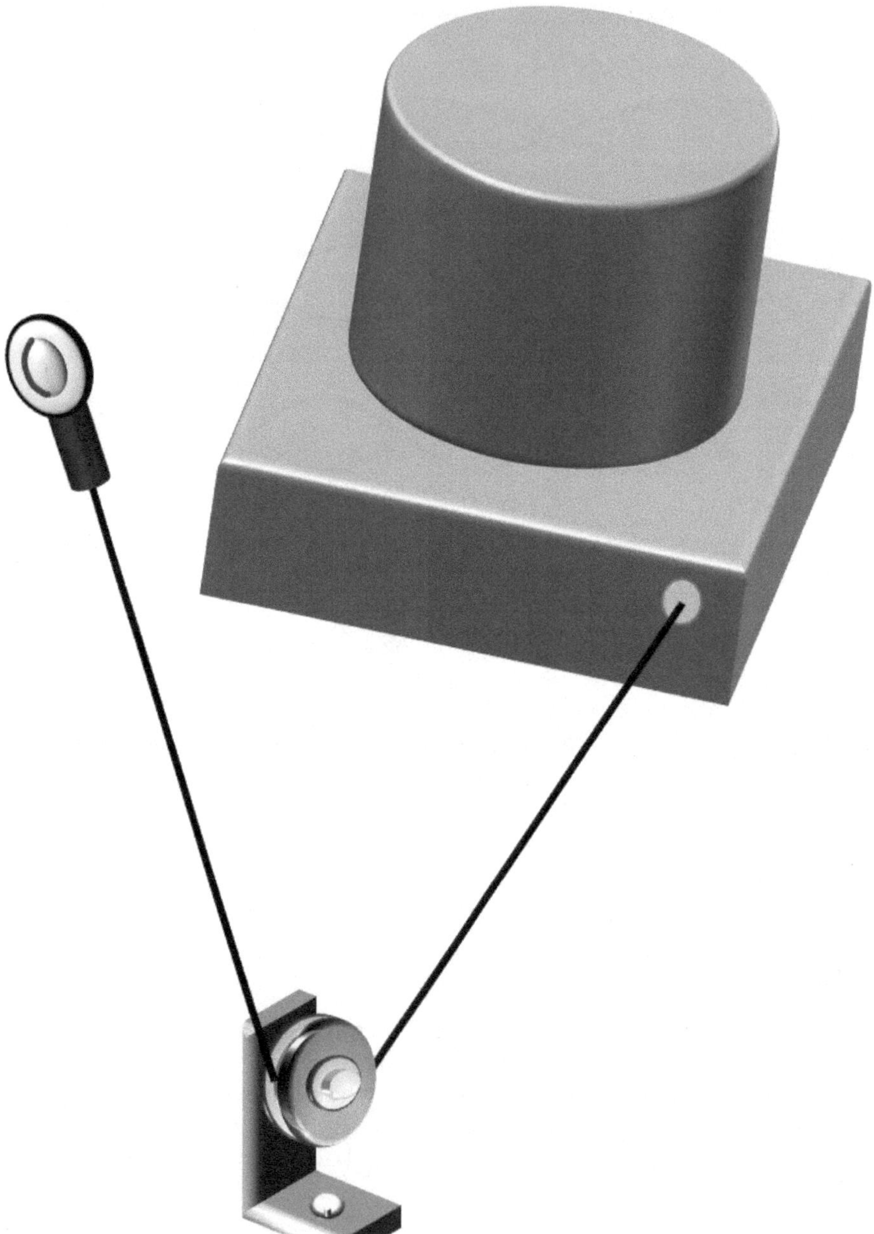

FIGURE 5.10 Adding one or more pulleys to direct the extension cable to a desired location.

The configuration shown in the figure can come in handy when fitting a CET into a tight space. In which case, the CET can be mounted where it happens to fit. Then the extension cable can be routed to the desired measurement area. If needed, the CET can be ordered so that some external cable is available when fully retracted. To maintain tension on the internal cable, a bead will be added at the point where the extra cable is added. Alternatively, the user can add more cable to the eyelet of a standard CET.

5.5 ADVANTAGES/DISADVANTAGES

CETs are a popular position-sensing choice because of their relatively low cost (some models), simple wiring, and because they are easy to understand. Here are some additional advantages that often may apply to CET applications:

- Extension cable can be pulled out straight or at an angle for linear measurement, or wound around a spindle or drum for rotation or angular measurement;
- Extension cable can be routed around barriers by pulleys;
- Easy mounting of the CET housing, as well as easy attachment of the cable extension end;
- Fast installation due to flexibility of routing the cable as desired;
- Wide range of measuring lengths from less than 50 mm to more than 60 m;
- Low power. Common application of 10 k pot at 5 VDC = 0.5 mA, = 2.5 mW;
- Wide operating temperature, some models operating over a range of –40 to 100°C;
- Relatively low nonlinearity and other errors, with nonlinearity typically in the range of ±0.1 to 0.25% of full range.

5.6 PERFORMANCE SPECIFICATIONS

5.6.1 Nonlinearity

The nonlinearity of a potentiometric CET is controlled by the nonlinearity of the potentiometer and the uniformity of winding the extension cable onto the drum (excluding sag and stretch). Overall resistance of the potentiometer is not as important as its nonlinearity, since it is most often used in a voltage divider circuit (even when further signal conditioning will take place in order to provide a different type of output).

When the wiper output is not buffered, a load resistance that is too low will increase nonlinearity, as explained in Chapter 4 on resistive position sensors.

5.6.2 Hysteresis

Hysteresis was explained in Chapter 1, and is defined similarly for a CET. It is the difference between two readings of the same measurement point, with one measurement being approached from downscale and the other being approached from upscale. In a CET, it is somewhat due to hysteresis in the potentiometer, but mainly due to hysteresis in the spring mechanism causing hysteresis in the cable tension, resulting in variable cable stretch when going upscale versus downscale. It is sometimes possible to account for this in the equipment that is using the CET output signal when the amount of hysteresis is known.

5.6.3 Sine Error

As the cable winds onto the drum, there can be a slight sinusoidal variation in data along the ideal nonlinearity line. This is mostly repeatable in a grooved drum, but not repeatable in a non-grooved drum.

5.7 TYPICAL SPECIFICATION

A typical set of specifications for a heavy industrial cable extension transducer with voltage divider output is shown in Table 5.1.

5.8 MANUFACTURERS

Some manufacturers of cable extension transducers include the following:

ASM	www.asm-sensor.com
Celesco	www.celesco.com
Fiama	www.fiama.it/en
Micro Epsilon	www.micro-epsilon.com
Midori America	www.midoriamerica.com
Opkon	www.opkon.com.tr/eng/default.asp

TABLE 5.1
Typical Specification of a Heavy Industrial CET with Voltage Divider Output

Full stroke:	0–75 inches, to 0–550 inches
Output signal:	voltage divider (potentiometer)
Nonlinearity:	±0.1% of full stroke
Repeatability:	±0.02% of full stroke
Resolution:	±0.002% of full stroke
Cable material:	stainless steel or thermoplastic
Potentiometer life:	≥250,000 cycles
Resistance element:	0.5, 1.0, 5.0, 10k ohms
Maximum voltage:	30 V (AC or DC)
Full stroke change:	94% of applied voltage, ±4%
Enclosure:	NEMA 4/4X/6, IP 67/68
Operating temperature:	–40 to 90°C
Max vibration:	10 g, at up to 2,000 Hz

Sauer-Danfoss	www.sauer-danfoss.com
Sensor Systems, SRL	www.sensorsystems.it (also www.baumer.com)
Sick	www.sick.com
SIKO	www.siko-global.com/en-us
SpaceAge Control	www.spaceagecontrol.com (also firstmarkcontrols.com)
Revolution Sensor (design)	www.rev.bz
TR Electronic, GmbH	www.trelectronic.com
Unimeasure	www.unimeasure.com
Wachendorf	www.wachendorff-automation.com/measuring_systems.html

5.9 QUESTIONS FOR REVIEW

1. **A ruggedized cable extension transducer will usually have a:**

    a. Rounded mounting foot
    b. Metal housing
    c. Fiber-optic cable
    d. Converter box
    e. Snap

2. **The fundamental frequency is also called the:**

    a. First harmonic
    b. Zero space
    c. Significant frequency
    d. Full scale
    e. Harmonica

3. **On a CET, force to retract the cable is usually provided by a:**

    a. Heavy weight
    b. Snap clip
    c. Retractor blade
    d. Ball bearing
    e. Torsion spring

4. **A difference between upscale and downscale readings of the same point is due to:**

    a. Nonlinearity
    b. Repeatability
    c. Sine error
    d. Hysteresis
    e. The repeal of Ohm's Law

5. A cable extension transducer is also commonly called a:

a. Draw-wire sensor
b. Snap cable
c. Line spool
d. Snot otter
e. Twisted pair

6. The extension cable of a string pot can be uniformly wound by using:

a. String theory
b. Elastic
c. A grooved drum
d. Adhesive
e. Braided line

7. As the extension cable is wound onto the drum, it must be maintained under:

a. Tension
b. Compression
c. Pressure
d. A 45 degree angle
e. An insulator

8. The resistance wire of a ten-turn wirewound potentiometer is usually wound onto a:

a. Thermoplastic cable
b. Cable core
c. Printed circuit
d. Metal shaft
e. Button shape

6 Capacitive Sensing

6.1 CAPACITIVE POSITION SENSORS

Position sensors based on capacitive sensing technology are very popular in the industrial world because they provide a relatively simple technique to implement a non-contact measurement. Some capacitive position sensors may have an actuating member that contacts the item to be measured, but the non-contact sensing element provides the opportunity for designing a sensor that will impose only minimal mechanical loading forces and sustain minimal wear. Other capacitive position sensors may have no contact at all with the measured target.

Capacitive sensors require a set of driving and conditioning electronics (signal conditioner), and therefore are inherently more complex than a (contact-type) resistive position sensor or cable extension transducer. A typical capacitive linear or angular position sensor comprises a variable capacitance sensing element, electronics, and suitable mechanical components to house them. The housing provides a means as may be necessary to maintain alignment of the movable elements while receiving the mechanical input of the measurand. One style of a completely non-contact capacitive position sensor is shown in Figure 6.1.

A capacitive position sensor as shown in the figure includes a sensing element, electronics, and housing with mounting and connection means. The electronics are contained within the housing of the sensor. This style of sensor can be installed by using a bracket having a hole through which the sensor housing is mounted and clamped.

Contrary to the capability of a linear potentiometer, the capacitive sensing element (transducer) does not produce a directly usable output. The transducer may comprise a sensing electrode, or electrode set, and driving electronics. In Figure 6.1, the sensing element would likely have a disc shape, and be located at the left end of the capacitive sensor. The driving electronics may be an oscillator having a frequency that changes in relation to the distance between the sensing element and a target, due to the effect of the target upon the sensing element capacitance. But other sensing element configurations are also used, including some with powered guard electrodes and powered or non-powered shield (more on this later). The basic concept is that the electronic circuit drives the sensing element with an alternating current, the sensing element changes capacitance due to changes in the measurand, and the resulting signal is a frequency indicative of the measurand.

In order to implement a complete sensor having a desirable type of output, signal conditioning electronics are added to measure or demodulate the signal, filter, amplify, and scale the output as needed. In addition, the circuit conditions the power supply and may convert the scaled signal into a digital form if desired. A housing is usually designed to contain all of the components, including the sensing element, electronics, cable strain relief or connector, actuator assembly (if used), and mounting features.

DOI: 10.1201/9781003368991-6

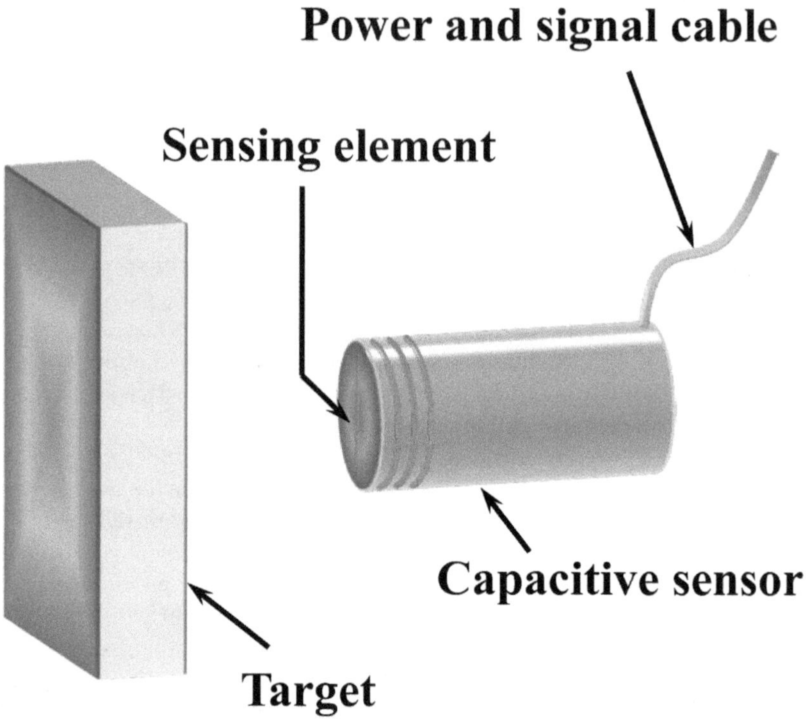

FIGURE 6.1 Capacitive position sensor and target.

6.2 CAPACITANCE

The amount of capacitance between two conductors is defined as the ratio of the electric charge Q on each conductor to the potential difference V between them. The SI unit of capacitance is the farad (F), which is equal to one coulomb per volt (1 C/V).

To understand and design capacitive sensors, it is helpful to be familiar with the nature of the electrical property of capacitance. The capacitance of a system is a measure of its capability to store electrostatic energy. An analogy can be drawn between an electrical circuit with capacitance, and a water system including a storage tank. In this analogy, a tank having a large diameter (large capacitance value) can hold many of liters of water (coulombs of electrons) for a given height of water (a given voltage). Voltage acts like water pressure, and water pressure is often measured by its height (for example, meters of water head, or pressure). A water tank extending up to a height of 20 meters above the ground will develop a pressure of about two atmospheres gauge pressure at ground level when the tank is full. If a pump (the electrical equivalent would be a generator or other power source) fills the tank at a certain flow rate in liters per second (electrical equivalent: charging a capacitor with a current in coulombs per second, or amperes), then the water level (voltage) will rise. If a smaller diameter pipe (series resistor of larger resistance value) is used to fill or drain the tank, the tank will fill (charge up) or drain (discharge) more slowly. The surface area

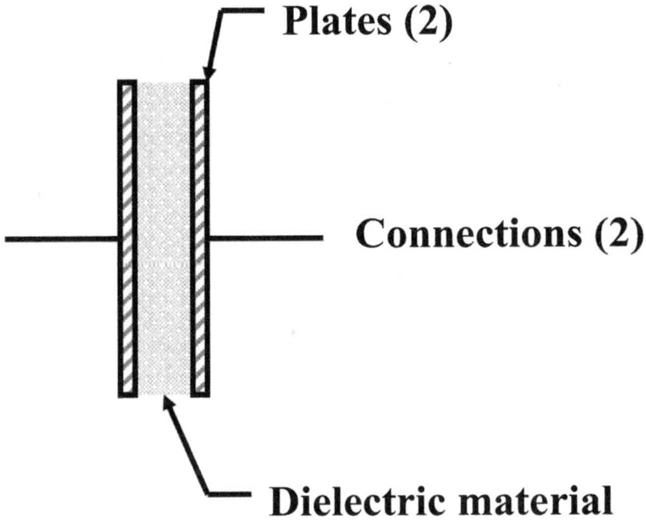

FIGURE 6.2 Construction of a simple parallel plate capacitor.

of a uniform diameter tank in square meters, for example, is analogous to the plate surface area and capacitance value (in farads) of a capacitor. So, a larger diameter tank (having greater capacitance) will fill or drain (change voltage) more slowly with a given flow rate (current) as compared with a smaller diameter tank. The product of the number of liters of water stored in the tank and the square of the height of the tank is proportional to the amount of energy stored. (See Equation 6.3.)

A simple capacitor (see Figure 6.2) is formed by two parallel electrically conductive plates in close proximity, separated by a dielectric material (an electrical insulator) by distance *d*. The dielectric material may be vacuum or air, plastic, fiberglass (as in a printed circuit board), or almost any non-conductor of electricity that is suitable for forming into the desired physical shape. The capacitance, *C*, would be given by:

$$C = \varepsilon_0 \varepsilon_r A / d \tag{6.1}$$

where ε_0 is the permittivity (permittivity is explained further in Section 6.3) of free space in farads/meter. ε_r is the relative permittivity of the dielectric material (a ratio, without units, also called the dielectric constant), *A* is the effective area of the plates in m^2, and *d* is the distance between the plates in meters.

Once the capacitance value is known, then the amount of charge stored in the capacitor can be determined when it is charged to a particular voltage. The charge, *Q*, is given by:

$$Q = CV \tag{6.2}$$

where *Q* is expressed in coulombs (the coulomb is the unit of charge), *C* is the capacitance in farads, and *V* is the voltage to which the capacitor is charged. By taking

Equation 6.2 and dividing both sides by *V*, capacitance can also be seen as equal to *Q*/*V*. Accordingly, one farad of capacitance, *C*, is equal to one coulomb of charge per volt (or *Q*/*V*). One coulomb is the equivalent charge of 6.24×10^{18} electrons. One ampere of electrical current flow is defined as the transfer of one coulomb of charge per second. The amount of electrostatic energy, W_E, that is stored in a capacitor is equal to:

$$W_E = \tfrac{1}{2} CV^2 \tag{6.3}$$

where W_E is energy in joules (or watt-seconds), *C* is in farads, and *V* is the applied voltage in volts. The stored energy has the letter *W* because energy is also known as the ability to do work. (It is interesting to note that some other formulas also have the same form, for example: kinetic energy of a mass at a velocity = $\tfrac{1}{2} MV^2$, kinetic energy of the magnetic field surrounding a coil having an inductance and a current = $\tfrac{1}{2} Li^2$, and so on. The form of many formulas is often the same in various areas of electricity, physics, mechanics, chemistry, and so on. This is the premise of the work shown by Olson [1].)

Stored energy is also known as potential energy. Stored electrical energy is analogous to the potential energy stored in the water of the elevated tank described earlier. When a given current, *I*, is applied to charge a capacitor, the rate of change of voltage, *dv*/*dt*, varies in direct proportion to the capacitance:

$$I = C\frac{dv}{dt} \tag{6.4}$$

As an example of a rudimentary capacitive position sensor, see Figure 6.3. According to Equation 6.1, the capacitance will vary inversely as the distance, *d*, between the parallel plates is varied (as plate distance decreases, capacitance increases).

When several capacitors are connected in parallel, the total capacitance, C_T, is the sum of the individual capacitances:

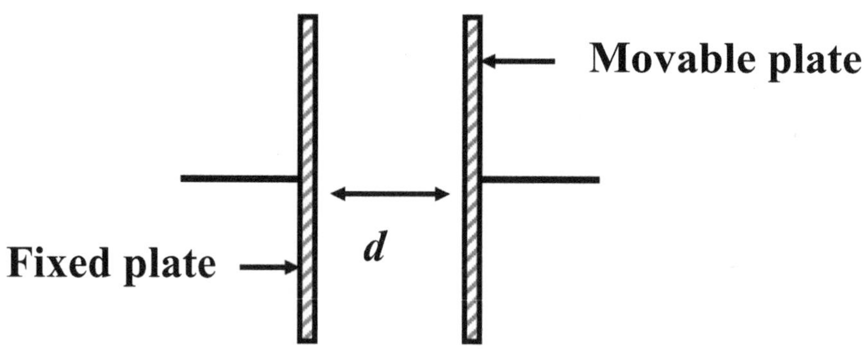

FIGURE 6.3 A rudimentary capacitive sensor—capacitance changes with plate separation distance, *d*.

$$C_T = C_1 + C_2 + C_3 \tag{6.5}$$

When several capacitors are connected in series, the total capacitance is equal to the reciprocal of the sum of the reciprocals of the individual capacitances:

$$C_T = \frac{1}{1/C_1 + 1/C_2 + 1/C_3} \tag{6.6}$$

The reader may have noticed that the formula for adding parallel-connected capacitors is similar to that for adding series-connected resistors (see the simple calculation method presented at the end of Section 4.2). Likewise, the formula for adding series-connected capacitors is similar to that for adding parallel-connected resistors.

6.3 DIELECTRIC CONSTANT

The permittivity of vacuum, ε_0, is 8.85×10^{-12} F/m (farads/meter). The relative permittivity of a material is equal to the ratio of the permittivity of the material to that of vacuum. Accordingly, relative permittivity is a ratio, dimensionless, and has no units [2, p 4]. The relative permittivity, ε_r, of vacuum is defined as 1. The ε_r of dry air is 1.0006, so it is almost the same as vacuum. The relative permittivity of a material is also called its dielectric constant, and this term is often used regarding the permittivity of dielectric materials of capacitors and capacitive sensors. The dielectric constant of a material is measured by using a standard capacitor, for example, a set of two parallel square plates, 1.00 cm on a side. The plates are made to sandwich a layer of the dielectric material to be measured, and the capacitance recorded as the first reading. Then another measurement is made with the dielectric material replaced by air or vacuum (since air and vacuum give almost identical results), with the same spacing between the capacitor plates as when the dielectric material in question was between them. This capacitance is recorded as the second reading. The first reading divided by the second reading is the dielectric constant (or relative permittivity). The dielectric constant of several common solid and liquid nonconductive materials is shown in Table 6.1 (as measured by the author, with vacuum as the reference of 1.0).

The dielectric constant of additional materials can be found at www.deltacnt.com/resources/level-and-flow/99-00032.

The higher the dielectric constant or permittivity of a material or space is, the slower an electric field will travel, and the larger will be the capacitance of a given size and spacing of parallel plates having the material or space between them. The velocity of electric field propagation is not a linear relationship with permittivity. The velocity of electric field propagation, v, in a material is inversely proportional to the square root of the permittivity and the permeability of the material:

$$v = \frac{1}{\sqrt{\varepsilon_a \mu_a}} \tag{6.7}$$

TABLE 6.1
Dielectric Constant of Several Common Insulating Materials

Material	Dielectric Constant
Vacuum	1
Air	1.0006
Paper	2 to 3
Nylon	4.5
Glass	4 to 9
Mica	6
FR-4 (PCB material)	4.8
Glycerin	60
Water	80
Gasoline	2
Kerosine	1.8
Mineral Oil	2.1

where ε_a is absolute permittivity, being the product of the permittivity of vacuum, ε_0, and the relative permittivity, ε_r of the material.

$$\varepsilon_a = \varepsilon_0 \varepsilon_r \tag{6.8}$$

and where μ_a is absolute permeability, being the product of the permeability of vacuum, μ_0, and the relative permeability, μ_r, of the material.

$$\mu_a = \mu_0 \mu_r \tag{6.9}$$

6.4 HISTORY OF CAPACITIVE POSITION SENSORS

Capacitive position sensors may also be called variable capacitance sensors. They must be combined with electronic circuits in order to operate, and are therefore not as old in the art as are the potentiometric position sensors. Early capacitive position sensors were built for a particular application, rather than being made as standard products for general use. Shorter stroke linear position sensors used variable spacing of the sensing plates, with air, another gas, or vacuum as the dielectric. In the 1970s, the author used a pressure sensor which incorporated an early capacitive position sensor. The capacitive sensor was coupled to a metal diaphragm that flexed in response to changes in pressure. The capacitive sensor measured the diaphragm deflection and produced an analog voltage output representing the applied pressure. This technology was sold as a standard line of pressure sensors.

The relatively large effort required to design electronic circuits to drive and condition the signal from capacitive sensors was substantially reduced when integrated circuits (ICs) for this use became widely available in the 1990s. These ICs enabled a reduction in the size and complexity of the circuitry, and included both the driving and the signal conditioning circuitry necessary for the operation of capacitive sensors. The lower cost also made the technology suitable for automotive use.

6.5 CAPACITIVE POSITION SENSOR DESIGN

The plates of a capacitive sensing element are typically constructed of one or more dielectric substrates onto which metallic layers are formed. In a two-plate sensor, the fixed plate can be a metal layer that is formed on a dielectric substrate by using standard electronic printed circuit techniques. The second plate would be a movable plate that is moved in response to the measurand. Other geometries are also possible, as explained later in this chapter. In addition to this basic technique, specialized circuitry, materials, and sensor configurations can be used to enhance resolution, signal to noise (S/N) ratio, sensitivity, stability, and temperature performance.

In accordance with Equation 6.1, the capacitance will change with a change in distance d between the plates, area A of the aligned area of the plates, and relative permittivity ε_r of the dielectric material. So, in practical position sensors, the dielectric material, the plate area, and/or the distance between the plates can be made to vary with the measurand. Figure 6.4 shows several representative diagrams of configurations that provide a changing capacitance in response to a change in linear position.

In Figure 6.4(a), capacitance between the movable and fixed plates increases with decreasing distance, d, between the plates. The plates are electrically conductive, such as being fabricated of metal sheet. The dielectric material between the plates is usually air. In Figure 6.4(b), the capacitance increases in proportion to the relative alignment between the movable and fixed plates, with maximum capacitance when

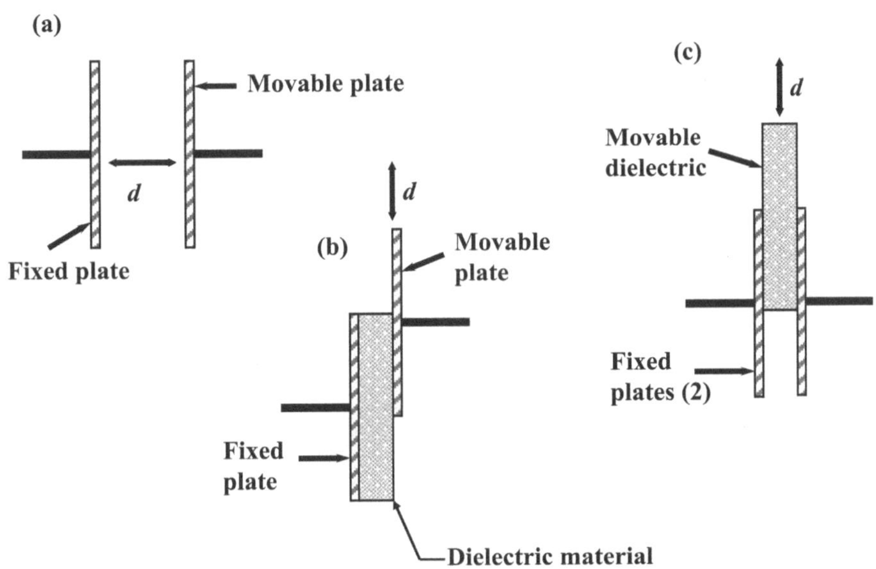

FIGURE 6.4 Several plate configurations for variable capacitance in response to a variation in linear position. (a) Variable spacing. (b) Variable area. (c) Variable dielectric.

the two plates are directly opposite one another. In Figure 6.4(c), the capacitance increases in proportion to the relative alignment of the dielectric material between two fixed plates. Assuming that the dielectric constant of the dielectric material is substantially greater than that of air, maximum capacitance is achieved when the dielectric material is fully within the space between the two plates. In Figures 6.4(b) and (c), the dielectric material is usually a plastic or insulative composite (such as printed circuit board fiberglass substrate material G10 or FR4) having a uniform thickness. G10 and FR4 fiberglass epoxy PC board materials are very similar to one another. The difference is a very slight change in the resin (epoxy). But the mechanical and electrical properties of interest are very similar. The name G10 was assigned by the National Electrical Manufacturers Association (NEMA) in the 1950s, designating a glass epoxy sheet composite. G10 originally stood for Glass (fiber) 10%, but may contain other percentages of glass. In 1968, NEMA designated FR4 as the flame retardant version. FR4 stands for Flame Retardant, and the 4 indicates that it comprises a woven glass reinforced epoxy resin. Since FR4 is flame retardant, it has mostly replaced G10. They are often written as G-10 or FR-4, and the spellings are acceptable with or without the hyphen.

Figure 6.5 shows two representative diagrams of configurations that provide a changing capacitance in response to a change in angular position (rotation).

In Figure 6.5(a), the rotor and stator (the movable and fixed plates, respectively) are each D-shaped (semi-circular) metal discs. The rotor is shown partially covering the stator, providing an intermediate amount of capacitance between the two plates. (The rotor is shown as being partially transparent, so that the stator can be seen. But in reality, the metal rotor would, of course, be opaque.) When aligned so that the rotor is not covering the stator at all, there is a minimum amount of capacitance. As shown, capacitance will increase as the rotor is rotated clockwise, and decrease if rotated anti-clockwise. The maximum amount of capacitance is provided when aligned so that the rotor is totally covering the stator. The total angular range that this configuration can measure is up to 180°.

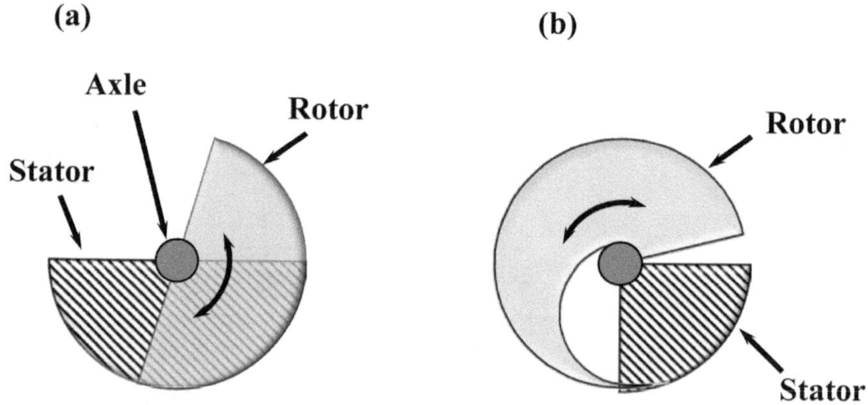

FIGURE 6.5 Rotor and stator configurations for variable capacitance in response to a variation in angular position. (a) 180° measuring range with semi-circular target. (b) 270° measuring range with spiral target.

In Figure 6.5(b), the stator is one-quarter part of a circular disc. The rotor has an approximately spiral shape. The rotor is shown slightly starting to cover the stator, providing a relatively small amount of capacitance between the two plates. As the rotor rotates anti-clockwise from the position shown, the capacitance will steadily increase. When the widest part of the rotor is completely aligned over the stator, the maximum capacitance will be reached. The total angular range that this configuration can measure is up to 270°.

There are trade-offs among cost, performance, and range of stroke length, that act to steer a designer toward choosing the best sensor configuration for a particular sensor design. Variable spacing, where the target being measured is a movable plate, is generally the least expensive approach. Longer stroke linear position sensors generally require variable area or variable dielectric designs. After choosing the sensing element configuration, a housing must be designed to provide alignment among the several components of the sensing element.

The type of capacitive linear position sensor that is most popular in industry works with a target supplied by the user. The sensor housing mounts near and parallel to the target. The target can be a metal plate that moves in an axis perpendicular to the sensing element within the sensor. A sensor of this configuration is shown in Figure 6.6, and has a sensing element similar to that shown in Figure 6.4(a).

In Figure 6.6, the linear distance to be measured is in the left and right direction of target position. This is called the x-axis, or sensitive axis. One error source however, is that the capacitance will also vary with motion at 90° to the x-axis. We'll call this the *y-axis*, or a *cross axis* (in the figure: into the paper, or up and down).

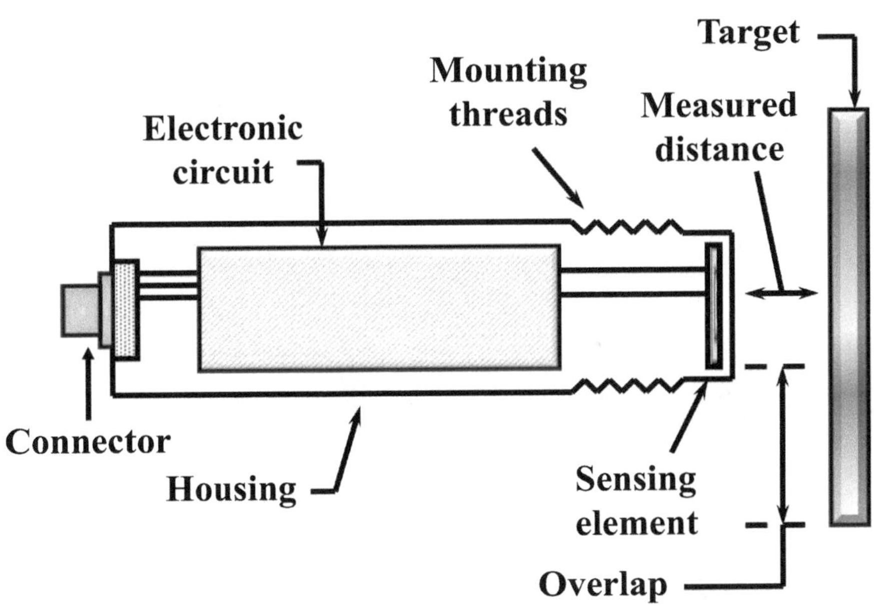

FIGURE 6.6 A capacitive linear position sensor, with user-supplied target, includes a sensing element, electronics, and a housing with mounting and connection means.

As shown in the variable capacitance distance position sensor of Figure 6.6, it is advantageous that the target be larger than the sensing element, by an amount called overlap. The overlap is pointed out in the figure for the overlap below the sensing element, but overlap can be present all around the main target area. This reduces cross-axis sensitivity, because (in the figure) moving the target slightly up or down will not change the measured capacitance.

The parallel-plate arrangement that was shown in Figure 6.4(b) is a variable-area-type capacitive sensor. In this case, the capacitance measured between the plates varies approximately with the percentage of the area of the movable plate that is aligned directly over the fixed plate. This assumes that the distance between the plates is small when compared with the dimensions along the sides of the plates. Overlap is also shown in the variable area position sensor of Figure 6.7.

Here, the movable upper plate is not as wide as the fixed lower plate. Thus, for a given position, there is little variation in capacitance with motion in the (y) cross-axis direction as long as the upper plate remains positioned over some part of the lower plate. A slight disadvantage is that the total capacitance is somewhat smaller due to the smaller width of the movable plate (as compared with a version having the top plate equally as wide as the bottom plate). There is a performance trade-off of accepting a lower signal level with the accompanying lower signal to noise (S/N) ratio in order to reduce the cross-axis sensitivity. This is an example of how the design of a sensing element can be optimized to fulfill an application requirement during the initial design and testing of a proposed sensor configuration. For this reason, and because each element of a sensor interacts with each other element, sensor design often includes several stages of iteration.

According to Equation 6.1, the capacitance will vary directly with the aligned area of the plates. Thus, in the arrangement of Figure 6.7, the capacitance will vary directly with the linear position of the movable plate as it moves in the sensitive axis.

Another consideration is that since the capacitance also varies inversely with the distance, *d*, between the plates, the thickness of the dielectric material must remain

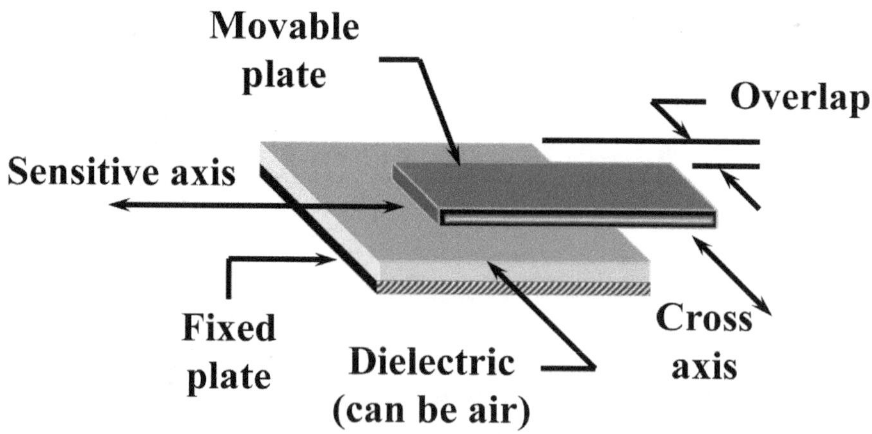

FIGURE 6.7 An improved plate design, with overlap, to reduce sensitivity to motion in the cross axis.

constant over time, and with temperature changes, and with motion in the sensitive axis. This can be accomplished in the case of an air dielectric by using mechanical means to ensure that the two plates will track in parallel, such as by fixing the bottom plate and having the upper plate ride within tracks along the two edges in the sensitive axis. An alternative way would be to use a plastic or ceramic dielectric spacer sandwiched between the electrode plates (the movable and fixed plates) instead of using air as the dielectric spacing. In addition to the fixed spacing between plates providing less cross-axis sensitivity in that direction (between plates), another advantage of this construction (with sandwiched dielectric plate) would be an increase in total capacitance due to the dielectric constant of the spacer being greater than air. The factor by which capacitance would increase by using a dielectric material other than air, according to Equation 6.1, is approximately equal to the relative permittivity of the dielectric plate material. With a greater overall capacitance, the absolute change in capacitance would also be greater. This may somewhat compensate for loss of signal level with a reduction in the movable plate size, due to overlap, as was shown in Figure 6.7.

Since most dielectric materials undergo some change in their permittivity with temperature change, this may need to be compensated in the mechanical design of the sensing element (such as having a reference section of the sensing element to determine dielectric changes), or in the electronics (such as having a temperature compensation program in the microcontroller).

An additional source of error in capacitive position sensors is the possible unwanted change in capacitance due to changes in the relative positions of nearby conductors. When an electrically conductive body moves within the vicinity of the sensor, it can provide additional capacitive coupling between the two plates of the sensor. Similarly, nearby electric fields can also be detected and thereby affect the capacitance reading. A shield can be added to surround the sensing area to reduce this problem. See Figure 6.8.

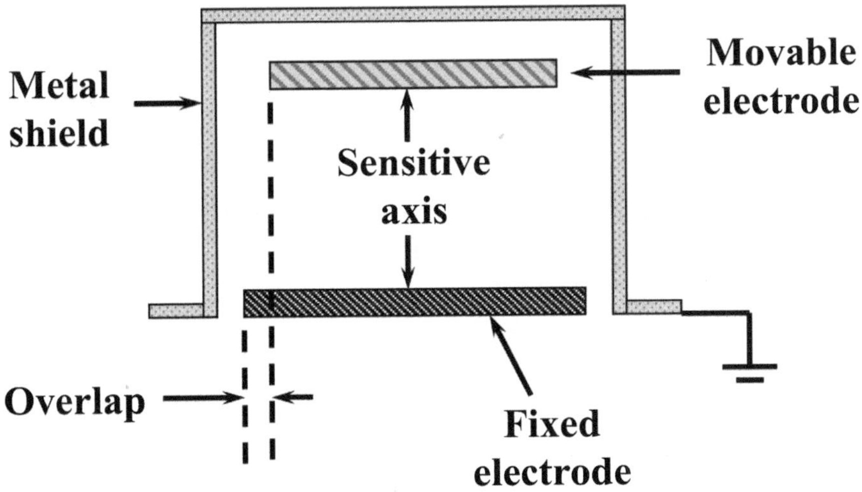

FIGURE 6.8 Adding a shield around a variable-spacing sensor to prevent unwanted error signals from nearby conductors or electric fields.

Overlap was explained relative to variable-area-type sensors with lateral motion to provide a variable area between the plates as shown in Figure 6.7, but the variable-spacing sensors of Figures 6.6 and 6.8 also benefit from overlap. According to Equation 6.1, the capacitance will vary inversely as the distance, *d*, between the parallel conductor plates. Like the variable-area design, though, there can be unwanted sensitivity in the cross axes in a variable-spacing sensor. Sensitivity to cross-axis motion in the y- or z-axes (if the sensitive axis is called x, in this case) is reduced by making the movable upper plate of smaller dimensions than the fixed lower plate. Again, the capacitance level is less, but this disadvantage is sometimes not as important as the advantage of reducing unwanted sensitivity in the cross axes.

6.6 ELECTRONIC CIRCUITS FOR CAPACITIVE SENSORS

The capacitive sensing element of a linear position sensor is driven with an AC or switching circuit. This allows the measurement of a variable frequency, phase shift, or amplitude change due to the change in capacitance in response to changes in the measurand. Typically, either an oscillator or a time period generator circuit would be used.

In a time period generating circuit, also called a mono-stable multivibrator, or one-shot, the length of the timed period depends upon the magnitude of the capacitance. See Figure 6.9.

A series of input pulses, usually at a fixed repetition rate (e.g. 100 kHz), are applied from an external source to the trigger input of the timer (such as a LMC555C integrated circuit). The timer generates a timed pulse width, having a time period depending upon the value of capacitance, *Cs*. As the capacitance, *Cs*, of the sensing element changes, the pulse width of the output waveform changes. If the output waveform is low-pass filtered to a DC voltage, such as by an RC filter or an active op amp filter, then a DC voltage indicative of the sensing element capacitance can be obtained.

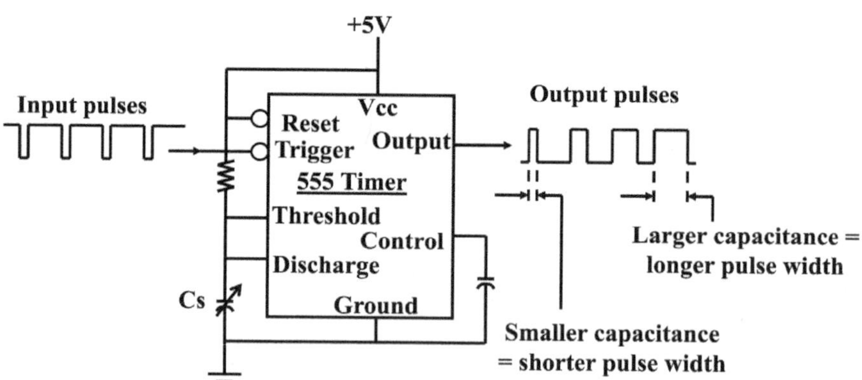

FIGURE 6.9 One-shot circuit in which the magnitude of a timed period depends upon the value of a variable sensor capacitance, C_S.

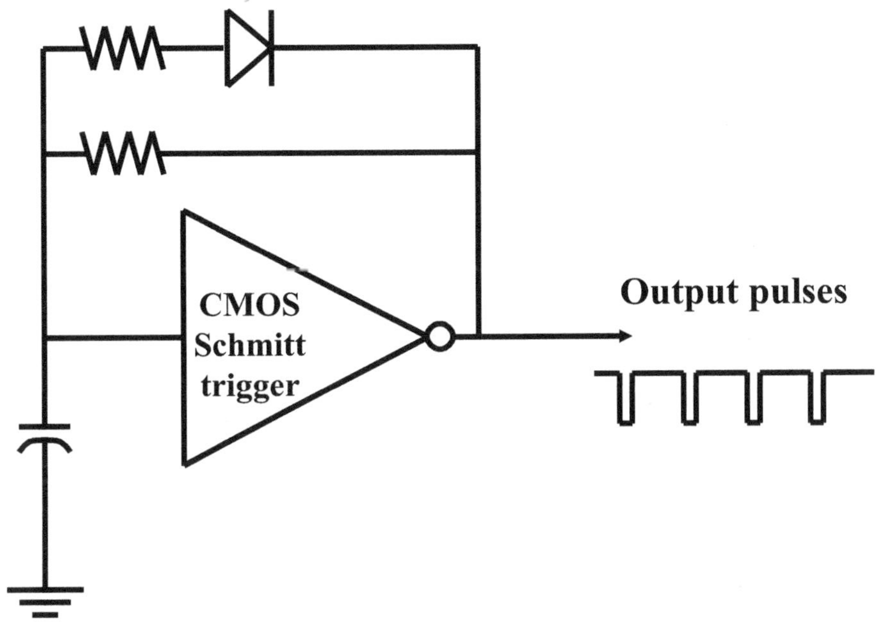

FIGURE 6.10 A free-running pulse generator to drive the one-shot circuit of **FIGURE 6.9** .

The input (trigger) pulses for the one-shot could come from a microcontroller circuit that sends the trigger pulse when a reading is desired, or the one-shot could be driven by a free-running pulse generator for continuous operation. A suitable free-running pulse generating circuit is shown in Figure 6.10.

The circuit of Figure 6.10 would generate a square wave output of approximately 50% duty cycle (± 5%) if the diode and its series resistor were not present. The frequency of the square wave would then depend on the RC time constant of the capacitor and the remaining resistor. For example, a resistance of 50k ohms and a capacitance of 100 pF would provide a time constant of about 5 μs, and the resulting square wave output would have a period of twice that (about 5 μs high time plus about 5 μs low time), or 10 μs, for a frequency of about 100 kHz. The exact frequency might be a little different from this calculation, depending on the type of CMOS circuit used. Adding the diode and its series resistor reduces the feedback resistance when the Schmitt trigger output is low. In the example, if the added resistor had a value of 2k ohms, then the low-time would depend on the parallel resistance of 50k and 2k ohms, and therefore be reduced to about 0.2 μs. So the total period of 5.2 μs would provide pulses at a rate of about 200 kHz. (Since the diode has a forward drop of about 0.6 volts, the actual low-time will be slightly larger than 0.2 μs.)

An alternative is to drive the capacitive sensing circuit with a free-running oscillator in which the frequency is determined by the sensor capacitance, C_S, and therefore by the measurand. See Figure 6.11.

The output frequency from the circuit of Figure 6.11 could then be fed to a circuit that produces a desired output signal from its variable frequency input. This could

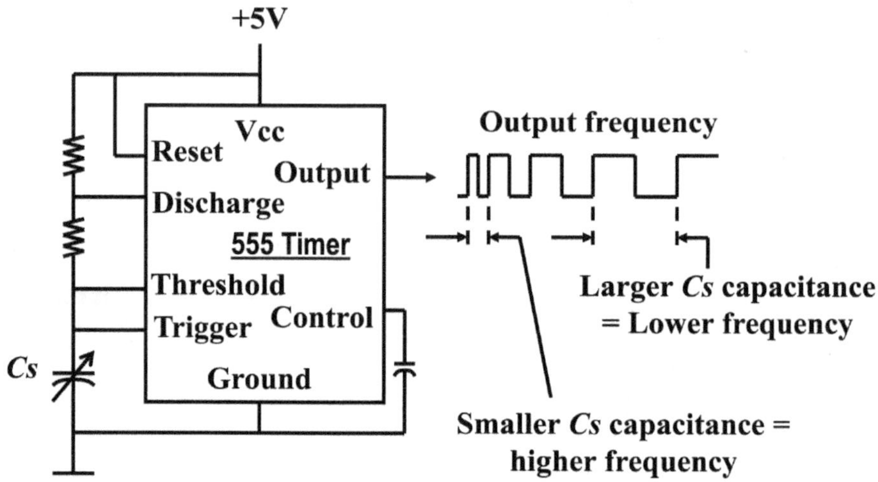

FIGURE 6.11 A free-running oscillator in which the sensing capacitance, C_S, determines the operating frequency.

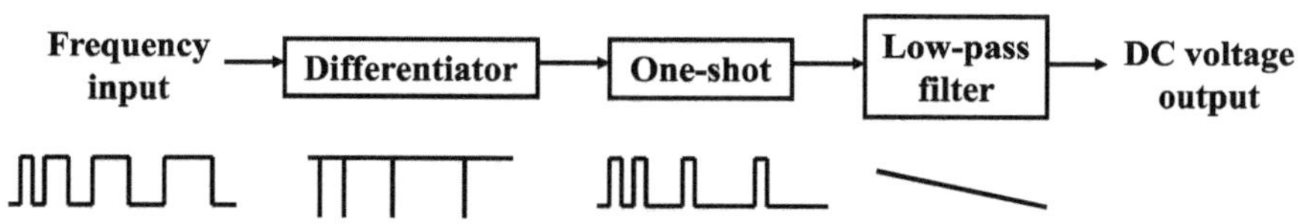

FIGURE 6.12 Frequency to voltage converter block diagram.

be a frequency to voltage converter circuit, if a voltage output is desired, a block diagram of which is shown in Figure 6.12.

Starting at the left of Figure 6.12, with a frequency input that varies with *Cs* (and with the measurand), a differentiator circuit provides very low-width pulses with each low-to-high transition of the input frequency. The differentiator circuit could be similar to that shown in Figure 1.11, but adjusted to produce only low-going output pulses, with high-going pulses prevented by a feedback diode. Alternatively, each high-to-low transition can be converted into a low-going pulse by a series capacitor, a pull-up resistor to V+, and a diode in parallel with the resistor as shown in Figure 6.13.

Next, the low-width pulses would next go to a one-shot, similar to that shown in Figure 6.9, but having a fixed capacitor in place of *Cs*. With a fixed capacitor, the output pulses from the one-shot would always have the same pulse width. So, there would be pulses of a fixed width, but how often the pulses occur would be controlled by the input frequency (and by *Cs*). Low-pass filtering of this pulse train would then yield a DC voltage signal indicative of the value of *Cs*.

Another way to obtain a variable DC voltage signal from an oscillator and variable capacitance is to develop an AC voltage that has an amplitude that varies with

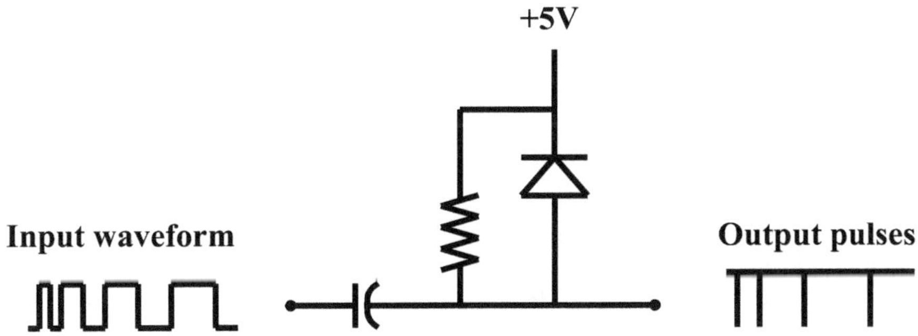

FIGURE 6.13 Using a series capacitor with resistor and diode to form a simple differentiator circuit.

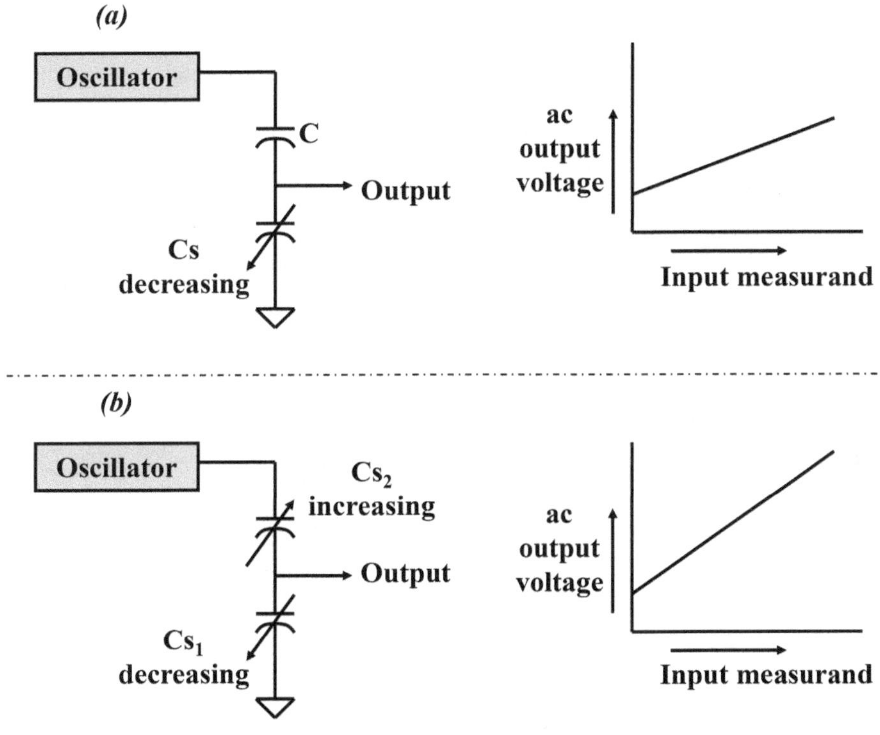

FIGURE 6.14 (a) Single-element circuit with variable voltage amplitude. (b) A dual-element circuit where one capacitance increases while the other decreases.

Cs (and the measurand), which can then be converted to a DC voltage. This is shown in Figures 6.14(a) and 6.14(b).

In Figure 6.14(a), fixed capacitance *C* and sensing capacitance C_S comprise a voltage divider circuit. The AC voltage across C_S increases as the capacitance decreases, because its capacitive reactance (impedance) increases. The capacitive reactance, X_C, is given by the equation:

$$X_C = \frac{1}{2\pi fC} \tag{6.10}$$

where X_C is the capacitive reactance in ohms, f is the frequency of operation, and C is the capacitance in farads (π has a value of approximately 3.1415926535).

With the increased capacitive reactance of C_S, the voltage drop is greater across C_S and less across C. In the dual-element sensor, one capacitance increases with the measurand, as the other capacitance decreases at the same time. This increases the output signal full-scale range and thus improves the S/N ratio. Once a variable AC voltage amplitude signal is obtained, it is then necessary to convert the voltage into a more usable form, such as a varying DC voltage. This can be done through the use of a rectifying or demodulating circuit. After a DC voltage is obtained, it can be converted to a digital signal by using an analog to digital converter, also called an A/D. Once digitized, a microcontroller can provide scaling and then produce the desired output signal.

Two methods to convert an AC voltage amplitude to a DC voltage include the use of either an asynchronous or a synchronous demodulator. An asynchronous demodulator often takes the form of a diode-type demodulator, and can be configured as in Figure 6.15. A single diode demodulator provides a variable voltage output with a DC offset, as shown in Figure 6.15(a).

The dual diode demodulator provides a differential voltage output as shown in Figure 6.15(b). (In this case, differential amplifier refers to the amplification of a

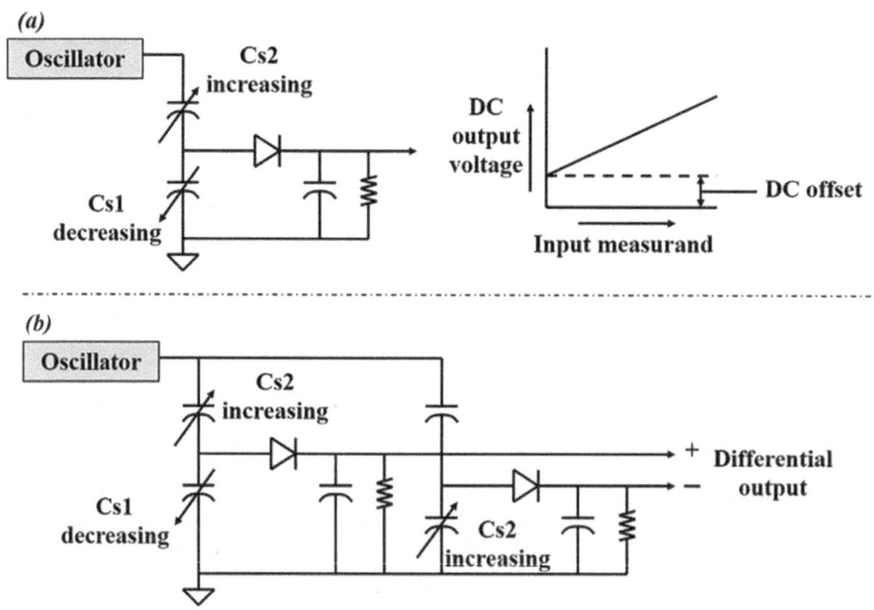

FIGURE 6.15 (a) A dual-sensing element with a single diode demodulator. (b) A dual-sensing element with a dual diode demodulator.

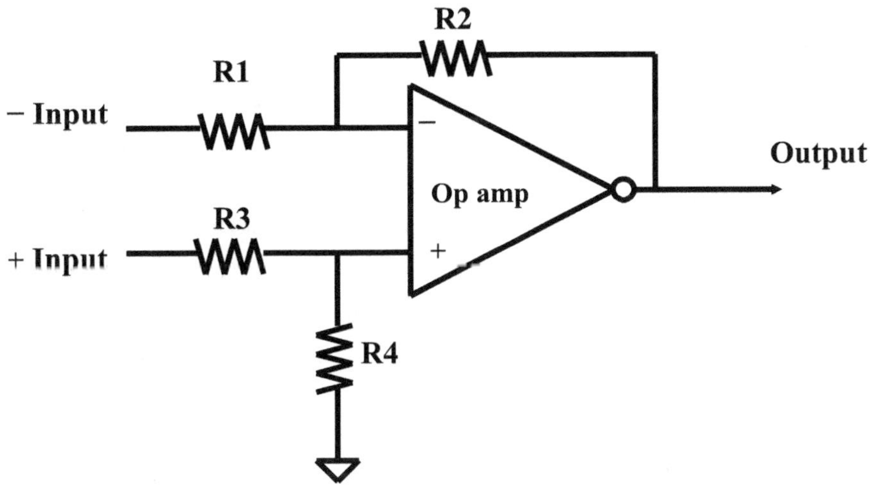

FIGURE 6.16 An op amp differential amplifier circuit.

voltage difference between the two inputs, not to a mathematical function of differentiation.) The differential voltage can be connected to the inputs of a differential amplifier similar to the one shown in Figure 6.16.

In the differential amplifier of Figure 6.16, input resistances *R1* and *R3* have equal values to each other, and should have resistance values high enough so they do not substantially load the circuit to which these inputs are connected. Values of 50k ohms are often used. *R2* and *R4* have equal values to each other, and the ratio of *R2* or *R4* to *R1* or *R3* defines the voltage gain of the amplifier. For example, if all four resistors are 50k ohms, the voltage gain is 1.

If the (+) output of the demodulator of Figure 6.15(b) is connected to the (+) input of a differential amplifier, and the (–) to the (–), then the amplifier output will center on zero volts when the two input voltages are the same as each other. For the single diode demodulator, a reference voltage or resistive voltage divider (not shown) can be set to equal the DC offset voltage of the demodulator and fed into the differential amplifier (–) input. The demodulator output would then be fed to the amplifier (+) input.

A diode demodulator is asynchronous, because there is no timing requirement between the signal phase and the operation of the diode. The diode just conducts when the anode voltage is higher than the cathode voltage, thus charging up the filter capacitor. With silicon diodes, the anode must be approximately 0.60 volts (depending on conduction current, from about 0.4 to 0.7 V) higher than the cathode for forward conduction of the diode to take place. This 0.60 volts is called the forward bias voltage of the diode. One problem is that the magnitude of the forward bias voltage changes with diode conduction current, temperature, and mechanical stress on the diode. The temperature sensitivity is about 2.2 mV per °C. The temperature sensitivity can be a major source of error in a diode demodulator. In a dual-diode demodulator, the temperature sensitivity of the two diodes approximately matches each other, but temperature compensation may be needed to obtain high accuracy signals. Using a matched diode pair molded into one case can help, but thermal gradient-induced stress can cause unpredictable shifts in forward bias voltage. One technique used by the author to reduce the thermally induced stress on the die within a diode with leads (not surface mount) is to form a loop in the leads before soldering them into the printed circuit board, as shown in Figure 6.17.

In a synchronous demodulator, an electronic switch is used instead of a diode. The switch closes when the signal voltage amplitude is higher and opens when it is lower. Closing the switch at the proper time allows the filter capacitor to charge, but without the errors of the forward bias voltage of a diode demodulator. For proper operation, the switch(es) must be opened and closed at the appropriate times, synchronous to the timing of the oscillator or to the signal itself. With a square wave oscillator driving a capacitive sensor, the switch can be operated by the oscillator output as shown in Figure 6.18.

In Figure 6.18, the switch can be a CMOS analog switch, such as a 4066 type. In the figure: when the oscillator output is high (for example, +3.3 or +5.0 volts), the switch is closed, while at other times it is open. As the capacitance of CS_1 increases and CS_2 decreases, the voltage divider formed of these two capacitances provides a higher voltage to the switch (because the impedance of CS_2 is becoming greater as the impedance of CS_1 is becoming smaller), and the RC filtered DC output will be greater. So, the DC output voltage changes with the sensing element capacitances. A small amount of ripple is shown in the output, to indicate charging the capacitor when the switch is closed (1–2), and some discharging during the switch open time (2–3). A larger value capacitor can reduce the amount of ripple.

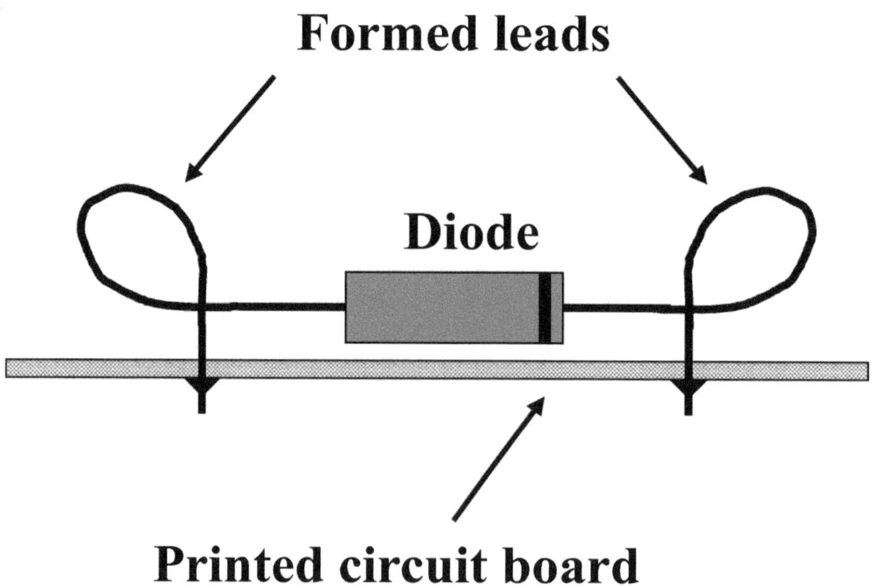

FIGURE 6.17 When using a demodulator diode provided with leads, thermally induced stress can be reduced by forming the leads into loops.

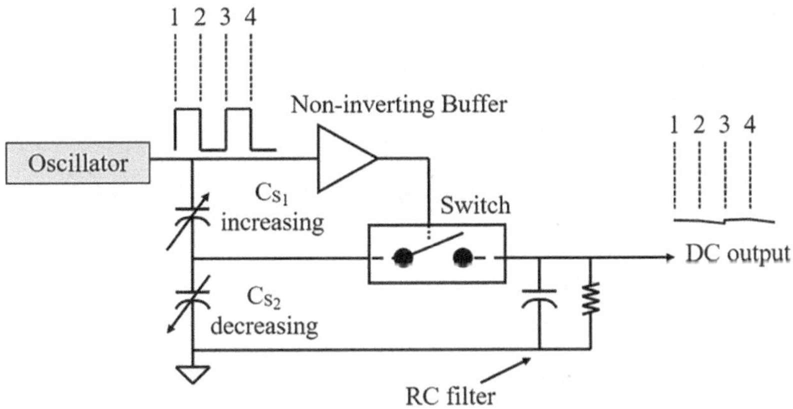

FIGURE 6.18 A synchronous demodulator derives a DC amplitude signal from a variable AC waveform by operating switches synchronously to the AC waveform.

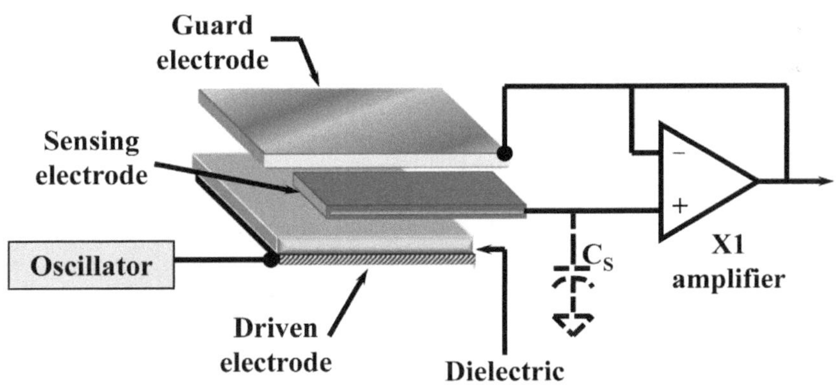

FIGURE 6.19 Driving a guard electrode with a voltage similar to that of the movable sensing plate reduces the sensitivity to electrical noise in the environment.

6.7 GUARD ELECTRODES

It is likely that a non-grounded movable plate of a capacitive sensor will pick up unwanted signals from nearby circuitry, as well as being affected by capacitive coupling with nearby electrical conductors. To minimize the errors caused by these sources, a guarding scheme is often used in conjunction with capacitive sensing elements and sensor circuits. The guard is a protective conductor plate, pattern, or other shape that surrounds, or is adjacent to, the movable sensing plate. The guard usually moves with the plate, and is electrically driven to a voltage equal to that of the plate. A shield could be implemented by connecting the guard to circuit common or ground (then it would be called a shield instead of a guard), but driving the guard as described prevents the guard capacitance from loading of the sensing capacitance. See Figure 6.19.

This is accomplished by driving the guard electrode from the low impedance output of a buffer amplifier. In the figure, the high impedance non-inverting input

(+) of the amplifier senses the AC voltage that appears on the sensing electrode. The amplifier is configured as a voltage follower, and has a gain of one (X1), so its output is the same voltage as the non-inverting input, but buffered so that it can drive the guard electrode without being affected by electrical noise of the environment. Because they are at the same voltage and phase (even though it is AC), there is no current flow between the sensing electrode (movable plate) and the guard (other than the input bias current of the op amp, but an op amp with a very low input bias current should be selected for this circuit).

6.8 EMI/RFI

An RC oscillator, with a capacitive sensing element comprising all or part of the frequency determining capacitance, has a tendency to synchronize or beat with other nearby oscillators and EMI sources that include energy at or near one of the harmonics of the frequency at which the capacitive sensing element is operating. (This is also true of most oscillators, not only RC oscillators.) As mentioned earlier, a harmonic of a particular frequency is that frequency, or any integer multiple of it. The particular frequency is the fundamental, or first harmonic. Two times that frequency is the second harmonic, three times is the third harmonic, and so on.

In the previous figures, the sensing capacitance was shown as being driven by a square wave oscillator. This was done for simplicity, but many sensors use a sine wave excitation instead of a square wave so that less electrical noise is radiated, and so that the signal can be more accurately demodulated. For synchronous demodulation, there would be a voltage (or an angle) along the sine wave at which the demodulation switch turns on (closes). When measuring the instantaneous voltage of a sine wave in order to determine the time to close or open the demodulator switch, the following conditions occur: as the capacitance of a capacitive sensing element is just starting to charge up during a given period of the oscillation, there is a larger voltage difference between the voltage across the capacitance being charged and the switching point of the demodulator circuit connected to the capacitance. This larger voltage difference means that it is less likely that it will be surpassed by noise voltage peaks, which could trigger the switching circuit prematurely. As the capacitor voltage nears the switching voltage, however, the amount of noise voltage needed to trigger the switch prematurely becomes much smaller. So the result is a variation of the circuit sensitivity to a noise spike, and the variation is regular with respect to the oscillator period. This can be avoided by producing the sine wave inside a microcontroller, so that the microcontroller always knows the angle of the sine wave generator. Then, the demodulator can be switched based upon the sine wave generator angle, rather than by measuring the sine wave voltage.

Also, when there are two oscillators in close proximity, a similar effect may occur, in which an electromagnetically coupled signal from a first oscillator is received in the circuitry of a second oscillator. The result is that the second oscillator frequency tends to track the interfering frequency of the first oscillator at a rate proportional to the phase difference between the two frequencies. This is called beating. The difference between the frequencies of the two oscillators is called the beat frequency. As an oscillator operating frequency beats with an interfering signal near one of its harmonic frequencies, an amplitude modulation of the oscillator output may occur. (One may notice this beat frequency phenomenon upon hearing two non-synchronized marine outboard engines operating at nearly the same RPM: the sound gets louder and softer at a regular interval. That interval is the period of the beat frequency, the *beat frequency* is the difference in frequency between the two engines.) An engine RPM synchronizer eliminates this by controlling the spark timing of one engine so that its RPM will exactly match that of the other engine. In a sensor, beating with an external energy source may be a cause of error in the sensor output, sometimes producing an unwanted amplitude modulation at the beat frequency. For example, if one sensor has an oscillator frequency of 100,000 Hz and a nearby sensor has an oscillator frequency of 100,005 Hz, each sensor may have an amplitude modulation of its output at a frequency of 5.0 Hz.

Even when there is not a sympathetic frequency relationship between the sensor electronics and a noise source, interference may still occur. Energy spikes in the noise level can induce error spikes in the sensor. Both of these effects can be reduced by shielding or guarding of the sensing element. Adequate bypassing of possible noise voltage on the power supply and other input/output lines can also help. Sometimes, when multiple sensors of the same type are operated near one another, it may become necessary to synchronize the frequencies of the several oscillators.

6.9 TYPICAL PERFORMANCE SPECIFICATIONS AND APPLICATION

The most popular type of capacitive linear position sensor is designed to detect the position of a metal target with respect to the sensor element, as was shown in Figure 6.6. A typical set of specifications for this position sensor configuration is shown next.

Operating range	0 to 8 mm from sensing face
Supply voltage	8 to 26 VDC
Output load	10k ohms minimum
Nonlinearity	±0.25% of full range
Repeatability	0.05% of full range
Resolution	0.01% of full range
Output ripple	≤5 mV RMS
Temperature coefficient	±0.01% of full range/°C
Operating temperature	−40 to 85°C
Storage temperature	−40 to 100°C
Housing dimensions	25 mm dia. × 75 mm long, with mounting thread

Industrial process systems often implement sensors of this type where non-contact measurement of roller position or other moving parts is desired. There is no maintenance required because there are no rubbing parts, although the full-scale range is limited when compared to distributed impedance, magnetostrictive sensors, or long stroke LVDTs. The long-term accuracy is about the same as that of an LVDT.

A representative specification of an angular (or rotary) capacitive position sensor is shown here.

Angular range	0 to 300°
Supply voltage	15 VDC ±10%, at 6 mA

Output	0 to 5 VDC
Output load	10k ohms minimum
Nonlinearity	±0.5% of full range
Repeatability	0.05% of full range
Resolution	0.01% of full range
Output ripple	≤5 mV RMS
Slew rate	6,000 RPM maximum
Temperature coefficient	±0.01% of full range/°C
Operating temperature	−40 to 85°C
Storage temperature	−40 to 100°C
Maximum starting torque	0.42 mNm (milli Newton meters)
Inertia	15 gcm^2

6.10 MANUFACTURERS

Some manufacturers of capacitive linear and angular/rotary position sensors include the following:

Balluff	www.balluff.com
Baumer	www.baumer.com
Capacitec	www.capacitec.com
Efector	www.ifm.com
Fargo Controls	www.fargocontrols.com
Loadstar Sensors	www.loadstarsensors.com
Locon Sensor Systems	www.locon.com
Micro-Epsilon	www.microepsilon.com
Microsense	www.microsense.net
Monitor Technologies	www.monitortech.com
RDP Electrosense	www.rdpe.com
Revolution Sensor Company	(design) www.rev.bz
Sick	www.sick.com
Turck	www.turck-usa.com

6.11 QUESTIONS FOR REVIEW

1. The amount of charge stored in a capacitor is measured in:

a. Ohms per square
b. Coulombs
c. Amperes/meter
d. Farads
e. A circle

2. Permittivity is normally measured in:

a. Vacuum
b. Farads per meter

c. A glass enclosure
d. Charge/m^2
e. Free space

3. Dielectric constant may also be called:

a. Conductance
b. Absolute resistivity
c. Irrational
d. Relative permittivity
e. Synchronous

4. Conductor driven to an equal voltage as a sensing electrode it surrounds is called a:

a. Ground plane
b. Common
c. Guard
d. Primary element
e. Secondary

5. When not timed relative to the signal phase, a demodulator is called:

a. Nonfunctional
b. Locked
c. Rectifying
d. Incremental
e. Asynchronous

6. With a capacitor operating at higher frequency, the value of capacitive reactance:

a. Is not affected
b. Is less
c. Is greater
d. Switches on
e. Is rectified

7. Relative permittivity is the ratio of the permittivity of a material to that of:

a. Glass
b. Water
c. Vacuum
d. A conductor
e. The earth

8. The transfer of one coulomb of charge per second through a conductor is:

a. One abampere
b. Very difficult
c. One ampere
d. Not possible
e. One henry

REFERENCES

[1] H. Olson, *Dynamical Analogies*. New York: D. Van Nostrand, 1943.
[2] P. Neelakanta, *Handbook of Electromagnetic Materials*. New York: CRC Press, 1995.

7 Inductive Sensing

7.1 INDUCTIVE POSITION SENSORS

The first thoughts in designing an inexpensive position sensor often involve an inductive type of sensing element. This is because they can be simple in theory, as the basic sensing element comprises one or more coils of wire, together with a movable core. See Figure 7.1. The transducers of inductive position sensors are sometimes called linear variable inductance transducers (LVITs), or rotary variable inductance transducers (RVITs). In some models, the transducer may be sold separately (i.e. the coil(s) contained within a housing, to be used with external electronics), and other models may comprise a complete sensor including the electronic circuit within the housing.

Though an inductive sensor seems simple, it may sometimes be difficult to achieve an acceptable trade-off among cost, performance, and measuring range. Still, linear versions have a wide range of practical applications where fairly accurate measurements are required over a relatively short measuring range (up to several tens of mm), or where medium accuracy over a medium range will suffice (up to a few hundred mm). And rotary or angular versions of inductive position sensors have advantages over a potentiometric type, due to their ruggedness and long life. Linear sensing models are generally not considered to be very practical over longer measuring ranges of more than about 500 mm due to the difficulty of maintaining uniformity along its length when winding the coil. The result is that longer ranges become more expensive than with other sensing techniques. Similarly, angular/rotary versions are often used where fairly accurate measurements are required over a relatively small angle of rotation, such as 0° to 90°. Or where medium accuracy over a somewhat greater angle of rotation measurement range is needed.

Complete sensor designs are shown in the figure, with electronics included within the housing for driving the sensing coil, demodulating, amplifying, filtering, and scaling of the signal.

For both linear and angular types, it is more difficult than with some other technologies (such as distributed impedance or magnetostrictive) to make each sensor perform the same as the next, due to the imperfect control possible in winding a coil of magnet wire. Like a capacitive position-sensing element, an inductive one does not produce a directly usable output. An electronic circuit is needed to provide an AC drive to the sensor, then demodulate and condition the signal to produce the desired output.

The outward appearance of the sensor housing is often similar to that of a Linear Variable Differential Transformer (LVDT) or Rotary Variable Differential Transformer (RVDT) (see Chapter 8 on the LVDT/RVDT). Since the sensing element is a non-contact device, the complete sensor can operate with very little wear or mechanical loading force. Typical inductive position sensors do make contact with the part being measured, however, through the use of a movable pushrod or a rotating

DOI: 10.1201/9781003368991-7

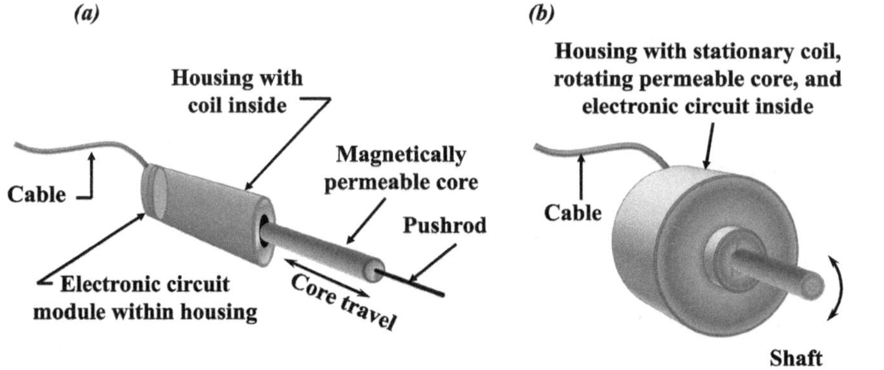

FIGURE 7.1 (a) Linear, (b) angular. An inductive position sensor includes a sensing coil, a magnetically permeable core, electronic circuit, and housing. The electronic circuit can be included within the housing.

shaft. On a linear sensor, one end of the pushrod is connected to or rides against the measured part, while its other end moves a magnetically permeable core piece within the sensing element. With a rotary sensor, the member to be measured is connected with the sensor's rotatable shaft. A somewhat flexible connection will provide allowance for any slight misalignment between the two rotation axes.

Inductive proximity sensors (prox sensors) are usually not called linear position sensors, so are not covered in this work, except for a short mention here. They are truly non-contact sensors, since they directly detect the magnetically permeable and/or electrically conductive member to be measured. Because they operate at relatively high frequencies, changes in the coil circuit due to the magnetic property of a ferromagnetic target (or by eddy currents dissipated in an electrically conductive target) can be detected. The signal from most prox sensors is a switched output that switches from low to high when a target comes within the calibrated range. Continuous functions are also possible, but are typically nonlinear, and affected by the shape and size of the target.

7.2 INDUCTANCE

Inductance is a property of all electrical conductors when conducting an electric current. This is because a current flowing in a conductor is always accompanied by a magnetic field. Electromagnetic energy is stored in the field. Inductance is a measure of the capability to store electromagnetic energy. In 1886, Oliver Heaviside first used the word inductance. An amount of inductance is indicated in henries (H), in honor of American physicist Joseph Henry, who independently discovered inductance at around the same time as British scientist Michael Faraday (the discoveries took place around 1831–32). The symbol L is used to indicate an inductor or inductance, in honor of Russian physicist Heinrich Lenz. For example, an inductor L_1 may have an inductance of 1.0 H: $L_1 = 1.0\ H$. (The plural is henries.)

An analogy can be drawn between an electrical circuit with current flowing (in amperes, i.e. coulombs/second) through an inductance, and a water system with water flowing (e.g. in liters/second) in a pipe when the pipe wall has an elastic quality.

For the purpose of this analogy, it is easier to envision a pipe made from rubber or an elastomer (such as a long, thin balloon), although a metal pipe is also elastic; but the elasticity of a metal pipe is too small to observe visually. When water is forced to flow in such an elastic pipe, the forcing pressure tends to expand the elastic pipe as the flowing water progresses. During this time, when the pipe is still expanding, energy is being stored as a volume of water at a pressure due to the pipe elasticity. When the pipe has been fully expanded for the given forcing pressure, the water will be flowing out of the pipe at a steady rate equal to that flowing into the pipe, as long as the amount of water flowing into the input to the pipe is constant. Under this condition, there is a steady pressure drop across the length of the pipe (from one end to the other) which depends on the flow rate and the pipe diameter. This pressure drop is similar to a voltage drop in an electrical circuit. If the force being applied to the water at one end of the pipe is increased or decreased, the total flow rate at the other end will tend to remain the same for a while because the elastic pipe will make up the difference. For example, with an abrupt increase in the input pressure, the pipe will further expand (building up more energy), thus tending to delay an increase in the output flow rate for a while. For an abrupt decrease in the input pressure, the pipe will reduce its diameter, thus tending to support the flow at its previous rate for a while. Additionally, if a valve is placed in the output line and the opening is suddenly reduced in size, the inertia of the water will cause a pressure increase in the pipe. For a while, this increase in pressure will act to delay the output flow rate from changing by as much as the smaller opening size would otherwise dictate.

The pressure drop from one end of the pipe to the other is analogous to the voltage drop across an inductor coil. A function of the inertia or mass of the moving water, together with the elasticity of the pipe, is analogous to the inductance of an electrical inductor. (The inertia and elasticity have similar effects as the core permeability and the number of turns of an inductor.) The flow rate (e.g. in liters of water per second, for example) is analogous to the electrical current through the inductance (i.e. coulombs per second). An inductance in a circuit tends to impede a change in current flow, just as the elastic pipe and the inertia of the water work together to impede the change of water flow rate. An inductance, likewise, depends on both the core permeability and the number of turns of wire in the coil creating the inductance. A simple inductor is formed of a coil of wire, with a material in the center of the coil being called the core, as shown in Figure 7.2.

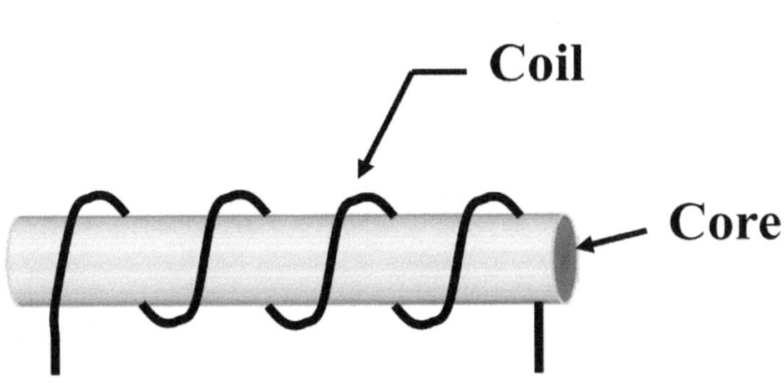

FIGURE 7.2 Construction of a simple inductor, with coil and core.

A coil with an air core does produce an inductance, but a higher inductance can be formed when utilizing a core having a higher magnetic permeability. When several inductors are connected in series, the total inductance, L_T, is the sum of the individual inductances:

$$L_T = L_1 + L_2 + L_3 \tag{7.1}$$

When several inductors are connected in parallel, the total inductance is equal to the reciprocal of the sum of the reciprocals of the individual inductances:

$$L_T = \frac{1}{1/L_1 + 1/L_2 + 1/L_3} \tag{7.2}$$

The formula for adding series-connected inductors is similar to that for adding series-connected resistors or parallel-connected capacitors. Likewise, the formula for adding parallel-connected inductors is similar to that for adding parallel-connected resistors or series-connected capacitors.

The coil in Figure 7.2 is, for simplicity, shown as a single layer with spacing between each turn and the next turn, although real coils are usually made with many layers of wire, and with the turns one next to one-another. A single-layered or multilayered coil that is wound onto a straight form is called a solenoid-wound coil. The core material may be air, but instead, it is often made from a material having a higher magnetic permeability than air, such as the alloys or mixtures containing iron, nickel, and/or cobalt, with iron or nickel usually being the highest percentage ingredient of the alloy.

A multilayered solenoid-wound coil is normally used in the design of an inductive linear position sensor, and with a movable core made of a ferromagnetic alloy. An angular or rotary inductive sensor can use an appropriately shaped core and a cylindrical coil, as shown in Figure 7.3.

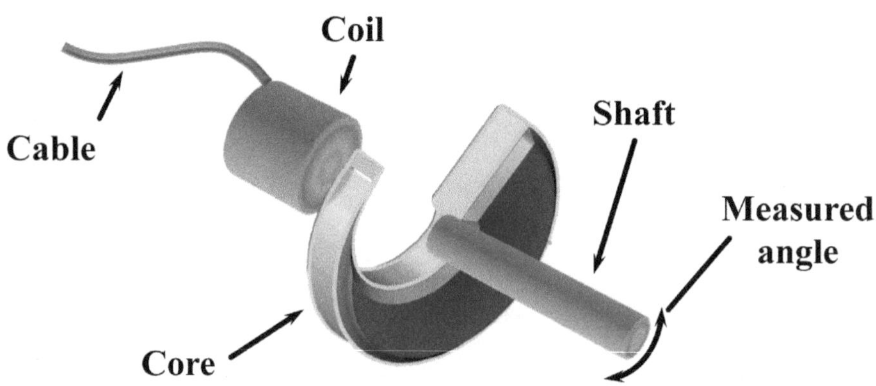

FIGURE 7.3 One of many possible configurations for an inductive angular position sensor. The shaft and core are mechanically coupled to rotate together.

In the figure, a magnetically permeable core is designed to incorporate a nonuniform cross-section so that the volume of core material in close proximity with the coil varies with rotation of the core axle or shaft. Alternatively, a coil could have a large-diameter bore, with the core passing through the coil, or the coil could be of a different shape.

With a higher permeability core, a higher inductance is achieved for given coil dimensions and number of turns. A coil assembly comprising a coil and a permeable core, with that coil assembly being affected by a rotating core passing near the coil assembly, will provide a greater change inductance change as compared to a similar coil if wound with an air core. This is true whether the position sensor is linear or angular. The formula for finding the inductance, L, of a coil varies with the shape of the coil and core, but for a solenoid-wound coil on a straight cylindrical core, where the core length is much greater than the core diameter, and the coil length is approximately the same as the core length, the inductance is approximately:

$$L = \frac{N^2 \mu_a A}{l} \tag{7.3}$$

where the unit for inductance is the henry, N is the number of turns of wire that are wound around the core, A is the cross-sectional area of the coil in meters^2, μ_a is the absolute permeability of the core in henries/m, and l is the core length in meters. (Absolute permeability, μ_a, is the product of the relative permeability of the material, μ_r, and the permeability of vacuum, μ_0; see Section 7.3).

The magnitude of electromagnetic energy, W, stored in an inductor is directly proportional to the inductance, L, and varies with the square of the current I, flowing through the coil.

$$W = \tfrac{1}{2} L I^2 \tag{7.4}$$

where W is the energy expressed in joules, or watt-seconds. W remains constant when the current in the inductor is constant.

Inductance in an electrical circuit opposes a change in current by generating what is called a "back EMF". *Electromotive force* (EMF) is measured in volts, and is the theoretical agent that tends to produce or maintain an electric current in a circuit. It is the electrical energy per unit charge:

$$E = W / Q \tag{7.5}$$

where E is EMF in volts, W is energy or work in joules, and Q is charge in coulombs. The direction of an electromotive force causing a charge to move in a portion of a circuit is the direction in which a positive charge would be forced to move in a circuit containing only this one source of EMF. When several sources of EMF exist in a circuit, the resulting EMF of the circuit is the arithmetic sum of all of the sources of EMF in the circuit. Those acting in one direction being called positive and those in the opposite direction being called negative. Those acting in a direction opposite to the resultant EMF are called back EMF.

When current through a coil having inductance, *L*, is changing, the EMF (*E*) across the coil is proportional to the rate of change of current (*di/dt*):

$$E = L\frac{di}{dt} \tag{7.6}$$

where the unit for inductance is the henry. The inductance of a conductor can be increased by forming it into a coil, having cross-sectional area *A*, length *l*, and *N* number of turns. The resulting inductance of the coil is:

$$L = \frac{N^2}{\Re} \tag{7.7}$$

where $\Re$ is magnetic reluctance.

For a coil having a uniform length *l*, which is much larger than its cross-sectional dimension, and core material with relative permeability μ_r, the reluctance is:

$$\Re = \frac{l}{\mu_0 \mu_r A} \tag{7.8}$$

where μ_0 is the magnetic permeability of vacuum ($4\pi \times 10^{-7}$ H/m), and μ_r is the relative permeability of the core material on which the coil is wound. Relative permeability of a material is the ratio of its permeability to that of vacuum (μ_r is near 1 for most materials, except that it is much higher for ferromagnetic materials such as iron, nickel, and cobalt). So then, for the case where the length, *l*, is much larger than the coil cross-section, formulas 7.7 and 7.8 can be combined to find the coil inductance, which is approximately:

$$L = \frac{N^2 \mu_0 \mu_r A}{l} \tag{7.9}$$

This is the derivation of formula 7.3. In addition to the inductance, a real coil has a parallel capacitance due to the adjacent turns of the wire. There is also a series resistance resulting from the resistance of the length of wire used to form the coil. A circuit approximation of a real inductor is shown in Figure 7.4.

The parallel capacitance, *C*, is usually shown across the series combination of inductance and resistance, as in the figure, but may sometimes be shown across only the inductance. The capacitance and resistance are each actually distributed along the length of the inductance among the various turns of the coil. Because of the parallel capacitance, there exists a frequency where the capacitive reactance is equal to inductive reactance. This frequency is called the self-resonant frequency of the

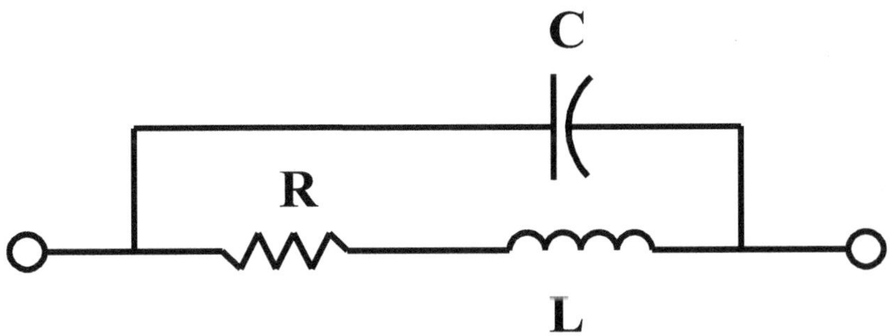

FIGURE 7.4 Circuit representation of the inductance, parallel capacitance, and series resistance of a real inductor.

coil, or simply the resonant frequency of the coil. When the coil is energized at its self-resonant frequency, the impedance is resistive, since the capacitive and inductive reactances are equal and act in opposite phase directions. (Impedance, in ohms, is the vector sum of capacitive reactance, inductive reactance, and resistance. By convention, the imaginary part is positive for capacitive and negative for inductive.) Accordingly, when below the resonant frequency, the impedance is inductive. Above it, the impedance is capacitive. The formula for inductive reactance is:

$$X_L = 2\pi FL \tag{7.10}$$

where X_L is the inductive reactance in ohms, L is the inductance in henries, and F is the operating frequency. The formula for capacitive reactance is:

$$X_C = \frac{1}{2\pi FC} \tag{7.11}$$

where X_C is the capacitive reactance in ohms, and C is the capacitance in farads. So, it follows that the condition for resonance is when

$$2\pi FL = \frac{1}{2\pi FC} \tag{7.12}$$

When using a fixed inductor as a circuit component, the operating frequency of the circuit must be far below the inductor's self-resonant frequency, so that it may act purely (or at least, primarily) as an inductor.

7.3 PERMEABILITY

By convention, a magnetic field is often represented by lines of flux: the greater the magnetic flux density, B, the closer together the flux lines are drawn. The magnetic permeability, μ, of a material, is the ability of that material to support magnetic flux. In vacuum, the magnetic flux density, B, is related to the magnetic field strength, H, by:

$$B = \mu_0 H \tag{7.13}$$

where B is expressed in newtons per ampere meter (N/(A·m); H is the magnetic field strength, or magnetizing force, expressed in amperes per meter (A/m); and μ_0 is the permeability of vacuum.

The permeability of vacuum is $4\pi \times 10^{-7}$ H/m, or 1.257 μH/m (μH/m is pronounced "micro henries per meter"). Since the relative permeability of a material is a ratio of the permeability of the material to that of vacuum, it has no unit of measurement. The relative permeability, μ_r, of vacuum is defined as 1. The relative permeability of dry air is 1.0006, so it is almost the same as vacuum. Accordingly, magnetic flux density, B, and magnetic field strength, H, are seemingly similar when no magnetically permeable material is present. One may consider H as a basically theoretical quality for use in calculation, whereas B accounts for permeability of materials and pertains to a real system. But often, the term magnetic field strength (or magnetic field intensity) is used when one is actually referring to magnetic flux density, B.

All materials have magnetic properties to some extent. Those known as ferromagnetic materials have rather high relative permeabilities, on the order of 50–5,000, depending on the alloy used. These include iron, nickel, cobalt, and their alloys. An atom in a ferromagnetic material has a magnetic moment that tends to align itself with an applied magnetic field.

Non-ferromagnetic materials typically have a relative permeability very close to 1. Those with a relative permeability only slightly above that of vacuum, such as aluminum and oxygen, are called paramagnetic. Materials having a permeability less than that of vacuum, such as copper and nitrogen, are called diamagnetic.

The higher the relative permeability of a medium, the slower a magnetic field will travel in that medium compared with its speed in vacuum or air (consider that light travels slower in a material of high refractive index, like glass, compared with its speed in vacuum or air). The higher the relative permeability of a material, the larger will be the inductance of a coil wound on this material and having a given size and turns count. The velocity of magnetic field propagation is not a linear relationship with permeability. The velocity of magnetic field propagation, v, in a material is inversely proportional to the square root of the permittivity and the permeability of the material:

$$v = \frac{1}{\sqrt{\varepsilon_a \mu_a}} \tag{7.14}$$

where ε_a is the absolute permittivity of a material, being the product of the permittivity of vacuum, ε_0, and the relative permittivity, ε_r, of the material.

$$\varepsilon_a = \varepsilon_0 \varepsilon_r \tag{7.15}$$

and where μ_a is the absolute permeability of a material, being the product of the permeability of vacuum, μ_0, and the relative permeability, μ_r, of the material.

$$\mu_a = \mu_0 \mu_r \tag{7.16}$$

7.4 HISTORY OF INDUCTIVE POSITION SENSORS

The English scientist Michael Faraday discovered in 1831 that a changing magnetic field in one electric circuit induced a current into a nearby circuit. This is now called electromagnetic induction.

It was already known at that time that passing a current through a wire conductor caused the wire to be surrounded by a magnetic field, as indicated by its effect on a magnetic compass needle. Faraday showed that the reverse was also true: that, in addition to a current causing a magnetic field, a magnetic field also can cause a current. Faraday demonstrated this by moving a magnet within a coil, and detecting the current induced into the coil. The relationship between a changing magnetic field and the induced current is now called Faraday's law of induction (or simply, Faraday's law), stating that electromotive force (EMF) is equivalent to the rate of change of magnetic flux, as in Equation 7.17.

$$\varepsilon = -\frac{d\Phi_B}{dt} \tag{7.17}$$

where ε is the electromotive force, and Φ_B is the magnetic flux.

Faraday's law of induction is a fundamental principle for operation of transformers, inductors, motors, and so on. The direction of electromotive force is indicated by Lenz's law, named after physicist Emil Lenz, who formulated it in 1834. The law states that a changing magnetic field induces a current into a conductor in a direction such that a magnetic field formed by the induced current opposes the initial changing magnetic field. This can be more easily understood by considering an illustration of the resulting "right hand rule", as shown in Figure 7.5.

An American physicist, Joseph Henry, made similar discoveries at about the same time. As mentioned, the unit of inductance, the henry, is named for him. In addition, when a magnetic field is produced by a current flow through a coil, electromagnetic energy is stored in that magnetic field. This is called self-induction.

Inductive position sensors, also known as variable-inductance position sensors, have been widely used since the middle of the twentieth century. They were first used in the very beginning of the twentieth century. Being easy to fabricate, but difficult with which to obtain high accuracy, variable inductance sensors have mostly been incorporated into measurement systems where simplicity was more important than performance. Their capability to be used in non-contact measuring configurations also lends their use to high reliability applications or other areas where elimination of wearing components is one of the most important features of a sensor. The sensing element can be designed for use at high temperatures of over 150°C, although the electronics module may need to be mounted in a lower temperature area unless

FIGURE 7.5 The "right hand rule" relates magnetic field direction and a current, *I*. Current moving in the direction of the thumb forms a magnetic field in the direction of the curved fingers.

specially designed for high-temperature use. When considering the development of high-temperature electronics, a designer may consider the high-temperature semiconductors that have been built using silicon carbide (SiC), gallium arsenide (GaAs), silicon on insulator (SOS), and several dielectrically isolated CMOS processes.

Since the basic sensing element comprising a coil and a movable core is simple, most improvements in performance over the years have been accomplished by advances in the related circuitry. These improvements include higher temperature operation, lower temperature sensitivity, greater stability, smaller size, and lower power consumption. Incorporation of a microcontroller allows the possibility of implementing temperature and nonlinearity correction, as well as providing various output types and communication protocols.

7.5 INDUCTIVE POSITION SENSOR DESIGN

Some configurations of linear and angular inductive position transducers were pictorially shown in Figures 7.1 through 7.3. These can be combined with the appropriate electronic circuits to develop and condition an output signal, thus producing a complete position sensor.

A typical inductive position transducer comprises a movable core, made from a ferromagnetic material, and a coil of wire wound onto a bobbin (or sometimes, more than one coil may be used). When the core is mostly outside of the coil bobbin, the coil inductance is lower. As the core moves into the bobbin, the coil inductance increases according to Equation 7.3. In a practical position sensor, a non-ferromagnetic pushrod or shaft (aluminum or plastic, for example) is attached to the core to break the magnetic circuit between the core and the movable element that will be measured. By way of the non-ferromagnetic push rod or shaft, the core is then caused to move with changes in the measurand.

Since the core can move in and out of the coil, the total sensor length with the core inserted approaches the length of the coil on a linear position sensor, as was shown in Figure 7.1(a). In this configuration, the total sensor length when the core is nearly pulled out of the coil approaches the sum of the coil and core lengths. This additional length is one of the drawbacks to using an inductive sensor, although the overall length is not as great as that in an LVDT (Linear Variable Differential Transformer; see Chapter 8) for a given measurement range.

Angular and rotary position sensors usually have an appearance similar to that shown in Figure 7.1(b), but linear sensors often have external features added in order to facilitate mounting into various applications. The sensor shown in Figure 7.6 is a rod-end configuration, having a housing, a movable rod, and eyes for attachment at each end.

It may have an appearance somewhat similar to a shock absorber of a vehicle suspension system, except of a smaller diameter. The eyes allow mounting the sensor between two movable pins or bolts, as when mounted in parallel to a shock absorber

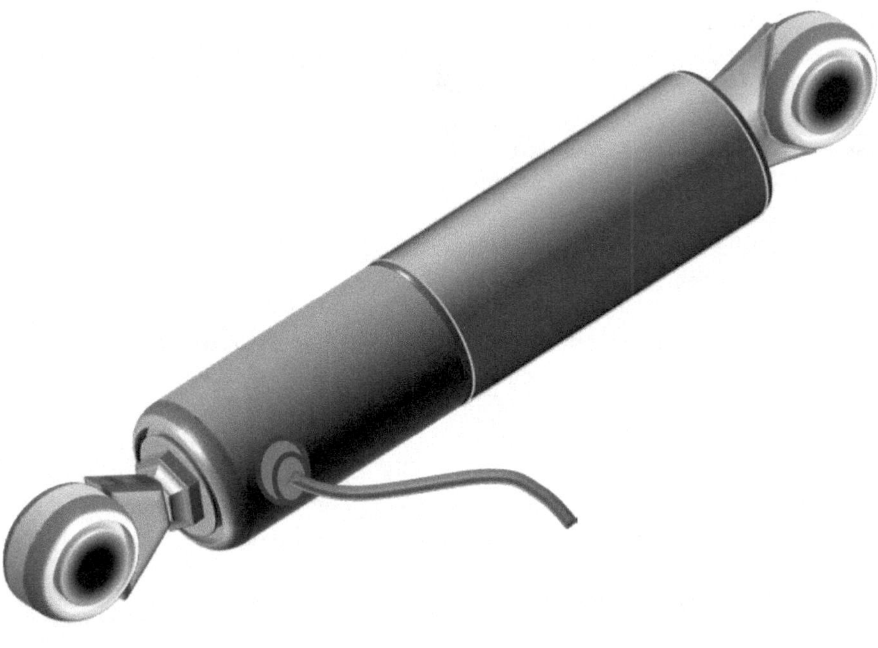

FIGURE 7.6 Inductive linear position sensor with rod-ends.

on a vehicle, so that each end is free to pivot around the mounting pins or bolts. Inside the sensor, the coil is rigidly mounted within the housing, while the core is attached to the rod that moves in and out. Rod guides and wipes are installed between the rod and the housing. The guides make sure that the rod moves freely without binding and help to extend the mechanical life of the sensor (life = number of motion cycles possible before wearing out the guides). The wipes prevent foreign material that may be stuck to the rod from entering the guides. Otherwise, the foreign material could damage the guides and reduce the life of the sensor. Also included within the housing is the electronic circuit.

7.6 THE COIL AND BOBBIN

It is important to carefully select the proper material for the bobbin that supports the coil, if a bobbin is to be used. Any warping of the bobbin during its manufacture or later, due to temperature variations, can produce a change in the coil inductance and result in an error in sensor output. A glass-filled high-temperature thermoset plastic material is often specified for molding of a bobbin in order to provide temperature stability. A thermoset plastic is cured (becomes hard, or set) when brought up to its processing temperature. When cured, it has its final shape, and cannot be re-melted for use again. A thermoplastic material softens when brought up to its processing temperature, and hardens when cooled. Thermoplastics may often be used over again by reheating and re-molding.

Most plastics have a relatively high coefficient of thermal expansion (t/c). Such plastics may be combined with glass, because glass has a low t/c. The resulting composite mixture has a lower t/c, and greater dimensional stability (less warping). Some plastics suitable for the molding of bobbins, when glass-filled, include: nylon 6/6 with 30 to 35% glass, polybutylene terephthalate (PBT) with 30% glass, polyethylene terephthalate (PET) with 30% glass, polyphenylene sulfide (PPS) with 40% glass, diallyl phthalate (DAP) with 30% glass, and liquid crystal polymer (LCP) with 30% glass. Variations of these materials are available under many trade names. A very helpful guide in designing a coil bobbin is listed here: www.cosmocorp.com/docs/en/cosmo-bcat-en-mdres.pdf.

It is also possible to wind a coil without using a bobbin. In that case, the magnet wire is wound onto a polished metal mandrel. The wire insulation outer layer material is one that can be softened with heat. The wire is heated during winding, so that the coils will be bonded together when cooled. Then, the finished coil is popped off from the mandrel after it has sufficiently cooled.

A coil-winding machine, coil supporting tools, and wire guide tool for winding the coil(s) of an inductive position sensor must be designed and calibrated to produce a repeatable winding profile, so that each sensor coil will be like the next. A spool of magnet wire is usually placed, with its axis in the vertical position, inside of a spool chamber, which is a tapered clear plastic tube with the top open. The top of the spool chamber is more narrow than the spool, while the bottom of the chamber is wider than the spool. See Figure 7.7.

A set of *whiskers* made from short, radially mounted strands of monofilament fishing line is mounted on top of the spool chamber. The assembly of a ring having

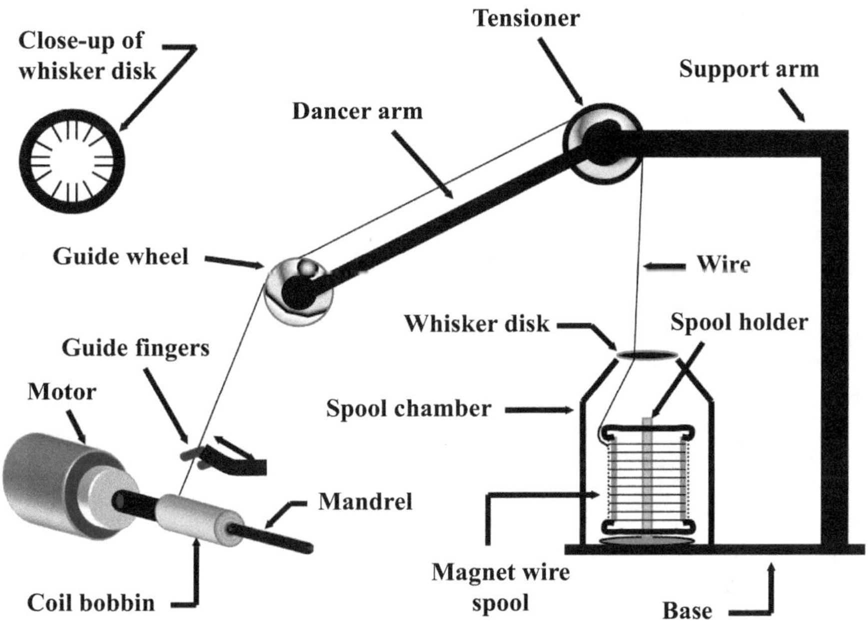

FIGURE 7.7 The basic elements of a single coil winder.

the whiskers disposed about the inside edge can be called a whisker disk, and a model is available commercially from Azonic Products and having the name *Wisker Disk*. The whiskers add some resistance against pulling the wire as it is stripped from the spool, helping to eliminate tangles. Then the wire goes to a tensioner and dancer arm assembly, followed by a guide wheel, before being wound onto the bobbin through the guide fingers. The tensioner comprises one or more parallel discs over which the magnet wire passes. The disks have an adjustable friction brake to control the rotation force of the disc(s) and provide the desired tension for winding. It is important to maintain a uniform tension on the wire as it is being wound onto the bobbin, so that the turns are placed smoothly and without digging into previous turns of wire or jumping up over adjacent turns. The dancer arm is a long spring that takes up the nonuniformity in wire feed rate as the machine speeds up and slows down. The tensioner and dancer arm also work together to prevent the magnet wire from breaking or excessively stretching during the winding operation. The winding machine itself has a mandrel to hold the bobbin, a traverse to hold and control the fingers, a motor to turn the mandrel, and a turns counter. The wire passes through the fingers close to the bobbin, to guide the wire onto the desired spot on the bobbin. There is usually a motor speed up and slow down cycle that engages when starting and stopping the winding operation. The first few turns are wound at slow speed to get the winding started properly. The last few turns are wound at slow speed in order to accurately count the correct number of turns before stopping.

After the coil is wound, the ends must be electrically and mechanically attached to lead wires, since the wire used in making a sensing element coil is normally of too small a diameter to be used as the leads for attachment to an electronic circuit. Such coils are often wound with AWG 42 to 46 wire, which can be easily broken. The wire ends from the coil are called *self leads*. The self leads are usually soldered to sturdy pins, and then a *lead-attach* operation solders lead wires to the same pins. The lead wires are of a larger diameter, for example, AWG 28, and are stranded and usually PVC or Teflon™ insulated. Before or after lead attach, varnish and/or adhesive tape is often used to stabilize the coil assembly before placing it into the coil housing. Sometimes the self leads are skeined before lead attach. The *skeining* of self leads means that the last few centimeters of the wire are doubled over one or more times and twisted together in order to increase the strength of the wire where it attaches to the pin. The lead-attach operation is often a manual one, so it can be one of the more expensive parts of the assembly process, but is very important. When a defect is found in an assembled coil, the problem is usually traceable to the lead-attach operation. If the self wire is stretched too tightly to its pin, flexing of the lead wires may move the pin enough to break the self wire. If the pin is not sturdy enough, or has been loosened by soldering for too long of a time period, the same result can occur. If the lead-attach area is potted, great care must be exercised in the design so that thermal cycling of the coil assembly will not break the connection from a self wire to its pin. This often dictates the use of a semi-rigid epoxy, rather than a rigid epoxy.

Besides providing mounting means and protection for the coil, the housing also provides magnetic shielding to prevent inductance changes due to nearby magnetic materials. To provide this magnetic shielding, the housing must be made from a ferromagnetic metal or alloy. A common selection is to use a (ferromagnetic) steel housing with a nickel plating. The nickel plating is also ferromagnetic, and adds corrosion protection and better appearance as well. The nickel plating is normally applied using an electroless process, where the part is immersed in a liquid that contains nickel ions, which plate onto the steel surface. The nickel thickness is determined by the liquid temperature and by the amount of time of immersion. The electroless nickel finish is smooth and shiny, as opposed to an electroplating process, where polishing would be required in order to yield a shiny surface.

After a coil is assembled into its case, a liquid potting material is often added, which then cures into a solid. This adds mechanical protection and seals against the entry of moisture.

The coil is usually wound with an enameled copper wire, called magnet wire. The insulating coating can be heat-strippable to aid in soldering the magnet wire to the coil leads. When using magnet wire having a heat-strippable coating, the wire can simply be wound around a pin without stripping the wire first. When the wire is heated by a soldering iron, and solder applied, the insulation melts and allows the soldering operation to be completed. If using standard magnet wire (not heat-strippable) an abrasive agent is used to remove a portion of the insulation before soldering.

One disadvantage in using copper wire is that the temperature coefficient of copper is about 3,900 parts per million per degree Celsius. If the sensor circuit is sensitive to resistance change, the change can be compensated by using either a separate compensating coil or a temperature-compensating resistor. The use of a compensating coil is explained in the signal conditioning section, 7.8. A temperature-compensating resistor can be one having a negative temperature coefficient (NTC) in series with the copper wire coil or can be located on the electronic circuit board. Alternatively, a linear positive temperature coefficient (PTC) resistor can be mounted on the electronic circuit board if utilized appropriately in the circuit to provide the desired compensation. The common NTC type of resistor is called a *thermistor*, and has a nonlinear temperature coefficient, where its resistance decreases as the temperature increases. Thermistors are available in a wide range of resistances, several resistance tolerances, and leaded as well as surface mount components. Once the resistance, package, and tolerance have been selected, the desired temperature coefficient characteristic is chosen from among usually ten or more variations. Consult the manufacturer's resistance versus temperature chart to select the desired resistance versus temperature characteristic.

A linear PTC thermistor is based on semiconductor material, or on the element nickel, and has a relatively linear positive temperature coefficient (not to be confused with a resettable fuse, also commonly called a PTC, but that should be called a PTC resettable fuse). The silicon version of a linear PTC thermistor is sometimes called a *silister* or a *silicon PTC thermistor*. They are only available in a very limited selection of resistances, and have a positive temperature coefficient of about 7,000 ppm/°C.

Alternatively, one can fabricate a PTC resistor by winding nickel wire onto a form (as the author did for compensating a pressure-sensing product designed in the 1970s, before PTC resistors were commonly available). Nickel wire has a positive temperature coefficient of resistance of about 5,900 ppm/°C. PTC or NTC resistors can also be used to compensate temperature shifts that are due to the electronic circuit, rather than due to the sensing element thermal characteristic. Alternatively, the temperature of a sensing coil can be measured, as well as the circuit board temperature, and corrections made within a microcontroller.

A way to remove the problem of temperature-induced changes in the resistance of a coil is to wind it with a wire that has a near-zero temperature coefficient of resistance (about 15 ppm/°C). This is called Manganin™ wire (invented by Edward Weston in 1892) and is an alloy of copper, manganese, and iron (Cu/Mn/Fe). It is available from magnet wire manufacturers in the same sizes and with the same insulating coatings as other magnet wires. The aforementioned information about coil winding, lead attach, potting, shielding, PTC/NTCs, Manganin wire, and so on, also applies to the manufacture of LVDTs, presented in Chapter 8.

7.7 CORE

The core material cross-section and magnetic permeability must be selected to provide the desired amount of inductance change as it moves into the coil. In addition to that, it is desirable that the core have some additional favorable properties. These properties include long-term stability, temperature stability, low-power loss at the operating frequency (relating to magnetic hysteresis), and some corrosion resistance.

Core materials are usually nickel-iron alloys, with some other materials added to tune the magnetic and mechanical properties. Some of the core materials that are the most common include Ni-Span C®, Hyperco 50B®, and various forms of Permalloy®. Ni-Span C has an advantage of very low permeability sensitivity to temperature change. Hyperco 50B has a higher absolute permeability. The manufacturer's data sheet must be consulted to define the proper annealing versus cold-working finished condition in order to obtain the desired magnetic properties. With most core materials, there is a final annealing process that relieves the mechanical stress from machining, and also makes the whole core magnetically uniform. If the core has an internal thread, or if it has undergone another machining operation, the annealing process must be applied after machining. In the usual annealing procedure, a batch of cores is brought to an elevated temperature for a set period of time, then allowed to cool at a somewhat-controlled rate. The recommended annealing schedule is available from the manufacturer of the alloy to be used. During the annealing process, oxygen is excluded in order to prevent oxidation of the metal, and a reducing atmosphere is usually established. The reducing gas can be hydrogen (with near-zero oxygen), which must be used with a method to prevent explosion (explained in Section 8.5). Alternatively, a forming gas can be used that has near-zero oxygen, and which has insufficient hydrogen concentration to support an explosion, but does have sufficient hydrogen to chemically reduce the surface of the core material. Such a forming gas may comprise 3.0% hydrogen, balance argon, less than 2 ppm oxygen (confirm with the bottled gas supplier that the forming gas being supplied meets with the appropriate regulations as being non-flammable, and confirm the type of pressure regulator that should be used).

7.8 SIGNAL CONDITIONING

Since a change in the measurand results in a corresponding inductance change in a variable inductance position sensor, an electronic circuit is needed to measure that inductance change, and to provide a desired output signal that is representative of the measurand. Accordingly, the electronic circuit includes several functions to drive the sensing coil, measure its inductance, and to produce a voltage, current, or digital output signal. The sensing coil is normally energized with an AC driving voltage from a sine wave oscillator. The coil's inductance change can be indicated by a changing oscillator frequency or by a changing amplitude, depending on the circuit configuration. For compensating errors from temperature variations or other sources, two or more coils can be incorporated into the sensor and a comparison between them used to produce an output signal. Figure 7.8 shows a simple circuit with a single variable inductor producing a change in voltage amplitude corresponding to a change in inductance.

In this case, the voltage across the inductor is not zero when a zero output voltage is desired. So, a resistive divider is used to provide an offset equal to the same voltage as the inductor would have with a measurand of zero. A differential amplifier can be used to subtract these two voltages, together labeled as V_{OUT} in the figure, to obtain a zero-based output. The circuit of Figure 7.8 provides an AC output voltage of varying amplitude, but often a DC output is desired, with a positive voltage indicating displacement in one direction, while a negative voltage indicates displacement in the opposite direction. This can be accomplished by using a circuit similar to the one shown in Figure 7.9.

A voltage developed from the resistor divider in the figure is demodulated by the lower diode and filtered by a capacitor to provide a constant DC voltage (at the lower V_{OUT} line of the figure). The voltage developed from the resistor and variable

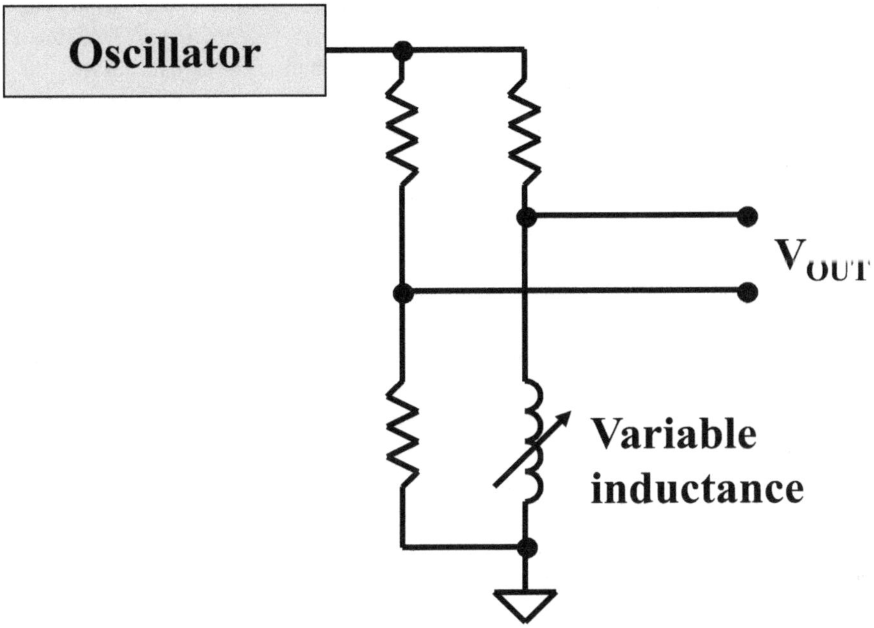

FIGURE 7.8 A change in the variable inductance, *L*, results in a change in the amplitude of the differential output voltage.

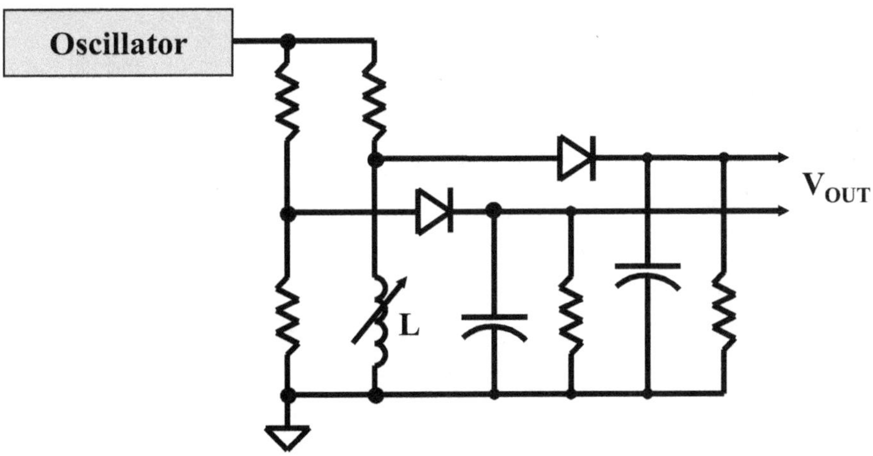

FIGURE 7.9 Variable inductance sensor circuit with a DC output voltage.

inductance divider is demodulated and filtered to provide a DC voltage (at the upper V_{OUT} line) that varies with the inductance value. The resistor shown across each capacitor is there to allow the voltage across the capacitor to reduce as the signal amplitude reduces, because the diode would not provide this function, since the diode would be reverse-biased. The RC time constant defines the response time of the signal, but the R must be high enough to allow the C to provide the filtering function. The C must be high enough to provide filtering at the operating frequency of the oscillator.

The circuit of Figure 7.9 is assumed to be connected with a high input impedance differential amplifier or instrumentation amplifier. If it is connected with a differential amplifier similar to that of Figure 6.16, the input resistances of the differential amplifier can provide the discharge means for the filter capacitors, and the two resistors shown across the capacitors of Figure 7.9 can be eliminated.

The same considerations regarding demodulation techniques (diode demodulator versus a demodulator circuit using a semiconductor switch) would apply as those presented in Chapter 6, and will not be repeated here. A circuit using a variable inductance, a non-variable inductance (for error compensation) and a diode demodulator is shown in Figure 7.10.

In Figure 7.10, if a temperature change, for example, causes a variation in the impedance of L_{sense} (sensing inductance), then a similar variation in L_{comp} (compensation inductance) will prevent any changes in the output voltage due to temperature sensitivity. This relies upon the two inductances being constructed in the same way, so that their temperature sensitivity characteristics will match, but the compensation inductance does not change with the measurand.

Greater sensitivity and measuring range can be obtained by using two coils arranged as shown in Figure 7.11, so that a core will move from one coil into the other coil. In this configuration, both coils change with the measurand such that the inductance of one coil increases as the other one decreases. When the core is centered and extending equally into each of the coils, the AC voltage across each of the two coils is the same. As the core moves more into one coil, the inductance of that coil increases, as the inductance of the other coil decreases.

In Figure 7.11, the coil resistances vary the same as each other, and should not need temperature compensation. The voltage at the connection between the two

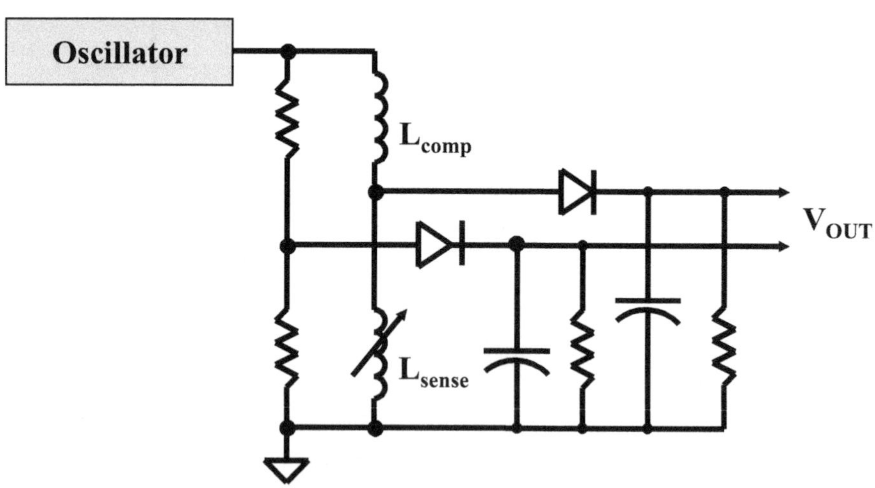

FIGURE 7.10 Sensing circuit with compensating coil and DC output voltage.

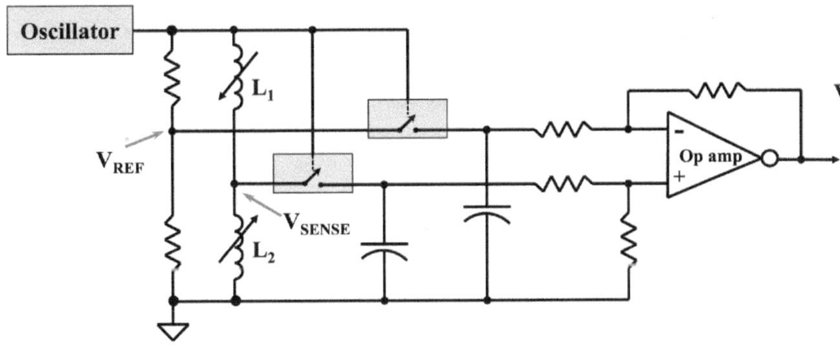

FIGURE 7.11 A dual coil variable inductance linear position sensor with semiconductor switches and DC output.

sensing coils (L_1 and L_2) can be called V_{SENSE}, and is the same as the voltage at the resistive voltage divider (V_{REF}) when the measurand is at mid-scale. V_{SENSE} is lower than V_{REF} when the measurand is at minimum. V_{SENSE} is higher than V_{REF} when the measurand is at maximum. A differential amplifier is added, to produce an output voltage swing that is centered on zero volts to provide a +/– DC output. Alternatively, a zero offset could be added for 0 to full-scale output voltage, if desired.

7.9 ADVANTAGES

A main advantage of variable inductance position sensors is their simplicity. This advantage is somewhat eroded by the need for an electronic circuit with a driver and demodulator to form a complete sensor, but it is still a popular sensor type.

Inductive sensors are typically robust, and can meet demanding physical requirements. Another advantage is that, when there is no need for supporting bushings, bearings, and so on, the sensor can be implemented as a non-contact device. Thus having a nearly infinite lifetime. In addition, typical inductive sensors with an analog output voltage have almost infinitely fine resolution, determined only by the ability to read the signal over whatever noise level may be present. Inductive sensors are also less affected by dust and humidity than are capacitive sensors.

A disadvantage is the relative lack of precision and stability that is attainable with several other technologies, such as with distributed impedance, magnetostrictive, and some encoders. Though inductive sensors tend to be physically robust, they can be bulky in some designs, and the carefully wound coils add to their manufacturing expense.

7.10 TYPICAL APPLICATION AND PERFORMANCE SPECIFICATIONS

Whereas distributed impedance sensors can provide errors of around 0.05% or less, magnetostrictive sensors in the range of 0.01 to 0.05%, and an LVDT in the range of 0.1% to 0.5%, inductive position sensors are generally limited to the accuracy range of having errors of about 0.2% to 1.0%, depending on the measuring range. This is not a problem for many applications where the most important aspect is to have a monotonic response with high resolution, such as in many position control loops. Some models may incorporate a microcontroller and nonlinearity correction. In that case, a nonlinearity of less than 0.2% may be possible.

Some inductive position sensors have the electronics mounted remotely from the sensing element, as shown in Figure 7.12. This is common when used in a hydraulic cylinder to minimize any head space requirement within the active portion of the cylinder.

The sensing element shown in Figure 7.12 uses a ferromagnetic core or *target tube* that is on the outside of the inductive sensing coil, rather than an internal core that was shown in Figure 7.1(a). This arrangement allows mounting this type of sensor within a hydraulic cylinder, in which the cylinder's rod is gun-drilled to allow the insertion of the inductive sensing element inside the rod. A gun-drilling machine is, obviously, the type of machine used to drill the bore for a gun barrel. It is capable of precision deep hole drilling by the utilization of a hollow drill bit through which a cooling liquid is applied at the cutting surface. A groove in the tool also allows the liquid to carry away the chips that are formed. Such machines are capable of producing extreme depth-to-diameter ratios.

The target tube shown in Figure 7.12 slides over the sensing element by a varying amount to be measured. But when a cylinder rod is gun-drilled, the drilled rod may serve to replace the target tube. As the target tube or gun-drilled rod (made of a magnetically permeable material) covers more and more of the inductive sensing element, the inductance increases.

Typical specifications for an inductive linear position sensor that includes a microcontroller to implement linearization and temperature compensation are shown in Table 7.1. As mentioned, one application of an inductive linear position sensor is the measurement of the amount of extension of smaller-sized hydraulic cylinders. The robustness of the

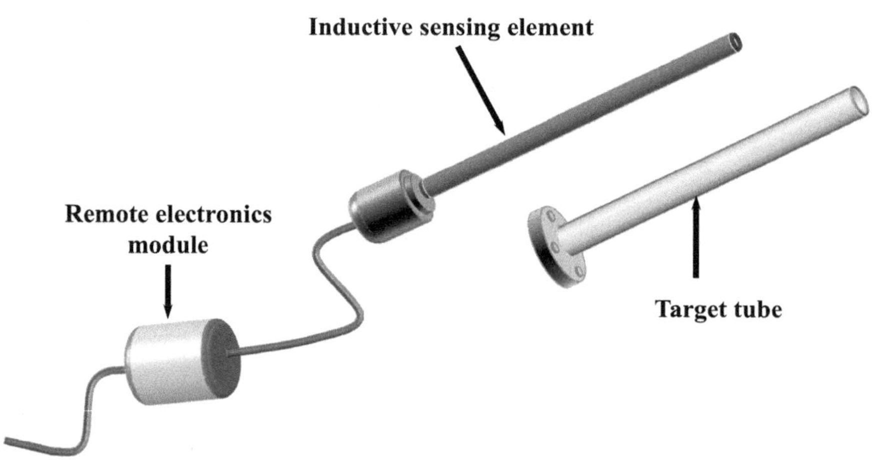

FIGURE 7.12 Inductive sensing element with target tube and remotely mounted electronics module. The target tube slides over the sensing element.

TABLE 7.1
Inductive Position Sensor Typical Specifications

Input Voltage:	± 12 to 15 VDC at 80 mA
Output:	± 10 VDC
Nonlinearity:	± 0.15% of full-range output
Resolution:	0.02% of full-range output
Operating Temperature Range:	−40 to 85 °C
Temperature Sensitivity:	± 0.02%/°C
Output Noise and Ripple:	0.01% of full-range output + 1 mV

inductive sensor lends itself well to this installation, but there is a limitation on maximum length, and there may be a limitation due to inaccuracy. Where a longer measuring range and/or higher accuracy is required, a distributed impedance or magnetostrictive sensor is often used. The magnetic coupling of a magnetostrictive sensor is sometimes desired, depending on the cylinder design. But often it may be easier to implement an inductive sensor or a distributed impedance sensor, as compared with a magnetostrictive sensor, since the cylinder rod must be gun-drilled in any case, even though a position magnet is used with a magnetostrictive sensor. Also, with a magnetostrictive sensor, it is important to use only non-magnetic materials in proximity to the position magnet.

Inductive sensors can be used successfully in many industrial and commercial applications. Valve positioners of many shapes and sizes for process control and on the factory floor are good candidates, as well as for measuring the drum position in a clothes washing machine, and other position-sensing requirements for consumer goods.

7.11 MANUFACTURERS

Some manufacturers of inductive linear and angular/rotary position sensors include the following:

Altheris Sensors & Controls	www.altheris.com
Balluff	www.balluff.com
Gill Sensors & Controls	http://gillsc.com
Revolution Sensor Company (design)	www.rev.bz
Sentech	sentechlvdt.com
LORD Microstrain	www.microstrain.com
Micro Epsilon	www.micro-epsilon.com
Zettlex	www.zettlex.com

7.12 QUESTIONS FOR REVIEW

1. **The core of an LVIT is usually made of this type of material:**

   a. Laminated
   b. Magnetically permeable
   c. Insulator

d. Non-ferrous
e. Thermoset plastic

2. This circuit can provide the AC waveform to drive an RVIT:

a. Exclusive OR gate
b. Flux capacitor
c. Zener diode
d. MOV
e. Oscillator

3. When the core is pushed further into an LVIT, its inductance:

a. Increases
b. Decreases
c. Oscillates
d. Rotates
e. Remains the same

4. An electric current flowing in a wire is always accompanied by a:

a. High resistance
b. Yagi
c. Chaperone current
d. Magnetic field
e. Sympathetic wave

5. Resistance to pulling wire from the spool of a winding machine may be added by a:

a. Brake
b. Roller
c. Whisker disk
d. Heat gun
e. Ohm meter

6. The ends of the magnet wire on a coil after winding are called:

a. Double-ended
b. Projections
c. Self leads
d. Electrodes
e. Pins

7. A reducing gas used in core annealing, but that is not flammable is called:

a. Hydrogen
b. Forming gas
c. Nitrogen
d. Argon
e. Halogen gas

8. PTC wire having a thermal coefficient of about 5,900 ppm/°C can be made of:

a. Manganin
b. Copper
c. Aluminum
d. Permalloy
e. Nickel

8 The LVDT and RVDT

8.1 LVDT AND RVDT POSITION SENSORS

The Linear Variable Differential Transformer (LVDT) and the Rotary Variable Differential Transformer (RVDT) are specialized types of transformers. In transformer theory, two or more coils are coupled by an alternating magnetic field. If a first voltage is impressed onto a first coil, a second voltage produced at a second coil is related to the magnitude of the first voltage by the ratio of the number of turns of the respective coils. The first coil, across which the voltage is impressed, is called the *primary*. The one or more coils that produce the relative outputs being called the *secondary* or *secondaries*. For example, a secondary with twice the number of turns as the primary will produce twice the voltage amplitude as compared with the voltage appearing across the primary, ignoring loading effects.

The LVDT and RVDT are position sensors that are non-contact, absolute reading, and have virtually infinite resolution. They each typically comprise three or more coils within which a magnetically permeable core moves to provide variable coupling between the primary coil and the secondary coils. Although the detection technique is non-contact, there is often a mechanical arrangement added to keep the core positioned within the coil throughout the stroke. Practical sensors can be designed with a nonlinearity of less than 0.2% and full-scale ranges (FSR) of less than 1 mm to over 100 mm. Resolution is virtually infinitely fine until the analog signal is digitized for use by a microcontroller for calibration, compensation, and so on. Popular applications include industrial machinery, such as metal forming machines, in-process dimensional verification, and valve positioners, as well as automotive and commercial products.

LVDTs and RVDTs require a set of driving and conditioning electronic circuits. A typical LVDT is sold as a sensing element with a core, the electronic circuit being supplied as a separate device. An RVDT product usually contains the movable core within a housing and with a rotatable shaft. An LVDT or RVDT that includes all of the required electronics within its housing is often called a *DC LVDT* or *DC RVDT* because they operate from a DC power supply and have a DC output, although the internal operation includes the normal AC driving, demodulation, and signal conditioning circuitry. A related product is the differential variable reluctance transducer (DVRT), or *half-bridge* LVDT.

8.2 HISTORY OF THE LVDT AND RVDT

Although variable differential transformers were constructed earlier, the Linear Variable Differential Transformer as a position sensor was described by G.B. Hoadley, and US patent 2,196,809 was issued to him in 1940 [1, pp 3–4]. Early uses were mostly military, because of the ruggedness possible with this type of sensor, and some quality assurance applications, because of the inherent high resolution of the

DOI: 10.1201/9781003368991-8

measurement. The LVDT started on its track to be coming more widely known, and utilized as an industrial sensor, when Herman Schaevitz published his paper "The Linear Variable Differential Transformer" in 1946. Also in 1946, Herman Schaevitz founded his development and manufacturing company, Schaevitz Engineering, based in Pennsauken, NJ. Additional improvements in construction and performance were made throughout the 1950s and 1960s. By then, the LVDT was widely used for industrial applications. US patent number 3,273,096, titled "Rotary Differential Transformer", was granted to Joseph Lipshutz in 1966. Other geometries of the coils and core for RVDT products soon followed. In 1976, Edward E. Herceg authored the book *Handbook of Measurement and Control*, published by Schaevitz Engineering. This book became a valuable reference in the field.

Further LVDT/RVDT improvements that came about in the 1970s and later were mostly in the driving and signal conditioning electronics. This included more accurate drivers, phase adjustment, and demodulation circuitry. Integrated circuits containing most of the complete LVDT driving and demodulation electronics functions became available as a standard product in the 1980s. DC LVDTs were available in the 1970s with "cordwood" construction: that is, axial-leaded parts were connected between two parallel printed circuit boards. But these were very difficult to design and manufacture. They had minimal functionality and performance characteristics, due to the lack of space to house a more complex circuit. With the ICs that became available in the 1980s, the DC LVDT became more practical to design and manufacture, with good performance resulting in the finished product. In the late 1970s, the author invented a low-powered LVDT circuit that provided LVDT excitation, synchronous demodulation, temperature compensation, and a filtered DC voltage output from a power supply of 6 to 24 VDC, while only using 1 mA of power supply current. A 4 to 20 mA version soon followed.

Available in 1987, the Signetics/Philips NE5520 integrated circuit was a very popular component due to its small size and low cost. It included circuitry for excitation, demodulation, and amplification, and was replaced by the similar NE5521 in 1988. The NE5521 was later discontinued sometime in the early 2000s, reportedly due to loss of the IC production mask in a fire. Although some LVDT integrated circuits became available from Analog Devices, they were much more expensive. The demise of the NE5521 caused manufacturers of LVDTs and RVDTs to develop their electronics modules without using an all-in-one integrated circuit. Since the late 1990s, microcontrollers had been added for calibration and other functions. In 2004, the author invented the first set of electronics to be contained within an LVDT housing while incorporating a microcontroller to form the excitation sine wave, control the signal demodulation, and provide calibration and linearization. The calibration and linearization functions were performed through a single wire connection in the cable or connector. By 2005, other microcontroller features and functions were incorporated, including temperature compensation. Later designs made further improvements in performance, and added other output types and industry standard communication protocols. In 2004, the author (Nyce) wrote the book *Linear Position Sensors, Theory and Application*, published by John Wiley and Sons. The chapter on LVDTs presented further information that extended the published knowledge base after the cited Herceg book.

8.3 LVDT AND RVDT POSITION SENSOR DESIGN

A basic LVDT or RVDT comprises one primary coil, two secondary coils, and a movable core. The coils of an LVDT are wound onto a tubular coil form, the interior of which is called the bore. The coil form is called a *bobbin*, and also includes flanges to separate the several coils. The flanges are like discs molded as part of the tube that forms the bore. See Figure 8.1 for a cutaway view and a side view of a typical LVDT.

The bobbin of Figure 8.1 includes four flanges, shown as being perpendicular to the bore. The two outer flanges are covered by end plates after being installed into the housing. The coupling between the moving (core) and stationary (coils) parts of the LVDT is achieved by means of a magnetic field that is generated in the primary coil by an AC excitation current. Due to the magnetic coupling, and no rubbing mechanical parts, any number of position changes can be made without incurring wear to the sensor parts. When using an LVDT to measure distance to a surface of a formed metal part, for example, the LVDT core is linked to a stylus that contacts the surface of the formed metal part. Although there may be wearing of the stylus due to continued use, there is no wear of the actual sensing element. The core of an LVDT is normally a cylinder of permeable material and provides inductive coupling between the primary coil and the secondary coils. The core of an RVDT has a more complex shape, as will be shown in Figure 8.4. The core moves within the bore of the LVDT and does not touch the walls of the bore. As the core moves along within the bore, the distance from the center between the secondaries to the center of the core length is read electronically. See Figure 8.2.

Figure 8.2 shows how the coils of an LVDT are drawn on a circuit schematic. Although the core actually passes within the center bore of all of the coils, the coils are drawn on a schematic as being next to the core, for simplicity. The core typically has a hole drilled through its center, and is tapped with a fine machine thread so that a core extension can be threaded into it. The core extension must be non-magnetic, and aluminum is the most common material for a core extension, but plastic or non-magnetic stainless steel may be used. If using a non-magnetic stainless steel core extension, it should be annealed after being threaded so that it remains non-magnetic.

The primary coil must be excited by an alternating voltage. Accordingly, an oscillator circuit is used that has sufficient driving current capability to easily drive the

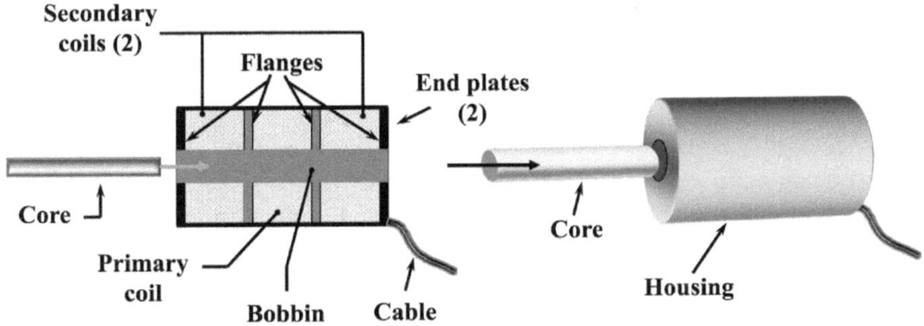

FIGURE 8.1 Cutaway view, left, and side view, right, of an LVDT.

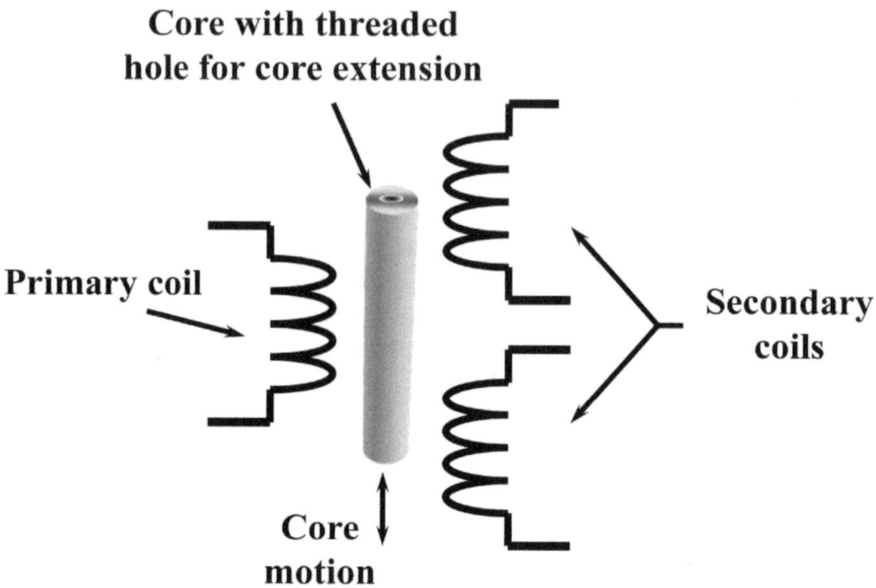

FIGURE 8.2 Pictorial representation of the coil configuration of a typical LVDT.

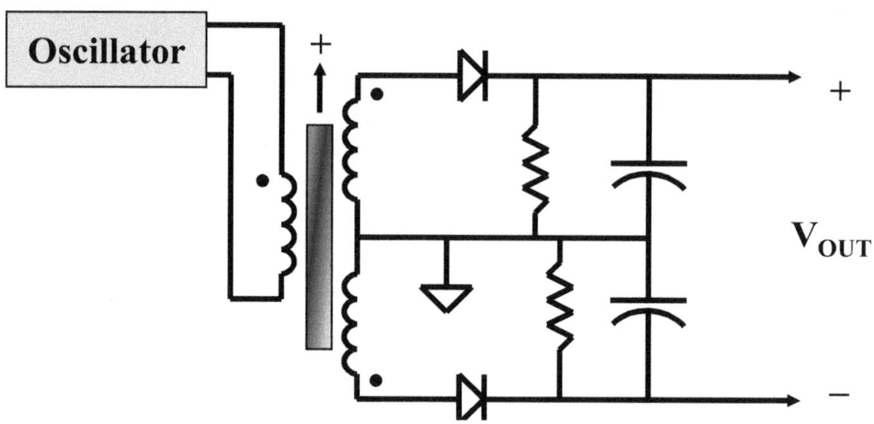

FIGURE 8.3 A simple LVDT signal conditioner with diode demodulator.

impedance of the primary. Since the output from the secondaries is also AC, the signal must be demodulated in order to provide the DC signal voltage that is normally required. A simple diode type of demodulator circuit is shown in Figure 8.3.

With a configuration as shown in Figure 8.3, the differential output voltage, V_{OUT}, will become more positive when the core is moved up, and become more negative as the core is moved down. When the core is centered within the secondary coils, the output voltage differential will be zero volts. When the core is centered this way, it is called the *null* position, or is described as the LVDT being at null. The diode demodulator of Figure 8.3 is called a *passive demodulator*, and operates *asynchronously*, due to the fact that it is not synchronized with the phase of the primary excitation. Alternatively, an *active demodulator* uses transistors or analog switches instead of diodes to convert the AC signal to DC, and may do so asynchronously, or may be *synchronous* with the phase of the excitation voltage or current. Demodulation techniques are explained further in Section 8.7, on demodulation.

8.4 COILS

The three or more coils of an LVDT or RVDT are wound onto a bobbin. It is important to select the bobbin material for rigidity and stability. Any warping of the bobbin during its manufacture or later, due to temperature variations, can produce an error in the sensor output. A glass-filled high-temperature plastic is usually specified (some common choices include glass-filled nylon, Ryton™, and Torlon™).

A winding machine is used to wind the primary and secondary coils onto the bobbin. A motorized coil-winding machine can be controlled by hand, with a start switch to be pressed after the first few turns of magnet wire have been hand-wound onto the bobbin. The winding operation then automatically stops when a set number of turns have been applied for the particular coil. The secondaries may have a different number of turns from the primary. Also, sometimes there may be more than two secondaries used in order to extend the linear range of the LVDT, and these secondaries may have a different number of turns from one another.

Alternatively, a fully automated winding machine can be used when it is planned to make large quantities (usually more than 10,000 parts per year). On such a machine, there are usually multiple bobbins being wound at the same time. In addition to winding each coil, a fully automated winding machine also winds the wire ends onto their respective termination pins, and may also solder the pins. Since the coils are wound with light gauge magnet wire, it is important to closely control the wire tension as it is wound, in order to prevent stretching the wire. The average tension is controlled by an adjustable brake. The instantaneous tension is averaged out by using a spring-loaded guide called a *dancer arm* (see Figure 7.7 in the chapter on inductive position sensors).

After the coil is wound, the ends are soldered or welded to end pins or lead wires (called the lead-attach operation), which are heavier and more durable than the wire used to wind the coils. When the magnet wire is of a very small diameter, less than about American Wire Gauge (AWG) 38, stripping the insulating varnish coating from the wire surface may be difficult to accomplish by an abrasive process without breaking the wire. In this case, it is advantageous to specify the use of a heat-strippable coating instead of a normal type of varnish on the magnet wire. There are several of these coatings available on the market, some of which are Beldsol™, Thermaleze™, and Solidon™. This coating can be soldered directly, without stripping. This coating acts as a solder flux, and melts away at soldering temperatures. It is available at several operating temperature levels (and will not melt at these operating temperatures). Care must be taken to specify a heat-strippable coating that will operate satisfactorily at the high end of the LVDT or RVDT operating temperature, while being able to be soldered at a temperature that will not excessively melt the plastic surrounding the end pins. On LVDTs and RVDTs rated for high-temperature use (above 100°C) this may not be possible. In that case, chemical or abrasive stripping must be used. When the magnet wire (called self wire) is too thin for reliable lead attach to the end pins, skeining may be used. This is the folding of two or more short lengths of the magnet wire one over the other, to increase strength. The several strands of magnet wire may be twisted together before soldering to the pin.

Varnish and/or adhesive tape are often applied over the coils to stabilize them. The adhesive tape is usually Kapton™ or Mylar™. This ensures that the coil wires will not be damaged during the remaining process steps, provides more uniform heating and cooling, and improves the ruggedness of the completed LVDT or RVDT in the field. The finished wound bobbin assembly is inserted into a housing and usually potted with epoxy to seal it and make it extremely rugged. The potting material should be one that cures semi-rigid, rather than rigid or fully hard, in order to protect the windings and connections from breakage due to thermal variations during use.

The housing is usually made of a material that will shield the coils magnetically. This is accomplished by making the housing parts of a magnetically permeable material, such as nickel-plated steel or 17-7 PH™ stainless steel. In addition to the housing tube, end plates of the same material are added to complete the shield. This magnetic shielding prevents interference from nearby electromagnetic fields and reduces effects from other magnetically permeable materials that may move in the LVDT proximity.

The temperature coefficient of the copper wire in the secondaries does not cause a noticeable error, since the secondary coils work together and their resistances track each other. But the increase in resistance of copper wire with an increase in temperature does have an effect on the current in the primary coil. As the wire resistance increases, the primary current decreases, causing a decrease in the primary current and a decrease in the output signal from the secondaries. This is usually on the order of about 1.5 to 2% of full range over 100°C. This can sometimes be avoided by driving the primary with a constant current rather than a constant voltage. Another way is to apply temperature compensation in the electronics module. A third way is to wind the primary coil with Manganin wire. This type of wire is an alloy of copper, manganese, and nickel, having a very low thermal coefficient of resistance (but is more expensive than copper magnet wire).

Since the LVDT and RVDT are transformers, the output voltage is proportional to the ratio of the number of turns of wire in a secondary coil to that of the primary coil. This is true as long as the core is not magnetically saturated, and sufficient energy is induced from the primary to support the output power in the secondaries. To design a lower-powered LVDT or RVDT usually requires a greater number of turns in the primary, so that it has a higher impedance. Then more turns are required in the secondaries to maintain the same output voltage level. Such a higher impedance design also requires a high input impedance of an amplifier connected with the LVDT or RVDT, so that the secondaries are not loaded so much that the signal level will be decreased.

A DVRT (half-bridge) uses only two coils, similar in physical arrangement to the secondaries of an LVDT or RVDT. The coils are connected in series with a center tap. Power is applied across the two ends of the series-connected set of coils, and the signal is taken from the center tap. As the core moves more into one coil, the impedance of that coil increases and the voltage across it increases. At the same time, the core is moving more out of the other coil, so its impedance and the voltage across it decreases. This signal can be demodulated and conditioned in a similar way as with an LVDT or RVDT.

8.5 CORE

The core material of an LVDT is usually a cylindrical or tubular-shaped component made of a nickel-iron alloy. It is often threaded at each end to enable attaching it to the element that is to be measured. The core of an RVDT is usually shaped differently, but also includes means for attachment or mounting within the RVDT housing. After the core is in its final shape, size, and is threaded or otherwise machined, it is then annealed. This assures that the magnetic properties are uniform throughout the core material, after those properties may have been disturbed through forming, drilling/tapping, and so on. The annealing process includes bringing the core to a high temperature (somewhere between 500 and 900°C) for a specified amount of time in order to remove mechanical stress and to make the permeability more uniform along the length of the core. This aids the completed LVDT or RVDT to exhibit a low null voltage, improve unit-to-unit similarity, provide lower nonlinearity, and yield better temperature performance. During the annealing process, a reducing gas is required in order to prevent oxidation. The annealing gas is usually hydrogen or a gas mixture including hydrogen. The hydrogen helps to prevent oxidation when the core is at an elevated temperature. With the presence of hydrogen, steps need to be taken to prevent a fire or explosion, such as keeping the hydrogen percentage below that needed for combustion. A nonexplosive mixture called *forming gas* may be used. A forming gas mixture could contain 3% hydrogen, for example, with the balance being argon. If instead, a high concentration or pure hydrogen gas is used, the processing operation must be performed within an explosion-protected room, and all of the electric and electronic equipment and other energy-generating or energy-storing equipment must be either explosion proof or intrinsically safe (see Section 2.18). The floor, walls, and ceiling of an explosion-protected room will usually be made of concrete. The room should have a sloped ceiling to direct any rising hydrogen, leading up to a vent at the peak to allow the escape of any hydrogen being given off from the process. Since hydrogen is lighter than air, it will rise to the ceiling and be vented. The room may incorporate a weak wall or other means to limit the pressure in the event of hydrogen ignition. The amount of hydrogen that could be vented from the process should be minimized by using tightly sealed process equipment, with the hydrogen flow rate indicated on a flow meter and the oxygen concentration in the process monitored by an oxygen meter (such as an electrochemical fuel cell) to keep it below about 20 parts per million (ppm). The oxygen concentration in the room should be monitored by an intrinsically safe oxygen monitor to assure that a high-enough oxygen concentration is available for the worker to breathe. Normal atmospheric oxygen concentration is about 20.94% (the balance is mostly nitrogen). The minimum oxygen concentration normally recommended for workers is 19.5% (check any local rules and those of the Occupational Safety and Health Administration, OSHA). A flammable gas monitor should also be installed in the room to warn if there is a process leak that is causing the hydrogen concentration to rise above the lower explosive limit (LEL; see Section 2.18). All intrinsically safe and explosion-proof equipment should be rated for hydrogen use (in the United States, this is group B).

Some alloys that are commonly used as core materials include Ni-Span C™, Hyperco 50B™, and various forms of Permalloy™. Ni-Span C has an advantage of exhibiting very low sensitivity of the modulus of elasticity and magnetic properties to temperature changes in the normal range of temperatures for industrial use (−40 to 85°C). Hyperco 50B has an advantage of higher permeability than Ni-Span C, but has a somewhat greater sensitivity of permeability to temperature changes.

When installing an LVDT core, the core should be mechanically connected to the member being measured by a non-ferromagnetic pushrod. The pushrod, or actuator rod, is screwed into the internal thread of the core. This enables movement of the core without affecting the core permeability. The connection is secured by using a set of jam nuts, or applying some nonpermanent thread-locking compound (such as Loctite™ 243). The pushrod can be made of aluminum, plastic, or non-magnetic stainless steel. However, a non-magnetic stainless steel (such as the ASM 300 series) must be annealed after machining because the cold working imparted during the machining process can align the magnetic domains of the alloy and cause the stainless steel to become more magnetically permeable.

The core of an RVDT must be of a shape that allows variable coupling among the primary and secondary coils as the core is rotated by rotation of a shaft. See Figure 8.4.

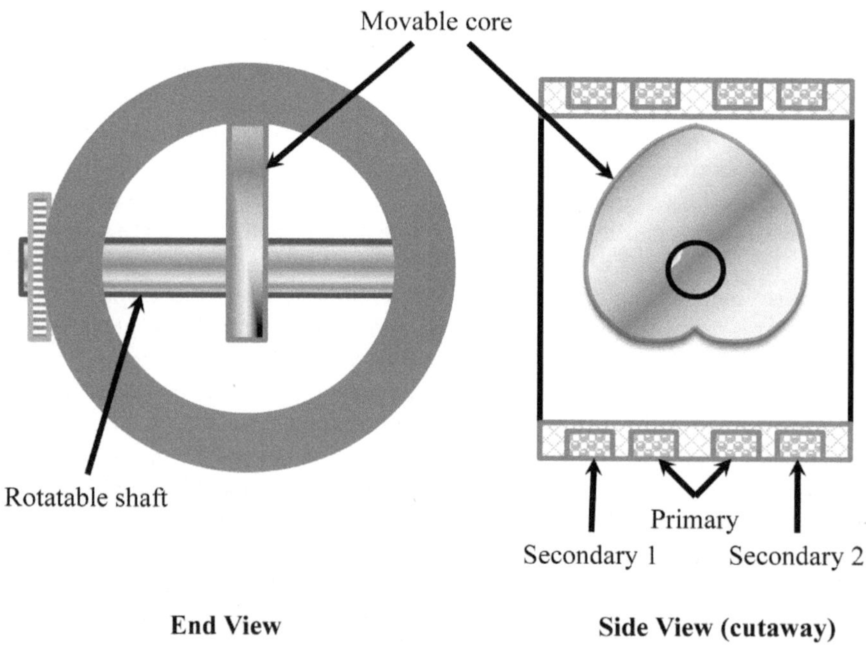

FIGURE 8.4 One type of core and coil configuration that can be used in an RVDT.

As can be seen in Figure 8.4, the movable core must be of a different shape from that of an LVDT. The core of an RVDT can also be called the *rotor*. The rotor is shown as being a relatively thin plate, and the housing as a cylindrical shape, for ease of illustration. An actual RVDT would be designed to maximize coupling between the rotor and coils.

The figure shows a knurled shaft connected to the rotor. This shaft is turned through an angular range as the measurand varies. Other types of shafts and couplings may also be used. The configuration shown is for an angular range of less than 180 degrees of rotation, such as zero to 120°, or ± 60°. When rotated past this amount, the rotor will again come within range of the coils on the other side of center, and again indicate the same range of angle. But usually, only one side of this range is used. Sometimes mechanical stops are included to prevent operation within the second range. As can be seen in the figure, a portion around the center of the rotor is always within the primary coil. As the rotor is rotated, more of the rotor volume is positioned within one coil, as less of it is positioned within the other coil. This is analogous to the movement of a cylindrical core within the secondaries of an LVDT.

When determining the dimensions of the core to be used with a new LVDT design, the most efficient design path is to start with the core dimensions used with an LVDT currently on the market having somewhat similar characteristics to those desired. The core should extend partway into each secondary coil even when the core is positioned at either end, at full stroke. The core should also engage the full length of the primary coil throughout the full range of core position. The main criterion that determines core length is the need to exhibit the least amount of nonlinearity error (see Section 2.5). First, take a set of readings of the output versus the core position. Then plot or analyze the data to determine the nonlinearity curve shape. A straight line, of course, is desired. If the curve bows up, so that the readings in the middle of the stroke are higher than the ideal line, the core is too short. If the curve bows down, so that the readings in the middle of the stroke are lower than the ideal line, the core is too long. If the nonlinearity is more than 1%, make a coarse adjustment to the core length to get it closer to the ideal. When the nonlinearity is less than 1%, make a new set of five cores with length variations in approximately 1% increments. Test these cores for nonlinearity error curve shape in the same way. When an S-curve is obtained, where a slight positive bow near one end of the stroke is approximately equal to the slight negative bow near the other end, the core length is optimum for that particular arrangement of coils.

Design of a rotor for use in an RVDT is not so simple. But a similar iterative process starting from a known design is the most expedient way to go. Otherwise, calculating and testing a new design from scratch takes a much longer amount of time.

The LVDT or RVDT housing can also have a great effect on the nonlinearity error, since it is generally made of ferromagnetic material in order to provide magnetic shielding. The author found that the housing dimensions and placement of the coils within the housing are especially important in developing a gage head LVDT, as shown in Figure 8.18. To make the gage head as small as possible, the very small coils are sometimes wound as free-standing coils without a bobbin. Such a coil is held together by the wire coating, which is softened with heat during winding. The softened insulation bonds each wire turn to the next, and forms a unitized coil when cooled.

When winding a bobbin less coil this way, a heat gun having a carefully controlled temperature is aimed at the coil winding area to soften the wire insulation. As the coil is finished being wound, the heat gun changes to a cooling cycle (with room temperature air). Also, a drop of alcohol can be applied to rapidly cool the coil. When the coils are completed, they are inserted into the housing and held in place by epoxy. The author found in one particular gage head LVDT design that exact positioning of the coils within the housing made the difference between having a nonlinearity as high as 0.35% and as low as 0.15% (after moving the set of three coils by only about 1 mm). With optimum positioning of the coils, an uncorrected nonlinearity specification of less than 0.20% could be guaranteed in that design. This required the use of a fixture to accurately position the coils during the process of applying and curing the epoxy.

8.6 CARRIER FREQUENCY

Although a few LVDTs are designed for operation at 60 Hz, excitation frequencies of 200 Hz to 10 kHz are more typical, with 1 kHz to 5 kHz being the most common. Generally, a higher carrier frequency is desired in order to obtain a faster response to variation in the core position. The response time is proportional to the filtering frequency for a given order (number of poles) of a low-pass filter circuit. In a low-pass filter, frequencies higher than the filter's corner frequency are attenuated. For a simple analog filter of two or three poles, the filter's corner frequency must be one-tenth to one-fifth of the LVDT/RVDT carrier frequency. The carrier frequency may also be called the *excitation frequency*. Having the filter's corner frequency five to ten times lower than the carrier frequency enables the signal conditioner filtering circuits to achieve an acceptable low level of AC ripple in the DC output signal (such as less than 2 mV AC_{RMS} for a 0 to 5 volts DC output). Typically, the highest frequency of interest in the measured signal should be no greater than 0.1 times the excitation frequency [2, p 278].

To assure an accurate LVDT signal, it is absolutely imperative that the excitation must have a duty cycle of exactly 50%. Otherwise, the demodulation will be inaccurate, since the positive and negative half-cycles would be uneven. This can be achieved by starting with a square wave that is double the desired frequency, then passing through a divide-by-two flip flop, followed by the low-pass filter to convert to a sine wave. The divide-by-two flip flop provides the exactly 50% duty cycle. Likewise, excitation of inductive and capacitive sensors of the previous chapters also benefit from excitation with a duty cycle of exactly 50%.

In most cases, the current required to drive the primary at a given voltage is lower when the frequency is higher, due to the increased inductive reactance of the primary coil. The impedance of the secondaries is also greater. A limiting factor, however, is that an excessively high excitation frequency (for example, >10 kHz) leads to eddy current loss in the core and results in lower output signal level, increased power dissipation, and greater temperature sensitivity, unless the core is very small, with a small wall thickness.

So there is a trade-off between frequency response and other performance parameters when determining the carrier frequency that will be specified. The most common LVDT carrier frequencies are between 2.5 kHz and 5.0 kHz, and RVDTs from 5.0 to 10.0 kHz. The ratio between carrier frequency and the effective low-pass filter frequency can be reduced, while maintaining a low ripple in the output voltage, by using a higher-order analog filter, or by using a combination of analog and digital filtering. The LVDT or RVDT signal can first pass through an anti-aliasing analog filter that is tuned relatively close to the carrier frequency, such as one half of the carrier frequency. Then the filtered signal can be digitized by an analog-to-digital converter (A/D, or A/D converter) and fed to a microcontroller. The microcontroller can execute a digital filtering program to reduce the signal ripple to the amount desired. If an analog voltage or current output is desired, the microcontroller can send the digital data to a digital-to-analog converter (D/A, or D/A converter). Further information on signal conditioning is presented in Section 8.8.

A digital filtering technique preferred by the author for use in position sensors utilizes a set of five registers configured as a first-in, first-out shift register (FIFO). Each new datum of measured position information (a data cycle) coming from the A/D converter is fed into the input (new datum) end of the register by the microcontroller, pushing out the oldest datum from the other end and discarding it. The microcontroller then reads the five data (four old plus one new), discards the one highest and one lowest readings, and takes the average of the remaining three. The resulting average then becomes the output datum. The result is that noise pulses are thrown away (likely to be the highest and/or lowest readings) and do not affect the reported measurement reading. Alternatively, instead of taking the average of the remaining three, the original five can be sorted low to high and the middle one selected as the output datum. A set of five registers, as described, is the smallest number that can be used effectively in this method; but more registers (>10) are often utilized. Using a larger number of registers yields a better (smoother) average, but increases the time to respond to a changing measurand. A second common method uses five registers (for example), but after operating similarly on one set of five data, throws them all away and takes a new set of five data. A disadvantage of this second method is that with the example of five data, it yields a lag time of five data cycles before a new update, whereas the first method has a lag time of only one or two data cycles before some change is indicated.

Zero-Phase Frequency (also *zero phase-shift frequency*)—a difference in phase (or a phase angle) that may exist between the primary excitation voltage of an LVDT and the signal voltage of the secondaries is called phase shift. The amount of this phase shift is related to the excitation frequency. And so there is usually a frequency at which the phase shift is zero, and this frequency is called the zero-phase frequency.

In many LVDTs and RVDTs, the lowest temperature sensitivity can be achieved by operating the excitation at the zero-phase frequency. Sensitivity to the amount of phase shift is mainly due to the effects of using a synchronous demodulator, in which a phase change results in a change in the demodulated signal. This can be mostly avoided by using asynchronous demodulation, but synchronous demodulation is often used for other reasons (see the following section about demodulation).

So, with most LVDTs, it is possible to adjust the excitation frequency in order to find the zero-phase frequency and thus reduce the temperature sensitivity of the signal amplitude. But even more importantly, it is best to confirm a frequency that provides the minimum *change* in phase shift due to temperature change. This could be different from the zero-phase frequency. By testing the LVDT or RVDT primary to secondary phase-angle relationship over temperature at several different excitation frequencies, an optimum excitation frequency can usually be found that minimizes the temperature-induced phase-angle sensitivity. Operating at this excitation frequency will provide the best temperature stability. When operating at other excitation frequencies, the temperature sensitivity of span may be greater. This should not be confused with the reduction in span as temperature increases that may occur due to the increase in primary impedance because of the copper wire thermal coefficient of resistance. And in fact, it is sometimes possible to operate at a excitation frequency that compensates for the temperature sensitivity of the primary wire resistance.

8.7 DEMODULATION

An LVDT or RVDT core is a magnetically permeable material, usually a nickel-iron alloy. When the core is centered on the primary and thus equally spaced relative to the secondaries, the voltage induced into each of the secondaries is approximately equal. The secondaries are usually connected into a circuit such that the voltages are *series bucking.* This means that the secondary voltages subtract and will have a total series-connected output of near-zero volts when the core is centered (at null). A dot on the schematic indicates the polarity of each coil, and each lead that is marked with a dot has the same phase. When winding a coil, the two ends are called the start and finish, indicating the end of the wire where the winding operation was started, and finished. The start is usually the end that is marked for phase or polarity. A series-bucking connection is shown in Figure 8.5. When the core is centered, it is called the *null position.*

The coils and their phase-marking dot are arranged in Figure 8.5 according to the way they are actually wound, rather than the way they were shown for simplicity in Figure 8.3. Either way is acceptable, as long as the phase dots are there to show the actual connection configuration. As mentioned, when winding a coil, the two opposite ends of the coil wire are called the start and finish, indicating when the

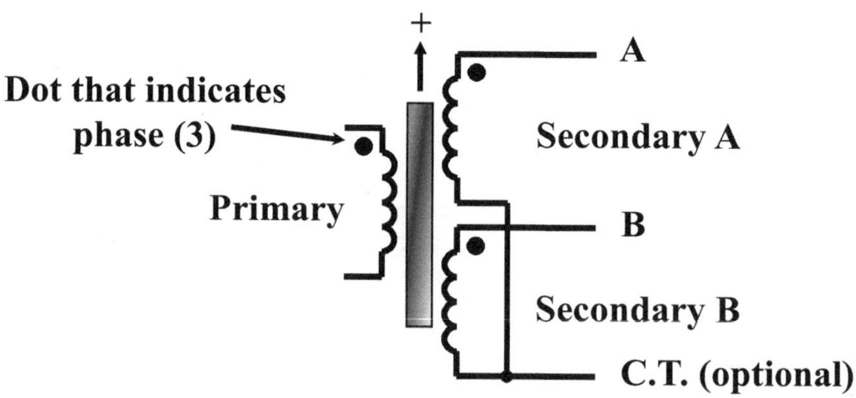

FIGURE 8.5 LVDT or RVDT series-bucking connection schematic.

winding operation was started (the innermost layer of magnet wire, wound right onto the surface of the bobbin) and finished (the outermost layer of magnet wire). It can be seen that the bottom end (the finish) of secondary A in the schematic of Figure 8.5 is connected with the bottom end (the finish) of secondary B.

When the core is not at null, the secondary that is more coupled to the primary, due to the core being closer to more of its coils than to the coils of the other secondary, will have a greater voltage output. So, the AC output of the two secondaries connected in series bucking will be at the minimum when at null, and increasing as the core moves away from null in either direction. To tell which direction the core has moved away from null, the secondary output is usually demodulated to a DC voltage, so the polarity of the DC voltage will indicate the direction of core travel. A three-coil LVDT or RVDT (without signal conditioning) can be supplied with three, four, five, or six lead wires. With six lead wires, all leads from the three coils are brought out. With five wires, two of the secondary wires are connected together internally and brought out as the *center tap* (C.T.). With four wires, two of the secondary wires are connected together internally but not brought out on a lead wire. With three wires, the center tap is brought out, but the primary is connected across the two secondaries. Or alternatively, there may be no primary coil and the two secondaries are powered at their ends, while the signal is found on the center tap. This last configuration is called a *half-bridge LVDT* or DVRT (differential variable reluctance transducer).

The simplest method to demodulate a DC output voltage from an LVDT or RVDT signal is to use a diode demodulator and differential amplifier circuit, similar to that shown in Figure 8.6.

The output signal across the diode cathodes is a varying DC differential voltage (the diodes are shown as a triangle with a line; the side with the line is the cathode). That is, if we call the voltage at one cathode the reference, then the voltage at the other cathode could vary, for example, from –100 mV to + 100 mV for core positions from – full stroke to + full stroke, respectively. This differential voltage is amplified by the differential amplifier to provide a desired output voltage, such as ± 5 volts DC or ± 10 volts DC. Circuits similar to this are commonly used when the LVDT or RVDT is a component part of another device, such as a pressure sensor, gage head, or valve positioner. A disadvantage of this circuit is that the *forward voltage drop*

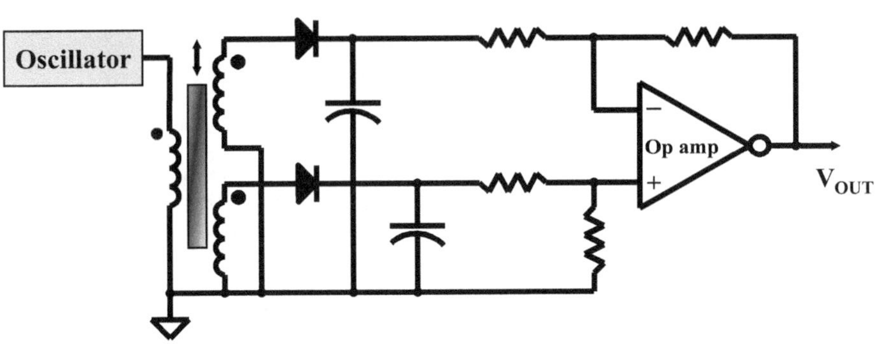

FIGURE 8.6 LVDT or RVDT with diode demodulator and differential amplifier circuit.

of the (usually silicon) diodes has a relatively strong temperature coefficient (about −2.2 mV/°C), and is also sensitive to mechanical stress. This is less of a problem when the demodulated DC signal is relatively high (≥ ±0.25 volts). But with lower signal levels, the problem is due primarily to variation in the voltage drop across one diode having a slightly different temperature coefficient as compared with the other diode. Besides variation in temperature coefficient from one diode to the next, there may also be a difference in mechanical stress exerted upon one diode as compared with the other. Mechanical stress also causes a change in the forward voltage drop of a diode. Mechanical stress on the semiconductor junction within a diode body can be caused by soldering while the leads are under stress, as well as from a difference in the thermal coefficient of expansion between the diode body and the circuit board. Using a pair of diodes contained within the same surface-mount package may provide better matching of the diode forward voltage drop thermal coefficient, as well as tending to exert the same stress on both of the internal diodes. For leaded diodes, a mounting method as was shown in Figure 6.17 may be used.

A more precise method for demodulating the LVDT signal is to use synchronous demodulation, utilizing active switches instead of diodes. With *synchronous demodulation*, a transistor, switch, or other device is activated to pass the signal at a desired time with respect to the phase of a primary or secondary voltage (usually the primary).

A simple method developed by the author in the 1970s for use in pressure sensors is shown in Figure 8.7. Bipolar junction transistors (BJT) or field effect transistors

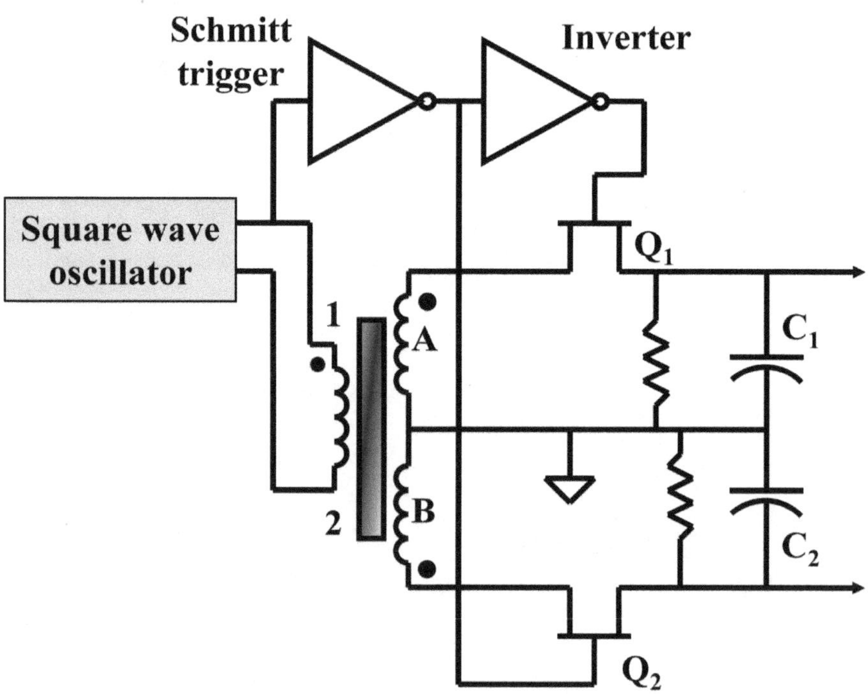

FIGURE 8.7 Square wave oscillator with synchronous demodulator using two transistors.

(FET) replace the diodes that were shown in Figure 8.6. They have a smaller forward voltage drop (between the collector and emitter on a BJT, or between the drain and source on a FET) than a silicon diode, with accompanying much smaller changes in the voltage drop over temperature. The voltage drop across each transistor when it is turned on is only a few millivolts, instead of the 500 to 600 mV of a silicon diode.

Although LVDTs and RVDTs have been traditionally driven by a sine wave, the author has found that a suitably designed LVDT can give good results when driven by an appropriate square wave and synchronously demodulated, as long as a specific LVDT is used so that the core can be adjusted to minimize nonlinearity, the exci tation frequency can be selected for the lowest sensitivity to temperature, and the leads between the LVDT and conditioning electronics are short (≤ 100 mm). Short leads can be used when the LVDT is included within another device, such as a pressure sensor, where the LVDT measures the travel of a pressure-sensitive diaphragm.

This simple circuit does not include phase adjustment, as in the demodulator shown in the next section, so this is one reason it is only suitable for use when the LVDT is located very close to the driver and demodulator circuits. Another reason that the leads must be kept short is that a square wave can allow the generation of more complex waveforms, with harmonics and their higher-frequency content, that can have a greater effect on the signal due to the reactance of a long cable. This could introduce unwanted signal variations if the cable is flexed or brought into close proximity with other dielectric or electrically conductive materials.

In Figure 8.7, the primary is driven by a square wave (typically 3.3 or 5 volts DC). This same square wave is used to switch on the demodulator transistors at the appropriate times. In the figure, the emitter of a BJT or the source of a FET connects with C_1 or C_2. In this example, when the primary is going positive at pin 1, transistor Q_1 conducts to charge capacitor C_1 up to the voltage of secondary A (this explanation is for an enhancement-mode N-channel FET or an NPN bipolar transistor, but the opposite polarity may be used with the appropriate configuration and transistor types). Similarly, when the primary is driven positive at pin 2, transistor Q_2 conducts to charge capacitor C_2 up to the voltage of secondary B. So, the voltage across C_1 represents the voltage of coil A, and the voltage across C_2 represents the voltage of coil B. When the voltages are equal, the differential output is zero. When the voltage across C_1 is greater than that across C_2, the differential output is positive, when the voltage across C_1 is less than that across C_2, the differential output is negative.

8.8 SIGNAL CONDITIONING

The circuitry that provides excitation to the primary and demodulates, amplifies, filters, and scales the signals from the secondaries is called a *signal conditioner*. The signal conditioner circuitry may be contained within the LVDT or RVDT housing, or may be sold as a separate product that connects with the LVDT or RVDT by a cable. When the electronics to provide a DC voltage output is inside the housing, it is called a DC LVDT or DC RVDT. When connected to an external signal conditioner by a cable, it is called an AC LVDT or AC RVDT. Alternatively, electronics within the LVDT housing, or in a separate device, may provide digital outputs of various protocols. The cable between the transducer and the signal conditioner may sometimes be longer than 500 m. When a signal conditioner is sold as a separate product, it normally has several adjustable features for use over a wider range of applications with a variety of LVDT or RVDT models. A signal conditioner product will typically include a sine wave generator with adjustable frequency, power amplifier with adjustable voltage output level, synchronous demodulator (sometimes with adjustable phase shift), low-pass filter (sometimes with adjustable filter frequency), circuits to provide the desired analog and/or digital signals, and power supply circuits as needed to serve the several types of analog and digital circuits. Accurate demodulation is an important function of the signal conditioner, but many other features can also be critical to good performance.

The power provided to the LVDT or RVDT primary coil is called the *primary excitation*, or excitation. It is usually in the form of a sine wave, and operates at the carrier frequency (or it can be called the excitation frequency). Some LVDTs and RVDTs only require less than 1 mA of excitation current, but some may require as much as 90 mA. So, when designing a standalone signal conditioner (that is, one that is made to connect with any AC LVDT or RVDT, rather than being directly coupled with a particular one), the signal conditioner should have a capability of powering the excitation at up to 90 mA. Most LVDTs and RVDTs can be powered with an excitation voltage of anywhere between 1 and 3.5 volts RMS (root mean square), but some signal conditioners have an adjustable excitation voltage.

The excitation frequency must be stable over time and with temperature changes, and have a total harmonic distortion (THD) level of 2% or less. There are many different circuits for generation of a sine wave. It is important to accurately control the sine wave amplitude and frequency, while also keeping distortion low. One popular circuit is called a Wien bridge oscillator, but the frequency and voltage stability, as well as distortion level are marginal for use in a signal conditioner. Another common circuit is the phase-shift oscillator, having less distortion and better frequency stability, but the voltage amplitude stability over temperature is still marginal for signal conditioner use. A third circuit configuration is the quadrature oscillator, having very similar results as with the phase-shift oscillator. A fourth type is to use an accurate square wave oscillator (such as a crystal-controlled one) and then use a low-pass filter to convert the waveform into a sine wave. This type can have good voltage amplitude stability if the filter capacitors are of a very temperature-stable type, such as the NP0/C0G type. Incidentally, those are zeroes and not the letter O. NP0 means negative and positive temperature coefficient of capacitance change are both zero (±30 ppm/°C). That same type of capacitor has an EIA code of C0G, indicating that the temperature coefficient is 0.0 (that's the C), the multiplier is −1 (that's the 0), and the tolerance is ±30 ppm/°C (that's the G). So, calling a capacitor type NP0 or calling it C0G is saying the same thing.

The circuit of Figure 8.8 was invented by the author as a way to produce a sine wave that is stable in amplitude and frequency, and having a low distortion level. It also provides the signals to drive a wide-range phase adjustment circuit.

Previous to this invention, phase adjustment was accomplished by analog circuitry, and the adjustment range was limited to ±45° (limited by the stability range of the analog amplifier). By using the circuit shown and adding a ±22.5° analog phase adjustment, a phase adjustment range of ±180° is easily obtained. The input of

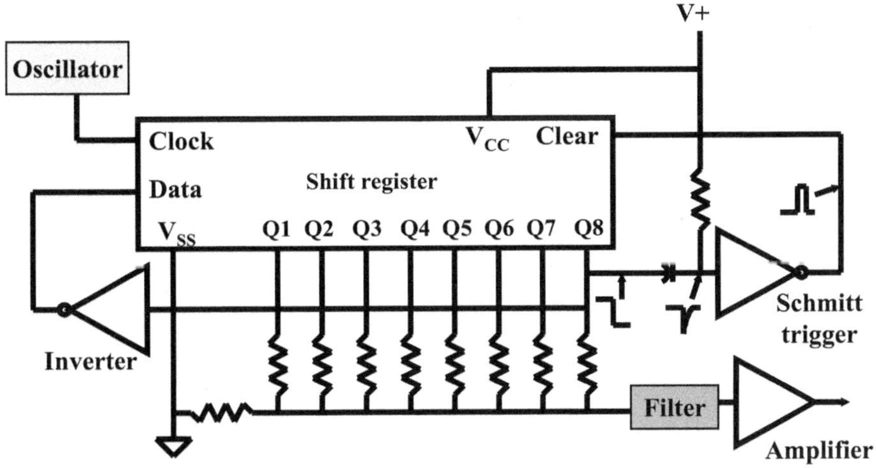

FIGURE 8.8 Frequency and amplitude stable low distortion sine wave generator, having phase-shifted outputs.

the analog phase adjustment circuit is simply switched among the Q outputs of the CMOS shift register, as needed to come within 22.5° for the final analog adjustment. In the signal conditioner product, the analog phase-shifter had a range of ±45° in order to allow overlap and allow easier adjustment.

The circuit shown in the figure synthesizes a sine wave digitally. This signal is derived from an easy-to-implement, stable DC power supply (usually 3.3 or 5 volts). Since the CMOS Q outputs are switched to within a few millivolts of 0 and V+, the waveform amplitude will be as stable as the V+ regulated supply. The input clock can be adjusted over a wide range while not affecting the accuracy of the sine wave shape. The values of the resistors are weighted to yield a voltage-divider output that varies according to the sine function. The output is then passed through a filter to smooth it and take away the switching spikes, leaving a clean sine wave (see Figure 8.9). If the input clock frequency is changed by a large amount, then the filter frequency should also be changed accordingly.

Because the sine wave is generated digitally, there is a square pulse for every 22.5 degrees of the full wave period available at the shift register Q outputs. A digital switch can select which 22.5 degree increment is selected, then the ± 45° adjustment is available from that starting point. This way, good resolution is obtained, with the adjustment only covering ± 45°, but the whole range of ± 180° can also be covered by using the digital switch.

An alternative way to generate a sine wave is to use a microcontroller that drives a D/A converter and amplifier. However, care must be taken to ensure that any subroutines used do not interfere with the continuous operation or frequency of the sine wave generator. The phase angle between the microcontroller-generated primary excitation voltage and the secondary demodulator switching circuit can then be adjusted in software to account for differences in phase from one LVDT or RVDT model to the next, and to allow for timing differences due to longer cable lengths.

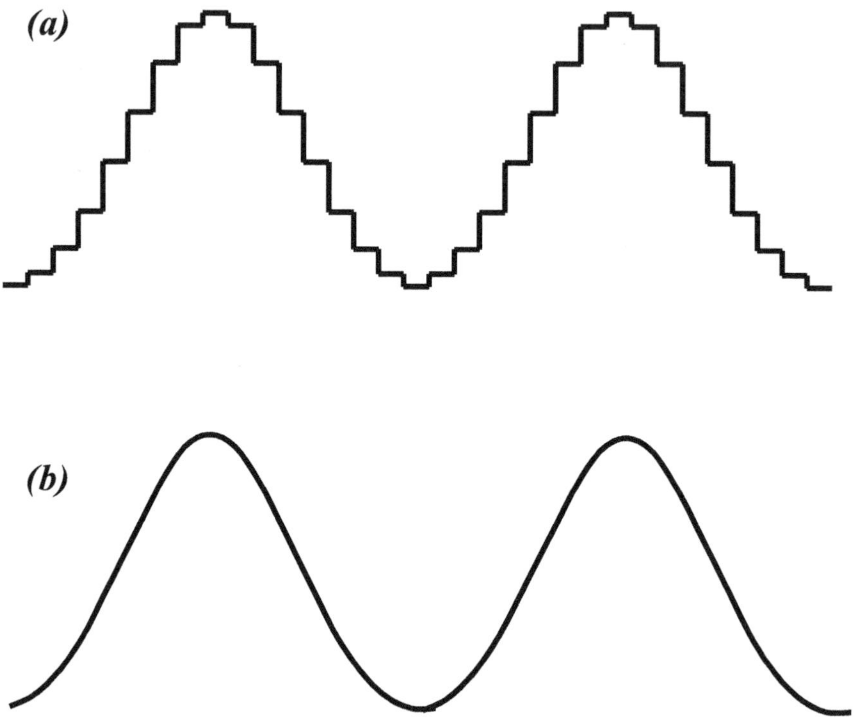

FIGURE 8.9 (a) The steps of the digitally generated staircase are weighted according to the sine function. (b) The filter circuit smooths the waveform.

The typical way to excite the primary of an LVDT or RVDT is to use a sine wave driver with constant voltage. An alternative way is to use constant current, in order to keep the current in the primary coil constant as the coil resistance increases with increasing temperature, and as inductance changes with core position. But the inductance changes with core position lead to nonlinearity in some LVDTs and RVDTs when excited by constant current. So, depending on the characteristic of the particular LVDT or RVDT used, one or the other (constant voltage or constant current) will provide the best performance over a temperature range. This is, of course, regarding RMS voltage or current, not instantaneous. Constant voltage is the most common.

Another method sometimes used instead of the difference between the output voltages of the two secondary coils, when a microprocessor is available, is to utilize the function:

$$Vout = \frac{Vcoil_A - Vcoil_B}{Vcoil_A + Vcoil_B} \tag{8.1}$$

where $Vcoil_A$ and $Vcoil_B$ are the output voltages of the two secondary coils, A and B. This formula tends to compensate for variation in the voltage of the sine wave drive, but assumes that the sum of the secondary voltages will remain constant as the core moves throughout its range. The assumption is true only if the core travel is somewhat limited to keep the core sufficiently within both secondary coils, so that the sum A + B remains relatively constant. This is the formula implemented in the integrated circuit AD598, but this IC is not suitable for use with LVDTs or RVDTs without either all four secondary wires or a center tap being available. The AD698 implements a similar formula, except the A and B voltage difference is divided by the primary voltage, and it can be used when only two secondary wires are available, without need of a center tap.

Figure 8.10 shows an LVDT with an internally mounted signal conditioner designed by the author. The signal conditioner uses a version of the digitized sine wave generator as described earlier and shown in Figures 8.8 and 8.9, but uses a phase adjusting algorithm in a microcontroller instead of in hardware. This way, the amount of phase difference between the primary excitation and the demodulator switches is adjustable in software.

Implementing phase adjustment in the microcontroller allows tuning of the excitation-to-demodulator phase relationship, and is accomplished through digital communication after the LVDT is fully welded with the signal conditioner inside its housing. This signal conditioner uses a switched-capacitor filter that enables changing the low-pass filter frequency through digital communication. The connector includes one pin that enables digital communication with an external digital device, such as a computer.

The separate signal conditioner of Figure 8.11 was designed by the author for laboratory or industrial use with any LVDT or RVDT. It also incorporates a digitized sine wave generator. To accommodate LVDTs and RVDTs having various cable lengths, the excitation-to-demodulator phase relationship is adjustable from the front panel.

The 4 ½ digit front panel display has a normal and a ×10 range, plus English and metric units, selectable by front panel switch. High and low alarms are set and visually indicated on the front panel, and electrically indicated by open collector outputs.

The LVDT/RVDT signal conditioner of Figure 8.12 was designed by the author as a position transmitter, with ultra-high stability, very low cost, easy installation, and simple calibration via the two ten-turn potentiometers for Zero and Span. The

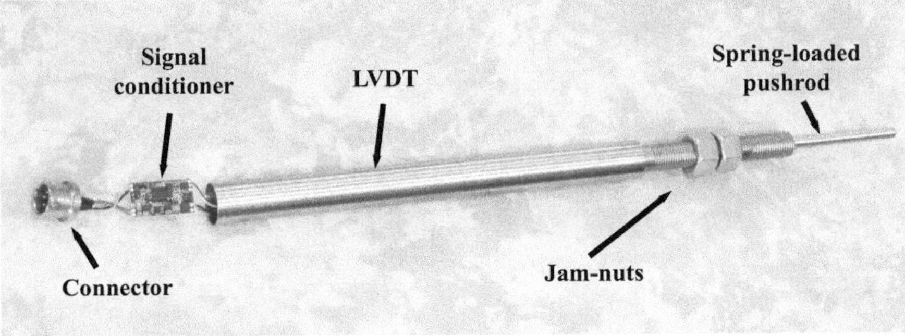

FIGURE 8.10 An LVDT with internal signal conditioner (disassembled for illustration).

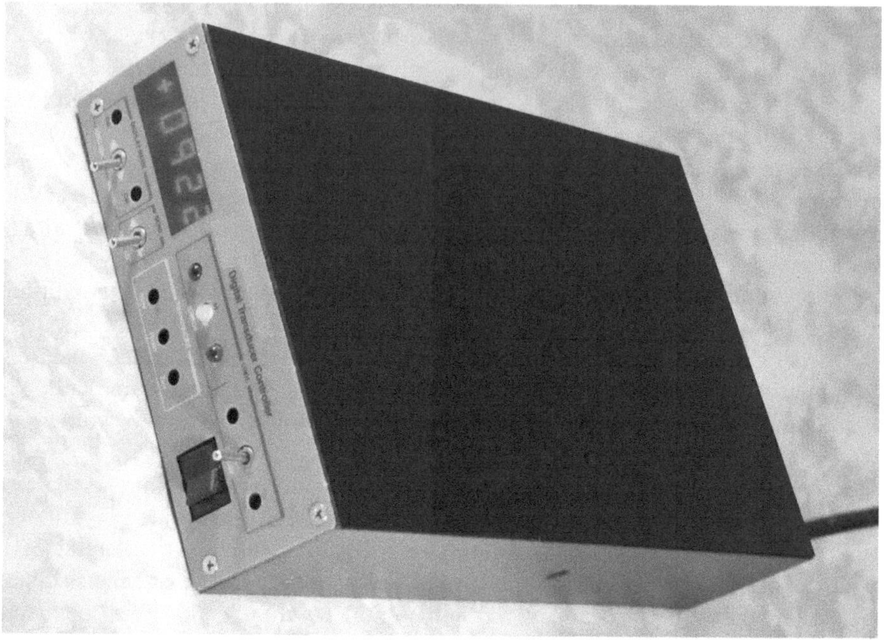

FIGURE 8.11 An external signal conditioner with digital display, analog output, phase adjust, and alarms.

FIGURE 8.12 A 4 to 20 mA two-wire current loop LVDT/RVDT signal conditioner, designed and manufactured by the author.

connectors are removable with the wiring intact, for easy replacement. The design is so accurate, that one transmitter can be switched with a replacement while only setting the potentiometers to the same number of rotations, thus maintaining the calibration (if the rotation-counting potentiometer option is installed).

It also includes a jumper selection for temperature compensation of the LVDT primary current. The removable screw-terminal connector is split into two parts: two terminals (plus shield) for the current loop connections, and five terminals for the LVDT connections. It can operate with any LVDT having three to six connections (for a 6-wire LVDT, two wires go to the same terminal for a center tap (C.T.)). It is called a transmitter due to its being configured as a 4 to 20 mA loop-powered device. It connects with a DC power supply, such as 12 to 24 volts DC, over two wires (a twisted pair). The amount of current drawn on these two wires indicates the measurand. A current of 4 mA is drawn when the measurand is at zero, and 20 mA when the measurand is at full scale. That means, for example, that the current draw would be 12 mA when at mid-scale. This way, the unit is powered by the pair of wires and also the signal is sent by using the same two wires. This makes installation easier, and any resistance of the twisted pair wires does not affect the reading. The circuitry includes a high stability voltage regulator, crystal oscillator, frequency divider, low-pass excitation filter, switched CMOS demodulator, non-interactive zero and span controls, low-pass signal filter, signal amplifier, and voltage to current converter. The aluminum spacers that are swaged onto the printed circuit board allow for easy mounting to a panel, using four bolts. Normally, the LVDT is utilized over its full range, including positive and negative excursion of the core from null. To calibrate accordingly, the LVDT connector is pulled out, and the zero potentiometer is set for a loop current of 12 MA (mid-scale between 4 and 20 mA). Then the connector is re-inserted, and the LVDT core position is adjusted to indicate 12 mA loop current (null). Last, the LVDT core position is adjusted to the positive full-scale position, and the span potentiometer is adjusted to indicate a loop current of 20 mA. The calibration is now complete. When the LVDT core position is adjusted to the negative full-scale position, the loop current will be 4 mA.

Figure 8.13 shows a microcontroller-based signal conditioner designed by the author, that mounts to a DIN rail. In the figure, it is already mounted to a DIN rail, and has all of the necessary connections to demonstrate its function.

This demonstration panel includes all of the components necessary to simulate a complete system. A digital micrometer and an LVDT are mounted into a fixture to simulate the LVDT measuring a variable position. Two batteries are powering the signal conditioner and the other battery powers the digital display. The digital micrometer provides a reference reading of the core of the LVDT, for comparison and evaluation of the LVDT and signal conditioner performance. The signal conditioner excites the LVDT with a digitally generated sine wave and conditions the LVDT signal output, which is indicated on a digital display. The signal conditioner has internal and external manual and visual interface capability (push buttons, switches, LEDs, and so on), as well as digital communication through an RS-485 communication port that may be connected with a computer or controller. Either the panel or the digital communication can be used to configure all of the signal conditioner parameters, including excitation frequency and voltage, analog output type, calibration, and other

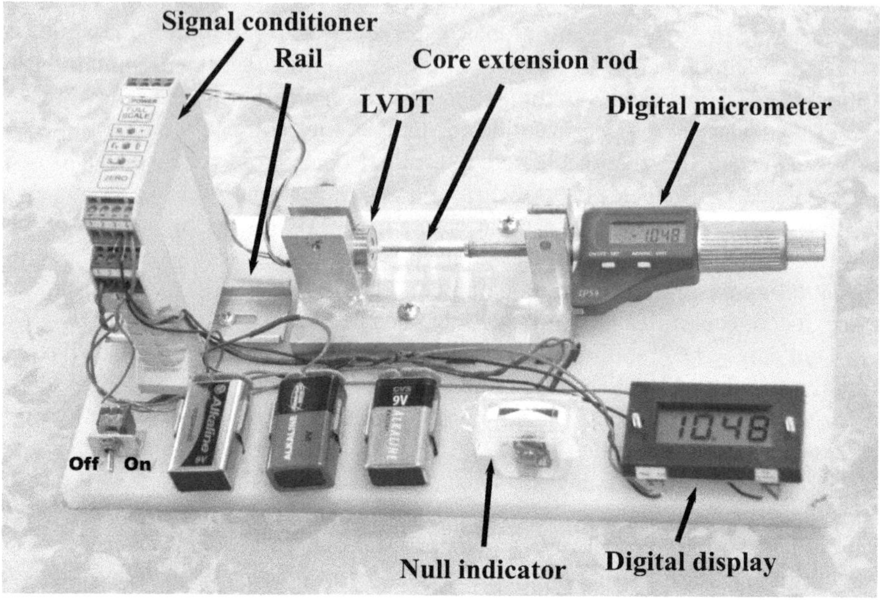

FIGURE 8.13 A DIN rail mounted signal conditioner, with RS-485 communication, analog and digital outputs, alarms, pushbutton calibration, and other features.

features. Selectable output types include several voltages and currents. When calibrating the LVDT or RVDT, two push buttons are used while the software algorithm leads the user through a simple automated calibration sequence. First, the user is instructed to move the core until the LEDs (and the null output) indicate the core is at the null position. Then the core is moved to positive and negative full scale, and then the calibration is complete. A single-wire connection means for synchronization among two or more signal conditioners is also provided.

The signal conditioner of Figure 8.14 was designed by the author to provide extremely low drift, such as for applications where periodic calibration checks might be difficult or impossible. It has selectable excitation frequency that is generated from a crystal oscillator, chopper-stabilized op amps, synchronous demodulation, a single-wire synchronization function for multiple units, and incorporates very high stability components.

Although LVDTs and RVDTs are already known as being very low-drift sensors, the author has invented a method and circuit that reduces that even further. A demodulation technique is used that provides low drift, and also reduces any sensitivity to cable length by a factor of two times lower than with the standard techniques. That is, the demodulation is synchronized with the actual voltage at the LVDT primary, rather than the oscillator voltage at the other end of the cable. In order to maintain the low drift when producing an analog output voltage, a fixed amplification factor is implemented, using very low drift and low-temperature coefficient resistors and capacitors. With a fixed amplification factor, final adjustment of the output signal to engineering units is accomplished within the receiving equipment. Two crystals are

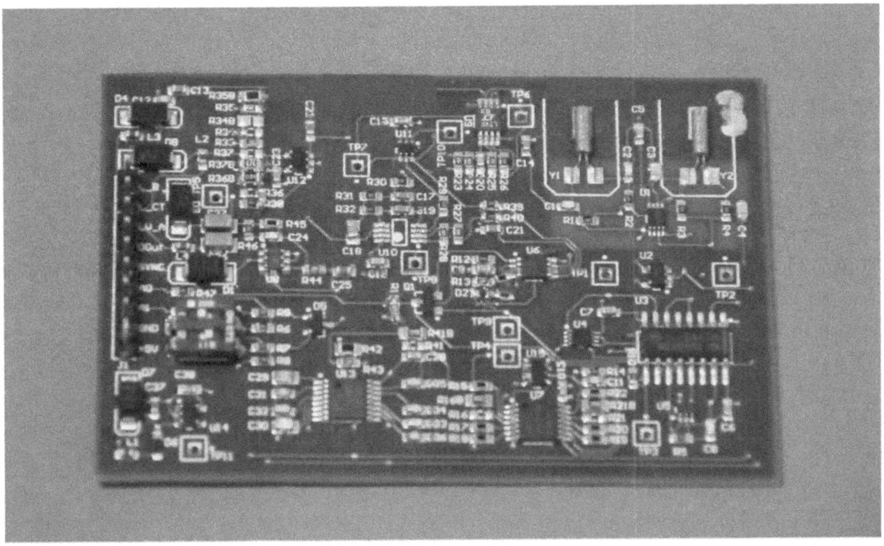

FIGURE 8.14 A high-stability signal conditioner with selectable excitation frequency and fixed amplification.

used with a digital selector circuit to provide a choice among four extremely accurate excitation frequencies, 1.5, 2.5, 3.0, or 5.0 kHz.

Synchronization among several of these signal conditioners mounted in the same location uses a method invented by the author that only requires the connection of one wire from each signal conditioner to a common point. See the following section on synchronization.

8.9 SYNCHRONIZATION

Previously, synchronization of a switched demodulator to the LVDT excitation was presented. In this section, synchronization among several signal conditioners in the same location is addressed. When two or more LVDTs and/or RVDTs are operating nearby one another at nearly the same or harmonically related frequencies, a beat frequency can be generated that interferes with the stability of the signal output. This is because, when selecting the "same" frequency among two or more devices, there is usually a slight difference in frequency from one to the next.

We suppose, for example, that two LVDTs and their respective signal conditioners are operating nearby one another, both having their excitation frequency set at 5.0 kHz, and are using very precise oscillators that are crystal-controlled. If the first LVDT is actually operating at a frequency of 4,999.75 Hz, and the second one is operating nearby at a frequency of 5,000.25 Hz, there is a frequency differential of 0.50 Hz. So, it may be stated that the beat frequency between the two oscillators is 0.50 Hz. If one observes the analog outputs from their respective signal conditioners on an oscilloscope, it is likely that a small recurring deviation of each signal will be present: their amplitudes will steadily rise and fall by a small amount at a rate of once every two seconds. The example used very accurate crystal-controlled oscillators, but works the same way if the frequencies are not as accurate. With RC oscillators, the difference in frequencies might be 2%, with one oscillator at 4,925 Hz and the other at 5,075 Hz, for example. The beat frequency might then be 100 Hz, and would be the frequency of the amplitude modulation of the output signal. This beat frequency phenomenon is the same that may be observed with two outboard marine engines that are not synchronized. In addition to the sounds of each engine, one can hear a continuous increase and decrease in their sound amplitude. That's why a beat frequency condition is also called motorboating. It can be prevented by ensuring that the two frequencies maintain exactly the same phase relationship with each other. With two motorboat engines, they may be synchronized by controlling the relative sparkplug timing. Synchronization has typically been accomplished in an LVDT signal conditioner by assigning one device to utilize its internal oscillator (the *leader*) and assigning the remaining devices (the *followers*) to maintain a fixed phase relationship with the lead oscillator. Figure 8.15 shows several ways in which multi-unit synchronization has been accomplished.

In Figure 8.15(a), the leftmost device is designated as the leader by wiring its leader (L) output to the follower (F) input of each other device. This scheme functions

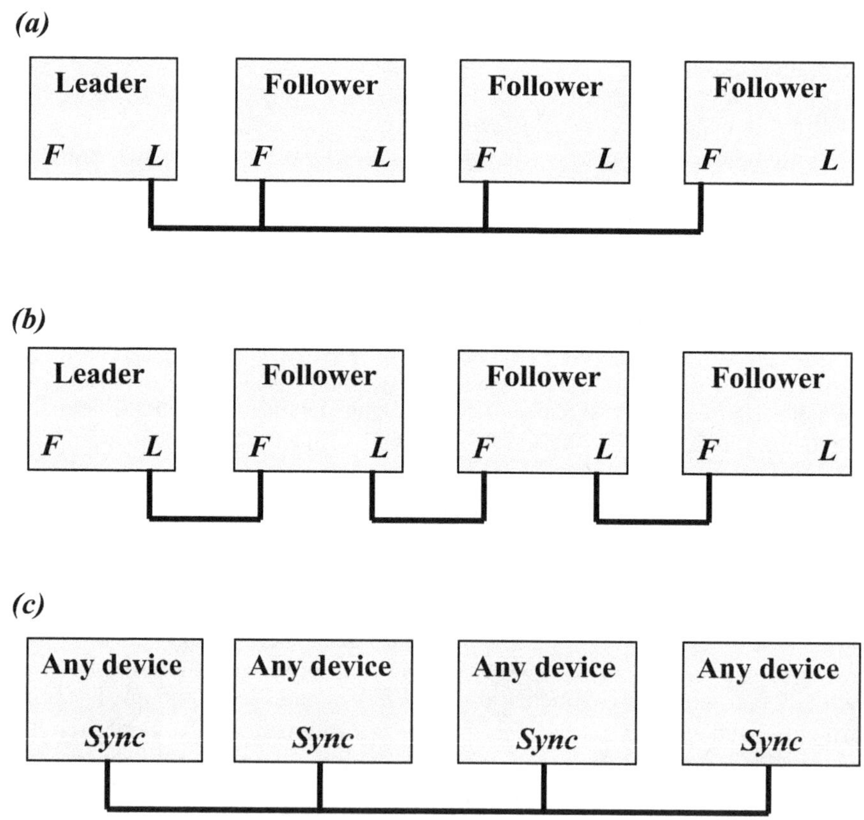

FIGURE 8.15 Various configurations of leader and follower oscillators to eliminate beat frequencies: (a) designated leader, (b) daisy chain, (c) one-wire sync.

well, except that a failure of the leader device excitation may cause the failure of all devices. But it is possible for some designs of follower devices to detect a failure at their follower input, and then switch to independent operation (but the beat frequency problem would then return).

In Figure 8.15(b), the leftmost device is designated as the leader by wiring its leader (*L*) output to the follower (*F*) input of the next device. Each next device drives the device coming after it. This configuration is called a daisy chain, and also functions well, except that a failure of any device may cause the failure of devices coming after it. It is possible for a device to detect its own failure, if otherwise still operational, and to short its follower input to its leader output, thereby preserving operation and synchronization of the remaining devices following it in the chain.

Figure 8.15(c) is a configuration invented by the author, called *one-wire sync*, and used in the product of Figure 8.14 and other products. Only one connection is used for synchronization, called the sync pin or sync wire. All of the sync pins of the various signal conditioners are connected together. One of the signal conditioners is automatically selected through an algorithm in its software to be the leader, and all others follow that excitation frequency and phase. If that leader loses its excitation function (or is removed from service), another leader is automatically selected. If the original leader is returned to service, it again participates in the leader selection algorithm. Synchronization among all operating devices is maintained.

8.10 CALIBRATION

When manufacturing a quantity of LVDTs and RVDTs, there are some differences among the many examples of a given product model number. These are due to small differences in the winding layers, the core size and permeability, welding of the housing, and other factors. Even though these parameters are precisely controlled during the normal manufacturing processes, there is still a limitation on that precision, and so slight differences from one unit to the next still remain. (This is essentially true for all transducers and sensors of any technology.) When an LVDT or RVDT is mated to its signal conditioner (either internal or external), a calibration procedure is implemented to standardize the output to the desired engineering units. With an LVDT signal conditioner having an analog voltage output of 0 to 5 VDC, for example, the calibration routine will provide adjustment to ensure that the output is 0 V when the measurand is at zero scale, and that the output is 5 V when the measurand is at full scale.

The first step in calibration is to set the desired excitation frequency, and if there is a coarse amplitude selection, to make that selection. Then the LVDT or RVDT core is moved to the null position. Some signal conditioners include a separate analog output for very accurate null determination with a handheld voltmeter. Some have a set of LEDs to indicate null fairly closely (both of these features, plus auto-cal, are implemented in the product of Figure 8.13). When the core is at null, the output is then adjusted to zero. This is done automatically in signal conditioners having auto-cal, and manually in others.

The next step in calibration is to move the core to the desired (usually positive) full-scale position. The gain of the output signal is adjusted by auto-cal or manually. If the LVDT will be used with the core on both sides of null, then the core is moved

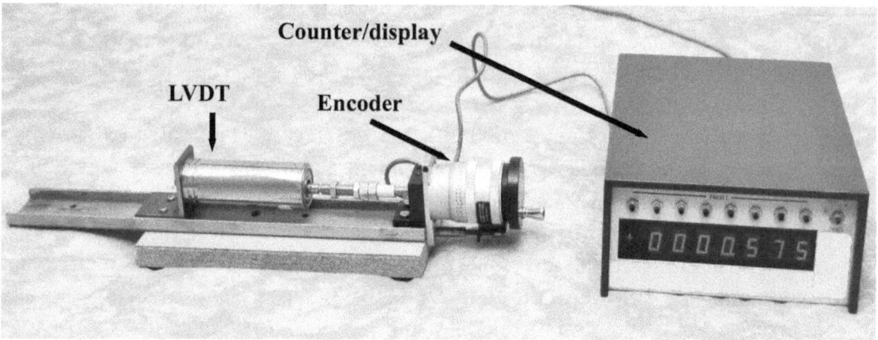

FIGURE 8.16 Fixture for accurately adjusting an LVDT core position.

to negative full scale and the gain of the output signal is adjusted by auto-cal or manually. Calibration would then be complete.

If the LVDT or RVDT is to be used with a long cable, calibration should be done with that cable in place. If this is not practical, then the signal conditioner phase adjustment should be used when the LVDT or RVDT is installed in the field to obtain the same output at zero or full scale as was set during calibration. But this step is not needed with the signal conditioner of Figure 8.14, since it is insensitive to differences in phase relationship between the excitation and demodulator.

Figure 8.16 shows one of the author's test fixtures for accurately adjusting the core position of an LVDT during calibration.

The figure shows a heavy-duty industrial LVDT having a larger, more robust housing. It is mounted securely into a fixture, along with a micrometer head that includes a rotary encoder. A handle on the micrometer is turned manually to change the position of the LVDT core, but automated fixtures can also be built that use a motorized system to move the core. As the micrometer is rotated, the encoder sends pulses to the counter/display unit to indicate the position of the core. The display has a zero pushbutton to zero the count when the core is at its null position.

Likewise, a fixture is needed to accurately position the core when calibrating an RVDT, as shown in Figure 8.17.

The author's test fixture of Figure 8.17 is a horizontal rotary table that is normally used with a milling machine, but is used here for calibration of RVDTs. Turning the crank causes the RVDT mount to rotate, and the angle of rotation can be read on a vernier scale along the degree marks. Or a rotary encoder can be added to supply a digital display. As with a linear test fixture, the rotary test fixture may also be automated by adding a motorized system.

8.11 ADVANTAGES

Since the LVDT or RVDT core does not touch the inside of the coil bobbin, the sensing element is non-contact. This means that many full stroke cycles can be endured without wear or degradation of the performance characteristics. In many applications, though, a bushing or ball bearing assembly is added to maintain the alignment

FIGURE 8.17 Fixture for accurately adjusting an RVDT core position.

between the core and the bore of the coil bobbin, as in a miniature gage head configuration designed by the author and shown in Figure 8.18.

As mentioned in Chapter 1, the spelling of gage head, as opposed to gauge head, has been the standard in the industry for many decades, and therefore *gage head* is the accepted spelling in the industry, and used in this text. This is similar to the accepted spelling of *strain gage*, another type of sensing element that is sometimes replaced by an LVDT gage head.

The gage head has a removable tip, so a suitable shape can be selected for the application. A rubber bellows prevents dust from entering the bearing that engages the plunger shaft. Some gage heads use an internal spring to return the plunger after it is pressed in. Others may use a pressure line to return the plunger with air pressure. When wear occurs in an LVDT gage head, there is little change to the LVDT accuracy, but there may be some additional force required from the measured element to drive the motion of the LVDT core.

Measurement resolution of an LVDT is virtually infinitely fine, while the signal remains analog. The main limitations imposed on resolution are due to noise, characteristics of the signal conditioning electronics, or limitations of the user's input circuitry. Higher excitation voltages are used in noisy environments to maintain a high signal to noise ratio (SNR, or S/N ratio). Quantizing error in the receiving electronics may limit resolution due to the analog to digital (A/D) converter that is often

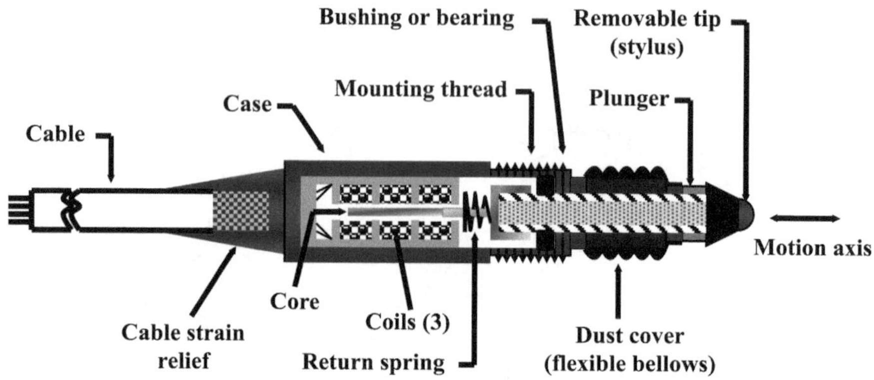

FIGURE 8.18 Cutaway view of a gage head LVDT.

incorporated in the signal conditioner or other digital electronics. Since the LVDT demodulated output is an analog signal, an A/D converter is needed to present the signal to a digital system such as a controller using a microcontroller. Use of a 16-bit A/D converter provides high resolution and reduces this limitation.

A hermetic seal is one that is gastight. Hermetic seals include welded, glass-to-metal, glass-to-ceramic, and may include high quality epoxy resin seals. A hermetically sealed LVDT or RVDT is a very rugged sensor. They can be used in high-humidity, high-vibration environments and over a wide temperature range. The metal end plates are typically tungsten inert gas (TIG) welded to the metal housing. One end includes a hermetically sealed connector. For TIG welding, a fixture is used to tightly press the end plates against a tube that is a part of the housing. The housing tube is clamped into a collet to hold it firmly while it is rotated by a geared-down motor. A tungsten electrode is brought close to the edge where an end plate joins the housing tube. An inert gas, usually argon, flows around the tungsten electrode to eliminate oxidation of the molten metal during the welding process. An electric arc is struck between the tungsten electrode and the joint as the LVDT housing is rotated by the motor. The heat of the arc melts some metal from the end plate and some from the housing tube so that a weld bead (a small puddle of molten metal) is formed at the seam. The metal solidifies after the tungsten electrode passes. When cooled, the resulting part is totally sealed against entry of water or gases. If a stainless steel material is used for the housing and end plates, a substantially corrosion-resistant LVDT or RVDT is produced. TIG welding is most common, although hermetic joining of metals can also be accomplished by laser or electron beam welding.

8.12 TYPICAL PERFORMANCE SPECIFICATIONS AND APPLICATION

Several styles and housing diameters are standard in the industry. An RVDT typically incorporates the rotor within a sealed housing. But a typical LVDT is supplied with a separate core. The core has internal threads on each end, by which a non-ferromagnetic actuation rod may be attached. The rod is usually aluminum, brass, or plastic, but can be non-magnetic stainless steel if it is annealed after machining so that it remains non-magnetic.

A second style of LVDT is the gage head or spring-loaded type, having a core mounted internally and with a plunger attached. The plunger is usually loaded by a spring, but can also be supplied with an air-actuated return mechanism instead of a spring.

Many LVDTs and RVDTs are AC operated, and must be used with an external signal conditioner. Some are DC operated, and that means a small signal conditioner is included within the LVDT or RVDT housing. For a DC unit, one typically supplies DC power in the range of 13.5 to 28 volts. That's to accommodate older systems operating on ±15 volts DC, as well as the more common industrial systems that operate at a nominal +24 volts DC.

Some LVDT models are available with a wider temperature range, and some are available for use in an ionizing radiation environment, such as X-rays or gamma rays at a nuclear power plant. The LVDT is a popular choice for high reliability military, aerospace, marine, and nuclear power plant applications. They are often used as the position feedback sensor in the actuator for control surfaces in airplanes. LVDT gage heads are used successfully throughout industry on the production floor as well as in quality assurance. An array of LVDT gage heads can be assembled to a template for inspection of an automobile door, for example. In this application, sample door panels are taken from the production line and placed against the template. Deviations from the standard profile are measured and used to determine if the tool making the door panels needs adjustment or re-working in order to keep the finished parts within the allowed tolerances. The author has designed sensors for this type of application, as well as a similar one that is used for auto windshields, to make sure that the correct contour is maintained during fabrication.

Incoming inspection tools use LVDTs to measure critical dimensions and read them into a computer that compares the data against the accepted tolerances. There is a resistance-welding tool on the market that uses an LVDT to measure the submicron motion that occurs in the mating parts as the welding progresses (this measurement is called the weld "set-down"). The author has used such a welder with set-down control in welding to the waveguide of a magnetostrictive position sensor (magnetostrictive position sensors are presented in Chapter 12). LVDTs and RVDTs have been used as components in many other products, including pressure sensors and valve positioners for process control.

Here are some typical performance specifications for an AC LVDT:

Input power:	2.5 kHz AC, 3.0 V_{RMS}
Supply current:	5 to 10 mA
Output:	differential AC signal
Nonlinearity:	≤ 0.2% to 0.5%, depending on stroke length and type
Resolution:	virtually infinitely fine
Repeatability error:	0.01% FSO
Hysteresis error:	0.01% FSO
Operating temperature range:	–55 to 105°C

Thermal coefficient of sensitivity:	−0.02% FSO/°C (nominal)
Thermal coefficient of zero:	±0.005% FSO/°C

Here are some typical performance specifications for a DC LVDT:

Input power:	13.5 to 26.5 VDC
Supply current:	30 mA
Output:	0 to 10 VDC
Nonlinearity:	≤ 0.10% FSO (full-scale output)
Resolution:	0.005% FSO
Repeatability error:	≤ 0.01% FSO
Hysteresis error:	0.01% FSO
Operating temperature range:	−40 to 85°C
Thermal coefficient of sensitivity:	−0.01% FSO/°C (nominal)
Thermal coefficient of zero:	±0.005% FSO/°C

Here are some typical performance specifications for an AC RVDT:

Input power:	2.5 kHz AC, 3.0 V_{RMS}
Supply current:	5 to 10 mA
Angular range:	±45°
Output:	differential AC signal
Nonlinearity:	≤ 0.5%, depending on stroke length and type
Resolution:	virtually infinite
Repeatability error:	0.05% FSO
Hysteresis error:	0.05% FSO
Operating temperature range:	−55 to 150°C
Thermal coefficient of sensitivity:	−0.036% FSO/°C (nominal)
Thermal coefficient of zero:	±0.01% FSO/°C
Shaft diameter:	4.75 mm
Moment of inertia:	1.8×10^{-6} Kg·cm^2

8.13 MANUFACTURERS

Some manufacturers of LVDT and RVDT position sensors include the following:

Active Sensors	www.activesensors.com
Alliance Sensors Group	www.alliancesensors.com
Kavlico	see Sensata Technologies
LORD/Microstrain	www.microstrain.com
Macro Sensors	see TE Connectivity
Measurement Specialties	see TE Connectivity
Micro Epsilon	www.micro-epsilon.com
Novotechnik	www.novotechnik.com
Penny & Giles	(Curtis-Wright) www.pennyandgiles.com
RDP Electrosense	www.rdpe.com

Revolution Sensor Company	(design) www.rev.bz
Schaevitz	see TE Connectivity
Sensata Technologies	www.sensata.com
Sensor Systems SRL (Baumer)	www.sensorsystems.it
Singer Instruments	www.singer-instruments.com
Sentech, Inc.	www.sentechlvdt.com
Solartron Metrology	www.solartron metrology.com
TE Connectivity	www.te.com
Trans-Tek Inc.	www.transtekinc.com

8.14 QUESTIONS FOR REVIEW

1. A simple LVDT most commonly has how many coils?

a. Three
b. Six
c. Four
d. Two
e. Five

2. An LVDT or RVDT core is usually made of alloys of what metals?

a. Copper, platinum
b. Iron, oxygen
c. Nickel, iron
d. Gold, cobalt
e. Silver, alumina

3. The best temperature stability is usually obtained by operating an LVDT at its:

a. Peak voltage
b. Demodulator null
c. Zero-phase frequency
d. Permalloy
e. FET

4. A half-bridge LVDT is also known as a:

a. Half-whit
b. Wheatstone
c. BJT
d. DVRT
e. HBVT

5. The form onto which one or more LVDT or RVDT coils are wound is called a:

a. Thermistor
b. Collateral
c. Gage head
d. Plastiform
e. Bobbin

6. The circuit to convert an RVDT or LVDT AC signal to a DC voltage is called a:

a. Voltage convoluter
b. Demodulator
c. Discombobulator
d. Carrier
e. Linearizer

7. Most LVDTs and RVDTs are traditionally driven by this type of waveform:

a. Triangle wave
b. Sound wave
c. Sine wave
d. Square wave
e. Sawtooth wave

8. A hermetic seal is one that is:

a. AC/DC
b. Gastight
c. Amphibious
d. Wet
e. Porous

REFERENCES

[1] E. Herceg, *Handbook of Measurement and Control*. Pennsauken: Schaevitz Engineering, 1976.
[2] W. J. Tomkins, *Interfacing Sensors to the IBM PC*. Upper Saddle River: Prentice Hall, 1988.

9 Distributed Impedance

9.1 DISTRIBUTED IMPEDANCE POSITION SENSORS

In Chapters 7 and 8, transducers and sensors having one or more variable inductance(s) were explained, including the LVIT (Linear Variable Inductive Transducer), RVIT (Rotary Variable Inductive Transducer), DVRT (Differential Variable Reluctance Transducer, also known as an LVDT "half-bridge"), LVDT (Linear Variable Differential Transformer), and RVDT (Rotary Variable Differential Transformer). A distributed impedance linear or angular position sensor, as presented in this chapter, is somewhat similar to these previously described inductive sensors, in that one parameter of the distributed impedance sensing technology for linear or angular position measurement typically comprises a variable inductance. All of the sensors of Chapters 7, 8, and 9 have a core or target that is movable in relation to some type of sensing coil(s), forming at least one variable inductance. But there are differences in construction, operation, and performance among these technologies. The distributed impedance technology has important differences, when compared with the inductive sensing technologies of Chapters 7 and 8, that can be advantageous in some applications. These differences are partly due to the distributed impedance sensor having both a capacitance and a variable inductance that are each distributed along the length of a linear sensing element, or around the rotation path of an angular sensor. Whereas the other types (of Chapters 7 and 8) have inductances that are simple coils of magnet wire, each coil having a lump capacitance in parallel with the inductance, because such coils of many layers have capacitance built up among the parallel layers.

A linear or rotary (angular) inductive sensor (LVIT, RVIT), as presented in Chapter 7, operates by the movement of a magnetically permeable core within the bore of an elongated or curved coil, or rotating a magnetically permeable flat target along the face of a coil. As a core or target moves further within, or increases alignment with a coil, the inductance of the coil increases. Such sensors are produced by Positek, Penny & Giles, Microstrain, and others. They typically have an operating frequency of around 50 to 100 kHz. This operating frequency is selected in order to provide optimum sensitivity, while minimizing eddy current loss into the movable core.

An LVDT or RVDT, as presented in Chapter 8, operates by movement of a magnetically permeable core within the bore of three coils arranged axially for an LVDT (some LVDTs may have a different number of coils, to provide the desired performance), or arranged to interact with a rotating core, for an RVDT. The core is energized by driving the primary coil (usually the center coil) with an AC excitation, and power is inductively coupled to produce a voltage across each of the secondaries in an amount varying according to the alignment of the core with each secondary coil. LVDTs typically have an operating frequency of 0.5 kHz to 10.0 kHz. This operating frequency is determined as needed to provide optimum position

DOI: 10.1201/9781003368991-9

measurement sensitivity and minimized temperature sensitivity, while minimizing eddy current loss into the movable core. A tightly wound coil, such as found in these inductive-type sensors of Chapters 7 and 8, has a parasitic capacitance, as mentioned earlier, that appears as a lump sum capacitance in parallel with the inductance of each coil. Some representative examples to illustrate this parasitic capacitance of the coils in an LVDT and an LVIT are shown in Figure 9.1.

A linear or angular position sensor based on distributed impedance, however, has capacitance as well as inductance distributed along its sensing axis. Figure 9.2 is a pictorial representation of the inductance and capacitance being distributed along the length of a distributed impedance position-sensing element. A small amount of resistance is also distributed along the length of the sensing element, but it is too small to have any noticeable effect on performance, and so is disregarded.

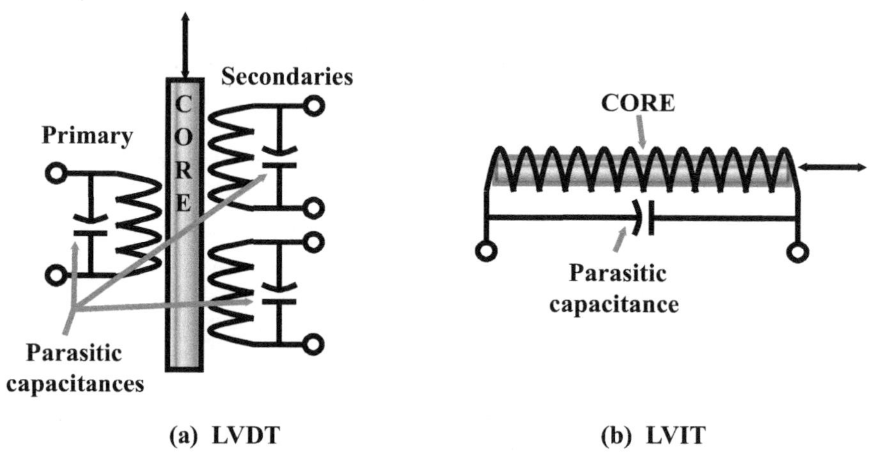

FIGURE 9.1 Inductive sensors have a lump capacitance in parallel with each coil: (a) shows a lump capacitance in parallel with each of the three coils of a typical LVDT, while (b) shows the lump capacitance in parallel with the coil of a single-coil LVIT (but some LVDTs and LVITs may have different numbers of coils).

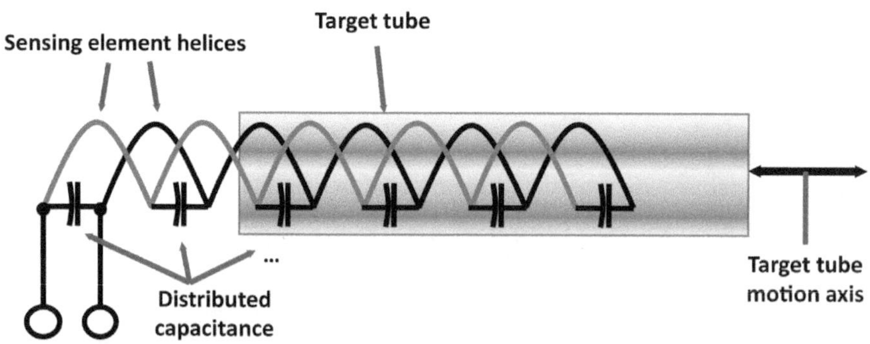

FIGURE 9.2 Distributed inductance and capacitance of a distributed impedance position-sensing element.

So, Figure 9.2 shows how the capacitance being distributed along the sensing element of a distributed impedance sensor is different from the lump capacitance of each coil of an LVDT, RVDT, DVRT, or LVIT, as was shown in Figure 9.1.

9.2 HISTORY

The distributed impedance sensing technology was developed by Revolution Sensor Company, in Apex, NC (now in New Bern, NC). US patent number 7,340,951 was granted to inventors David S. Nyce and Yuriy N. Pchelnikov in 2008. Both linear and angular position sensors were subsequently developed utilizing this sensing technology, with added circuitry and software to produce products for the industrial, commercial, military, medical, and other markets.

Distributed impedance sensors are also sometimes called NyceWave™ (a trademark that was coined by a user of the technology, and named after the inventor, and owner of Revolution Sensor Company, David S. Nyce). This user didn't think that the technology being named "Distributed Impedance Sensing" was suitable for marketing purposes, and wanted a non-technical name with a better sound. They suggested NyceWave, and permission to use that name was received. Subsequently, Revolution Sensor Company has become the owner of the NyceWave trademark.

Sometimes a distributed impedance (aka NyceWave) position sensor has been mistakenly called an LVIT, but that is incorrect because there are important differences in construction, operational theory, and performance between a distributed impedance position sensor and a linear variable inductive sensor, as are explained in this chapter.

9.3 OPERATIONAL THEORY

A distributed impedance position sensor typically operates at a much higher frequency, such as 2 MHz to 10 MHz, than an inductive position sensor operating at 10 kHz to 200 kHz. A sensing element based upon the distributed impedance technology employs two or more electrically conductive patterns, forming an electrodynamic element, having the conductive patterns positioned one with respect to another so that an electromagnetic wave travels along the length or arc that forms the measuring range (called the motion axis, sensing axis or sensitive axis) of the sensing element. The conductor patterns together (sometimes supported by a dielectric substrate) comprise a sensing element, having at least two terminals for electrical connection. At least two of the conductor patterns are similar, or can be identical, and are electrically insulated from one another along the operating length of the sensing element. One of the conductor patterns in a distributed impedance linear position sensor, taken individually, often may approximate a helical shape. Then, two of such conductors approximate a dual helix configuration, and may be wound onto a cylindrical dielectric substrate, such as a fiberglass rod. The resulting sensing element has an electrical impedance that can be measured between at least two of the terminals. See Figure 9.3.

The RLC (resistance, inductance, capacitance) circuit of the sensing element will have a resonant frequency. When an alternating current (AC) at the sensing element

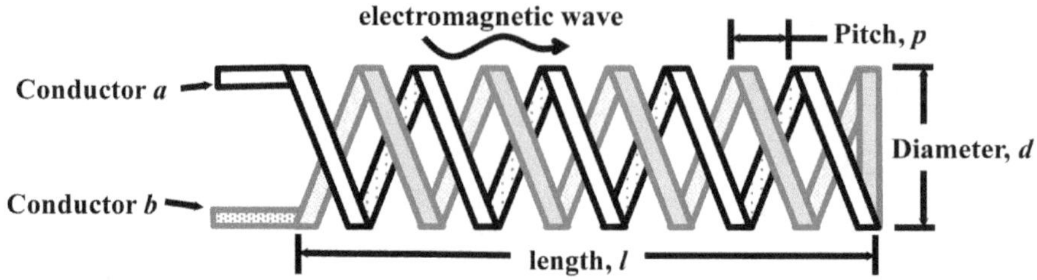

FIGURE 9.3 One configuration of a distributed impedance linear position sensor is a dual helix set of conductors that may be wound onto a dielectric cylindrical substrate material.

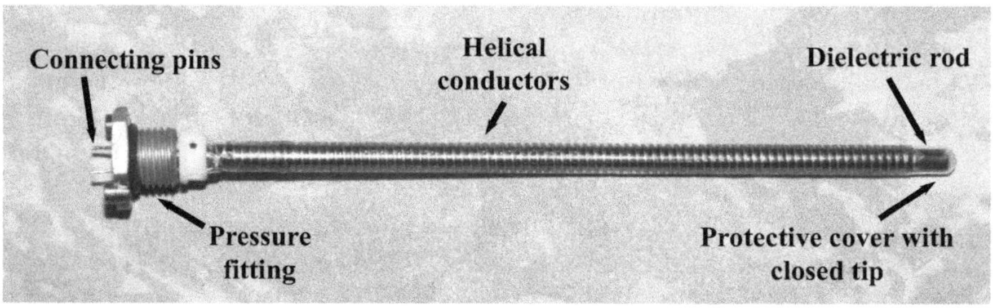

FIGURE 9.4 Photo of a distributed impedance linear position-sensing element (without target tube).

resonant frequency is applied through conductors *a* and *b*, an electromagnetic wave propagates along the length *l*, of the sensing element. The resonant frequency is determined by length *l*, pitch *p*, diameter *d*, the width of the conductors of the electrodynamic element pattern, permittivity of a dielectric support that may support the electrodynamic element, and other factors. The most important of these other factors is the position of a target tube (a target tube will be shown later in a photo of a completed position-sensing device). But for now, we will concentrate on the sensing element itself. Figure 9.4 is a photo of a distributed impedance linear position-sensing element (without a target tube. A target tube would cover a portion of the sensing element).

In Figure 9.4, the outer layer of the electrodynamic element can be seen, and forms one helix of the bifilar helices mentioned earlier. The other helix is beneath the visible one, and the two are separated by a gap *g*, of a dielectric material. This gap (or thickness of the dielectric material) is one of the determining factors of the amount of capacitance that is distributed along the length of the sensing element. The dual helices of the figure are wound onto a fiberglass rod, and a protective cover is applied over the helices. The protective cover can be nylon, polyethylene, Teflon, or other dielectric material. At one end is mounted a pressure fitting, which includes a hermetically sealed set of connecting pins.

The magnetic field, *H*, surrounding a coil of an inductive sensor differs from the magnetic field surrounding the electrodynamic element of a distributed impedance sensor. A pictorial example of this is shown in Figure 9.5.

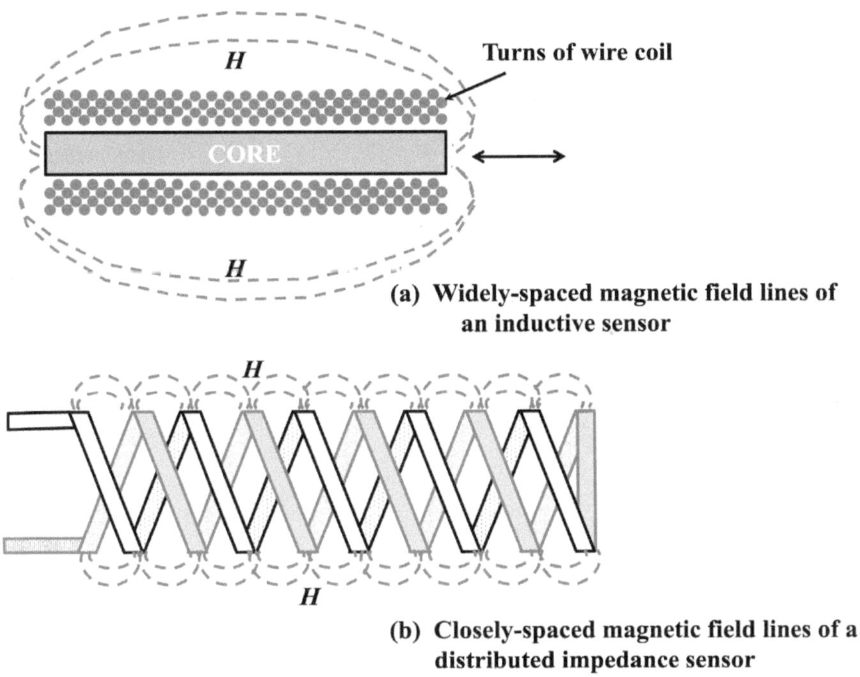

FIGURE 9.5 Magnetic field lines of inductive and distributed impedance sensors.

One of the attributes of a distributed impedance sensor is that the electromagnetic wave may be split into its component parts of electric wave and magnetic wave, so that they are each concentrated into their respective desired locations. The magnetic field strength (H) is indicated by dotted lines (lines of magnetic flux, or *flux lines*) in the figure. Figure 9.5(a) shows a single coil of a typical inductive sensor, having widely spaced flux lines that travel farther into space, due to the extended distance between the ends of the magnetized core. Figure 9.5(b) shows a distributed impedance sensing element, having closely spaced flux lines that travel less far into space. In this case, the magnetic field is concentrated close to the sensing element, due to the magnetic field being generated between adjacent poles of the electrodynamic conductor pattern. The electric field is concentrated between the two helix layers, and cannot easily be shown in this view. The location of the electric field can be better seen in Figure 9.6, of an electrodynamic element for a distributed impedance position sensor for measuring the distance to a flat metal target.

In Figure 9.6(a), two conductors of a spiral electrodynamic element are shown. The two conductors, upper and lower, are separated by a dielectric material (such as two flat spiral copper conductors, one on each side of a printed circuit board fiberglass dielectric material). Upper and lower conductors are separated by a gap, g (the thickness of the fiberglass PCB material). This configuration is suitable for use in sensing the distance to a flat conductive plate, as shown in Figure 9.6(b). The figure shows the separation of the electric (E) and magnetic (H) fields into different

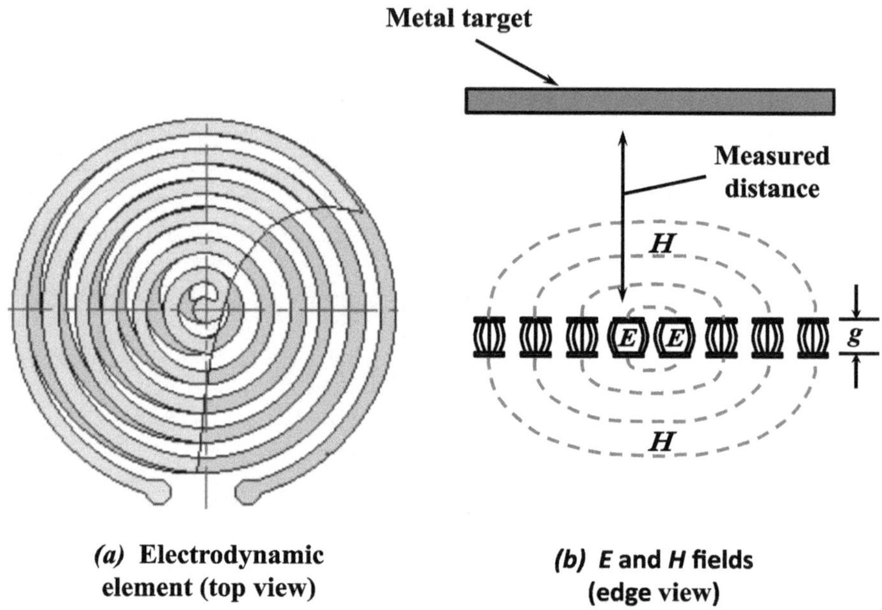

FIGURE 9.6 Electric and magnetic fields (*E*, *H*, respectively) of a flat configuration of a distributed impedance position sensor electrodynamic element.

locations. The magnetic field is disposed into the area adjacent to the electrodynamic element faces, so that the magnetic field can interact with a target that may be in that space. The electric field is concentrated between the two conductor patterns of the electrodynamic element. To implement an angular sensor, an electrodynamic element may cover less than 360 degrees, and more than one electrodynamic element may be disposed in a pattern, as in the photo of Figure 9.7.

In Figure 9.7, the sensing element on the left has four sets of conductor patterns, 1a and 2a on top, with 2a and 2b on the bottom (bottom not visible in the photo). Each approximately D-shaped pattern covers about 180 degrees of arc. The approximately D-shaped target is shown at the right. In use, the target would be placed directly above the conductor patterns, with a shaft passing through the hole of the target and the hole central to the conductor patterns. When the target is rotated so that its straight edge is approximately vertical (as it appears in the photo, but if moved over to match the shaft holes), the impedance of each coil would be the same. As the target rotates by up to 90° clockwise or anti-clockwise, the impedance of a conductor pattern would increase as the target becomes less aligned with it, and decrease as the target becomes more aligned with it. Suitable electronic circuitry can be used to determine the angle of the target and produce an electrical output that indicates the angle of shaft rotation. In the configuration of Figure 9.7, angles of up to 180° (or ±90°) may be measured. Other configurations can support angular measurements of up to 360°.

When measuring the position, velocity, or acceleration of a target as it moves along a motion axis, the sensing element is positioned so that its sensing axis is

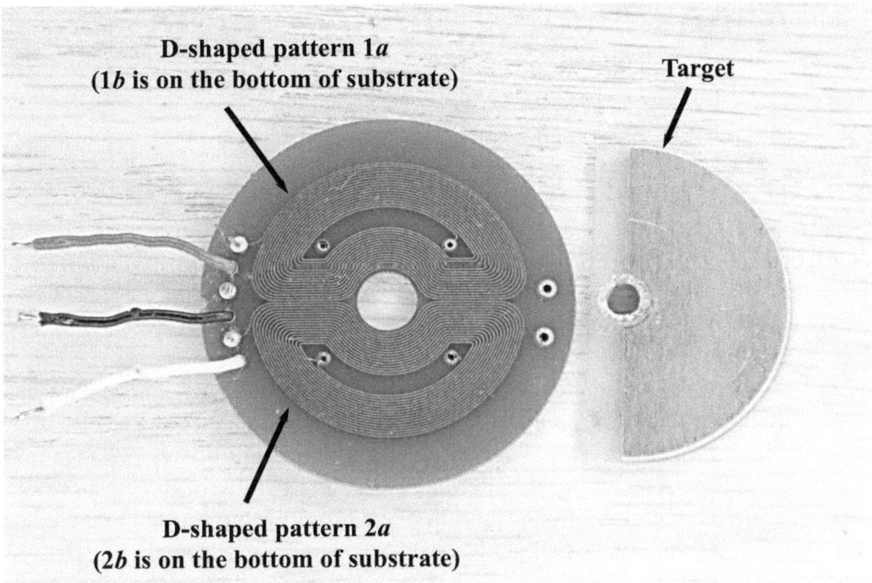

FIGURE 9.7 Angular sensing element (left) has electrodynamic conductor patterns on top and bottom of a fiberglass substrate. A metal target is shown to the right.

approximately aligned with the motion axis, and the target can be formed of any electrically conducting material, such as an aluminum tube or other shape (as opposed to an inductive position sensor having a core of magnetically permeable material).

At least two of the conductive members of the electrodynamic element are excited by an alternating-current oscillator in an anti-phase mode. This means that a first terminal can be connected to ground, while a second terminal is alternately energized by applying a positive and then a negative voltage. Alternatively, the first terminal can instead be connected to a first alternating voltage while the second terminal is connected to a second alternating voltage that is not in phase with the voltage at the first terminal. Such anti-phase energizing of the sensing element allows for the measurement of the sensing element AC impedance, and changes in the impedance that are caused by changes in the position of the target tube.

9.4 THE DISTRIBUTED IMPEDANCE SENSING ELEMENT AS A TRANSMISSION LINE

In the example of a distributed impedance electrodynamic element shown in Figure 9.3, there is an inductance formed by the number of turns of the electrical conductors, and there is a capacitance formed at each crossing of one conductor over the other. Thus, both the inductance and capacitance are distributed along the operating length of the sensing element. This configuration can be described in a way similar to a transmission line, as shown in Figure 9.8.

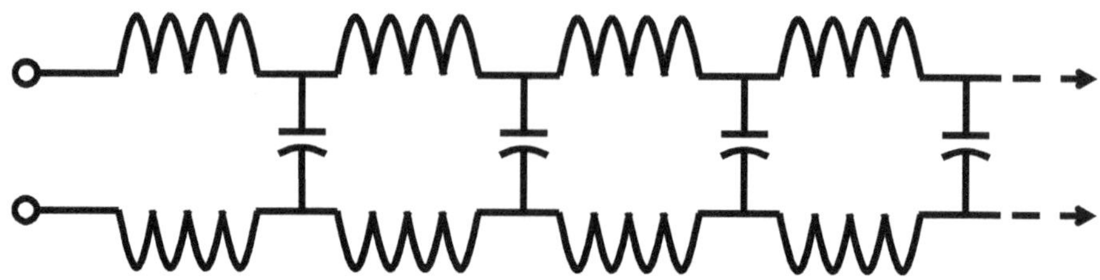

FIGURE 9.8 A transmission line, formed by a distributed impedance electrodynamic element, such as a bifilar helix, has capacitance as well as inductance distributed along its length.

Other distributed impedance sensor types (for example, the non-contact measurement of liquid level) may rely on a change in capacitance. But the distributed impedance linear and angular sensors described here rely mainly on a change in the sensing element inductance in relation to a change in position of one or more electrically conductive members (the target) in close proximity to the sensing element, while also relying on the distributed capacitance to help shape the electric and magnetic fields. If the position of a target affects the inductance of the transmission line, then a phase delay or variable resonant frequency of the sensing element can be used to indicate the target position, with one or both ends of the transmission line connected to a measuring circuit. The distributed impedance position sensors shown previously have electrodynamic elements in which one or more impedance conductors are coiled approximately in helical or spiral forms (but may also have other configurations as needed for a desired performance). The velocity of electromagnetic wave propagation in any transmission line is equal to $1/\sqrt{L_0C_0}$ where L_0 and C_0 are the specific inductance and specific capacitance, respectively. The values of L_0 and C_0 depend upon the configuration of the conductors, the distance between them, and associated dielectric materials. Coiling one or both conductors of a two-conductor transmission line increases its inductance and adds capacitance. This leads to a reduced velocity of an electromagnetic wave progressing along the length of the transmission line. The amount of velocity reduction is the ratio, n, of the velocity of light in vacuum, c, to the phase velocity of the wave v_p in the transmission line:

$$n = c / v_p = \sqrt{L_0C_0 / \varepsilon_0\mu_0} \tag{9.1}$$

where ε_0 and μ_0 are the permittivity and permeability, respectively, of vacuum.

In a simple, two-conductor transmission line, a decrease in the distance between conductors causes a similar increase in capacitance C_0 and a decrease in inductance L_0, allowing the wave velocity to remain equal to the velocity of light. But in the arrangement of conductors in a distributed impedance sensing element, the influence of the distance between conductors on L_0 and C_0 differs from that of a simple transmission line, making it possible to change the wave velocity, v_p, by changing the distance between conductors of the sensing element.

The wave impedance of the sensing element is proportional to the wave velocity reduction (n) of the electromagnetic wave. This enables the impedance to be changed over a wide range, from a few ohms to thousands of ohms.

The wave velocity reduction causes concentration of the electromagnetic field near the impedance conductors, reducing energy radiation and increasing sensitivity to a target disposed in this concentrated field. The thickness of the area of energy concentration is approximately equal to $\lambda/2\pi\, n$, where λ is the electromagnetic wave length in vacuum.

9.5 PERIODIC STRUCTURES

The individual electrodynamic elements used in distributed impedance sensors are periodic structures. Typical configurations for these structures include helices, spirals, and meander lines, but other shapes may also be used. A periodic structure is considered to be infinite, in theory. But in practice, a portion of a periodic structure is also called a periodic structure. When an electromagnetic oscillation is applied to a periodic structure, a wave may propagate along its length. The z-axis of the periodic structure is defined as being along the direction of wave propagation. Any component A of the electric and magnetic fields at a point along the periodic structure (at coordinate z), can be represented by the series:

$$A(x, y, z) = \sum_{m=-\infty}^{\infty} am(x, y) e^{j\omega t - \beta mz} \tag{9.2}$$

Each term of the series is called a space harmonic. a represents amplitude. am is the amplitude of the mth space harmonic, ω is the angular velocity of the oscillation, t is time, $\beta m = (\beta + 2\pi m/d)$, z is the wave number of the mth space harmonic, and j is an imaginary number [1, p 414].

9.6 HYBRID WAVES

Electromagnetic waves in a distributed impedance electrodynamic element, for example, a dual helix, are hybrid waves, comprising fields with electric (E) and magnetic (H) polarization. In many practical cases, such a wave can be represented as a sum of E and H waves, the transverse components of which are expressed in terms of the derivative of the longitudinal component of the electric or magnetic fields with respect to the transverse coordinate. Due to the configuration of the electrodynamic elements, the E and H waves exist together and have the same velocities as each other.

Although having the same velocities, the distribution of the electric and magnetic fields can be different from each other, due to their difference in polarization. For example, an E wave can be shielded by a longitudinally conducting screen, whereas an H wave can be shielded by a transversely conducting screen. Accordingly, considering the example of a dual helix electrodynamic element, sensitivity can be increased by separation of the E and H waves, and shaping them individually. Therefore, the electrical parameters of a medium or an object (for example, the conductivity, permittivity, and permeability) can have a different effect on the *E* wave as compared with their effect on the *H* wave. This can occur in accordance with the shape of the medium or object, and to the respective directional dependence of the *E* and *H* waves.

Due to the periodic structure of the impedance conductors, the *E* or *H* waves can be represented by the sum of space harmonics with different phase velocities. The velocity of the zero harmonic is defined by formula 9.1, the velocities of other harmonics decrease as their harmonic number increases. The amplitude of the zero space harmonic or, for some structures, the amplitude of the total field of *+1* and *−1* harmonics, sufficiently exceeds the amplitudes of the other harmonics to allow consideration of only the zero harmonic or only the *+1, −1* harmonics. The velocities of the *+1, −1* harmonics are much less than the velocity of the zero harmonic. This results in a significant decrease in the area in which the energy is concentrated, with a thickness approximately equal to $p/2\pi$, where p is the pitch (or period) of sensing element pattern. In a hybrid wave, the field of the *E* wave can be represented by the zero space harmonic, while the field of the *H* wave can be represented by the *+1, −1* harmonics, as, for example, occurs in dual helices due to the opposite directions of the currents in the adjacent conductors.

When an anti-phase wave is excited in a sensing element of a dual helix configuration, the *E* wave is concentrated between the impedance conductors and the *H* wave is concentrated outside of the conductors. Remembering that the electric energy is concentrated mainly in the *E* wave and the magnetic energy is concentrated mainly in the *H* wave, splitting of the electric and magnetic fields is therefore apparent. Such splitting increases the sensitivity of a sensing element to the desired target, and decreases sensitivity to surrounding eclectic fields, as well as limiting radiation of electric or magnetic energy.

9.7 DISTRIBUTED IMPEDANCE SENSOR DESIGN

The dual helix of a linear position-sensing element can be formed of an enameled copper magnet wire that has been flattened. This can be wound onto a stiff rod of dielectric material, such as fiberglass, using an adhesive to keep the helical coil in place. The flat wire is wound with a controlled pitch to obtain the desired resonant frequency of the finished sensing element. After winding the first helix, an insulating layer is placed over the winding. This provides the dielectric layer that will separate the two helices at the crossover points along the sensing element. The thickness and relative permittivity of the dielectric layer, along with the area of the parallel conductors at the crossover points, determine the amount of distributed capacitance. Then the second helix is wound over the first. Care should be taken to ensure that the termination of the two wire leads are stabilized to prevent any movement when in service. If one end of the sensing element is to be inserted into a metal housing (for example, a housing containing the electronics), the two conductors should be spaced from the metal housing so that the capacitance between the conductors and the housing is minimized. Hermetic glass-to-metal seals are recommended.

After the dual helix and terminations are completed, an outer protective jacket of dielectric material should be added in order to protect the sensing element from damage due to the target tube sliding over it. Such protection can be afforded by the application of a protective cover or jacket formed of nylon, polyethylene, PTFE, PFA, FEP, or other durable and corrosion-resistant dielectric material. After the jacket is formed tightly to the outside of the sensing element, the sensing element tip should be heat-formed so that the tip is also sealed.

When designing an angular sensor, it is important to make sure that the movable target is parallel with the sensing element, and that it is rigidly attached to the rotating shaft, so that vibration will not cause noticeable motion of the target. The shaft must be supported with fine bearings that will not allow the target to wobble. It is recommended that there be a counterbalance weight if the rotation rate can be high. This will reduce any tendency to wobble due to an unbalanced load on the shaft and bearing.

9.8 ELECTRONICS

Whether a distributed impedance sensing element is of a linear or an angular configuration, it can be driven by a relatively simple oscillator circuit. A typical oscillator circuit is shown in Figure 9.9.

In this Pierce oscillator configuration, the sensing element acts as a resonant feedback element, and controls the frequency of oscillation. The circuit of Figure 9.9 includes U_1 as the active element. It is usually a CMOS gate that can be a part of a hex inverter, if other inverters are used in the circuit, or it can be a single inverter. Single gates and other single logic element ICs are sometimes called "tiny logic", and are available from many integrated circuit manufacturers. R_1 is a feedback resistor

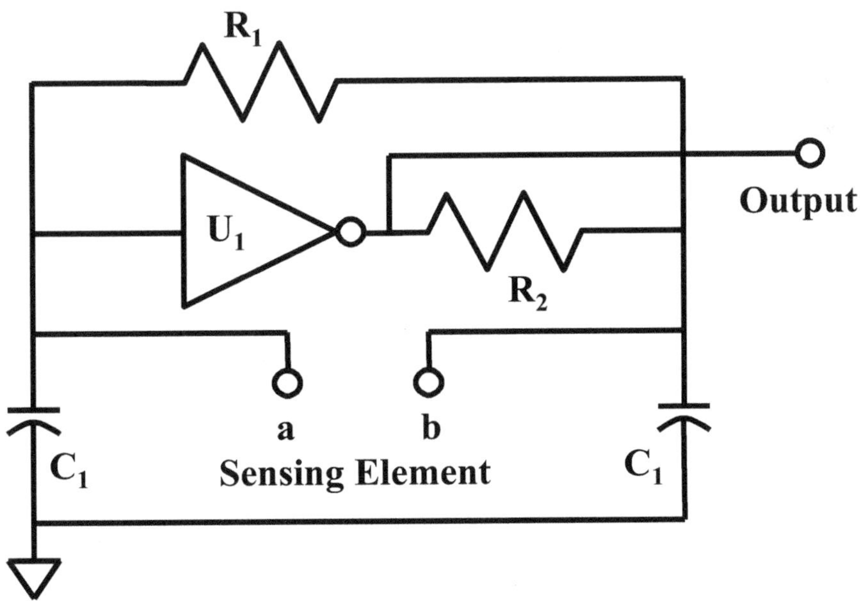

FIGURE 9.9 Pierce oscillator circuit to drive the sensing element.

to ensure oscillation, and can be in the range of 10k to 1 meg ohms, depending on the maximum resistance that still ensures the oscillator will start with a given sensing element design. R_2 limits the peak current pulse of the gate output, and is typically 200 to 500 ohms (this limits EMI radiation from the sensing element that could otherwise be due to higher switching current at sharp edges of the sensing element driving waveform). The resistors are typically a 1%, 100 ppm/°C, 1/8 watt type. C_1 and C_2 are usually equal to each other, and about 22 to 100 pF. They are typically 5% tolerance, 50 V, and of a ceramic, temperature stable, NP0/C0G type. The output of the oscillator is a square wave, having a frequency that varies with the position of a target tube that slides over the sensing element (for a linear position-sensing element, as shown in Figure 9.4), or a metal plate that moves relative to the sensing element (for an angular position-sensing element, as shown in Figure 9.7). As an alternative to operating the sensing element in a feedback loop to control oscillator frequency, one can use a fixed oscillator and the sensing element may be incorporated as a time delay, and compared with a time delay reference, or as a variable impedance with the signal amplitude being measured. Other circuit types may also be used, relying on the resonant frequency, time delay, or impedance of the sensing element.

Typically, some additional circuitry is used in order to condition the signal, scale it, and provide a desired output, as shown in Figure 9.10.

All of the electronics shown in the figure will fit onto a PCB that is about 1.5″ in diameter, or smaller if passive components smaller than 0603 size and leadless ICs are utilized (such as QFN, BGA, and so on). Input power for this type of sensor is usually 13.5 to 28 Volts DC. That's because modern industrial circuits mostly use 24 VDC, and some older analog systems use ±15 VDC. So, 13.5 to 28 V will accommodate both power supply voltages. Input power protection prevents damage from overvoltage or reverse connection. All input and output lines should have protection against ESD (electrostatic discharge), and should be standard practice on all sensor electronics designs. The voltage

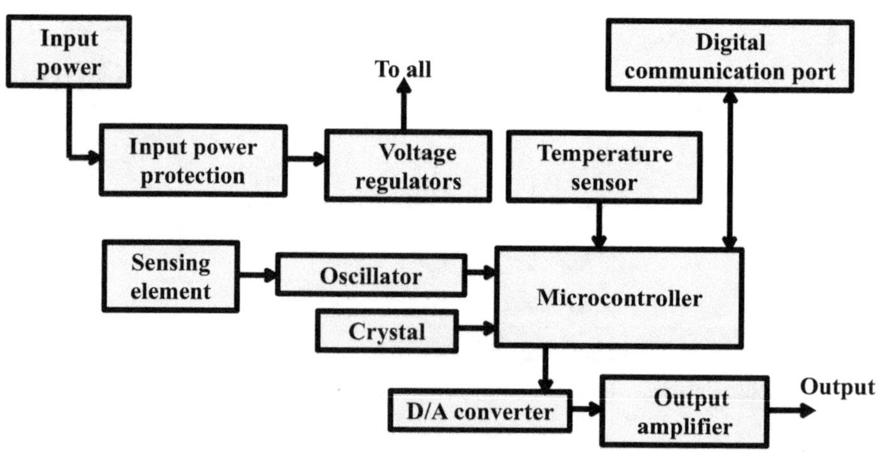

FIGURE 9.10 Block diagram of complete position sensor circuit to provide a conditioned output signal.

regulators usually supply 3.3 V (older versions used 5 V), plus any other voltages that may be needed by other circuits. The oscillator and sensing element will provide a sensor signal to the microcontroller, usually in a range from 2 to 10 MHz, and having about 30% change in frequency (or other parameter) in response to changing position. For example, if the resonant frequency is 3.0 MHz with the target removed (that is, at zero position), then the frequency could be 2.1 MHz with the target fully engaged (that is, at the full-scale position), representing a 30% change (span), since 2.1 MHz is about 70% of 3.0 MHz). A change of 30% is a relatively large amount of change, and will support accurate and stable measurements, as well as high resolution. (See Section 9.11, titled "Infinite Resolution?")

In Figure 9.10, a microcontroller in the sensor electronics measures the resonant frequency and stores a digital count representing the frequency. This count should be capable of 16-bits representation so that a high resolution is available. Calibration should be accomplished through the digital com port, interfaced with, for example, a PC. Firmware should include accommodation of setting zero, full scale, temperature compensation, linearization, filtering, and selecting the type of output desired. A good method for filtering the signal before going to the DAC (digital to analog converter, or D/A) is to use median filtering. For example, for a 5 reading median, sort five consecutive measurements according to value from lowest to highest. Select the middle number for forwarding to the DAC. Take one new measurement, throw out the oldest measurement, and sort again, then repeat.

The filtered data can be sent to a 16-bit DAC, and from there to a suitable amplifier to generate the analog voltage or current output signal. If sending a digital output, appropriate drivers are used instead of a D/A.

9.9 ADVANTAGES

Distributed impedance position sensors provide the virtually infinite resolution of an inductive sensor or LVDT, but are easily produced over a much wider range of measuring lengths, have a much lower temperature sensitivity, and are equally as rugged. They are more rugged than magnetostrictive sensors, because distributed impedance position sensors don't have a sensitive pickup device as needed in a magnetostrictive sensor. Distributed impedance sensors are also relatively inexpensive, and do not require a complex electronic circuit to operate. A magnetostrictive sensor has an equally high resolution at longer lengths, but has limitations on response time when longer than about 1 m, due to the travel time of the torsional wave (see Chapter 12 about magnetostrictive sensors).

9.10 TYPICAL PERFORMANCE SPECIFICATIONS AND APPLICATIONS

Distributed impedance linear position sensors have been built having measuring ranges from as low as 1 mm to more than 6 m in length. Angular sensors can measure over a full 360° of rotation, or have a limited scale, such as ±45, ±60, or ±90°, for example.

An example of a distributed impedance linear position sensor is shown in Figure 9.11.

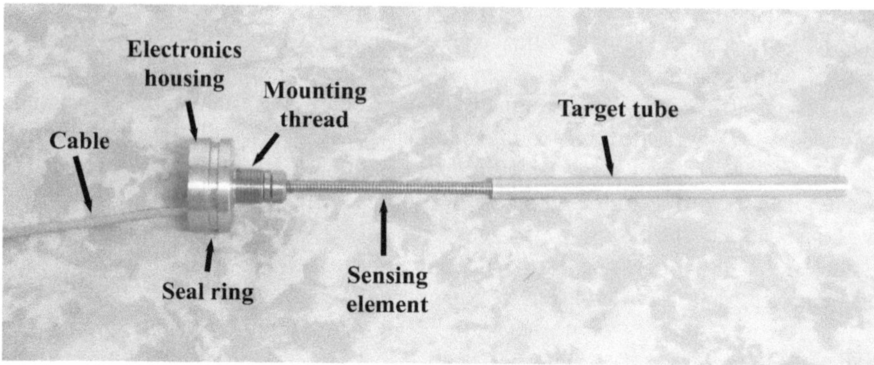

FIGURE 9.11 A distributed impedance linear position sensor, with stainless steel housing and aluminum target tube. (Sealing O-ring not installed.)

This linear position sensor has a non-magnetic 316 stainless steel housing, with a threaded nose to facilitate screwing into the end of a hydraulic cylinder, and a groove machined into the housing circumference for a sealing O-ring. An aluminum target tube is shown. The target tube can be of any electrically conductive material, but would typically be a bore that is gun-drilled into the piston rod in a hydraulic cylinder application, rather than using a separate target tube.

Distributed impedance linear position sensors have been used in many applications, including hydraulic and pneumatic cylinder positioning, vehicle suspension systems, vehicle steering, factory automation, aerospace, solar panel positioning, animatronics, and others.

The angular position sensor of Figure 9.12 has an anodized aluminum outer housing, and internal parts to support the sensing element and PCB (printed circuit board). The internal parts include metal and non-metal supports, PCB, sensing element, target, and high-performance bearings. The stainless steel shaft has a flat spot machined into one side (not shown in the figure) to reference the zero position, relative to a mark on the housing. The square plate on the right side is for mounting. The dimensions are standard for the mounting of servo motors. This particular model has an angular rotation of up to 120°.

Distributed impedance angular position sensors have been used in many applications, including valve positioners, factory automation, weighing systems, tensioners, throttle control, aircraft actuator position, and agricultural vehicles.

Some typical performance specifications for linear and angular distributed impedance sensors include:

Input power:	13.5 to 28 VDC
Full-scale range:	0–10 mm to 0–2,000 mm (linear) 0 to 30°, 0 to 180°, 0 to 340° (angular)
Operating temperature:	−40° to 85°C, or −40° to 125°C (wider ranges also possible)
Outputs:	DC voltage: 0 to 1 V, 0 to 5 V, 0 to 10 VDC, and others

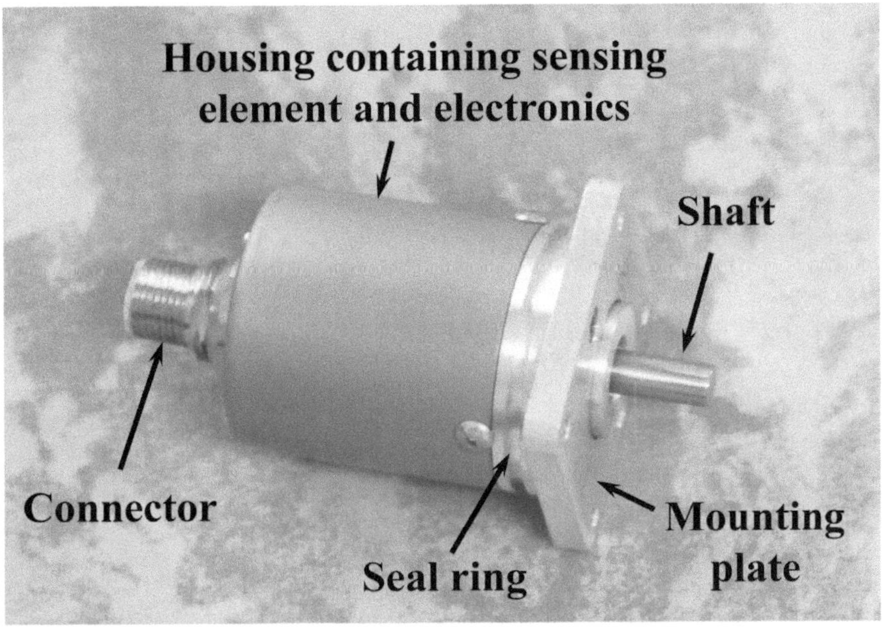

FIGURE 9.12 A distributed impedance angular position sensor, with anodized aluminum housing and mounting plate, with stainless steel shaft.

	Current 4 to 20 mA
	Digital: frequency (square wave), CANbus, SSI, and others
Static error band:	±0.1% (nonlinearity, hysteresis, repeatability)
Zero temperature sensitivity:	±50 ppm/°C
Span temperature sensitivity:	±50 ppm/°C
Resolution:	nearly infinite, limited by noise or digital conversion

9.11 INFINITE RESOLUTION?

In a dual helix type of distributed impedance linear position sensor, laboratory measurement of the frequency typically yields, for example, the stability of a 3 MHz reading to within 3.000000 MHz ± 0.000001 MHz (1 Hz) on a sensor having a measuring range of 250 mm. With a span of 0.9 MHz representing the 250 mm, this indicates an instability of no more than 0.28 μm. With very careful manual operation of the positioning fixture, position changes resulting in increments of 0.000001 MHz were possible, meaning that a resolution of 0.28 μm, or 0.00011%, was demonstrated. The equipment was not capable of any finer motion or measuring capability, and therefore the actual lower limit of resolution capability of the sensing element could not be determined. So, this can be considered to be as close to infinite resolution as could be easily determined.

See technology comparison Tables 1.1 and 12.1.

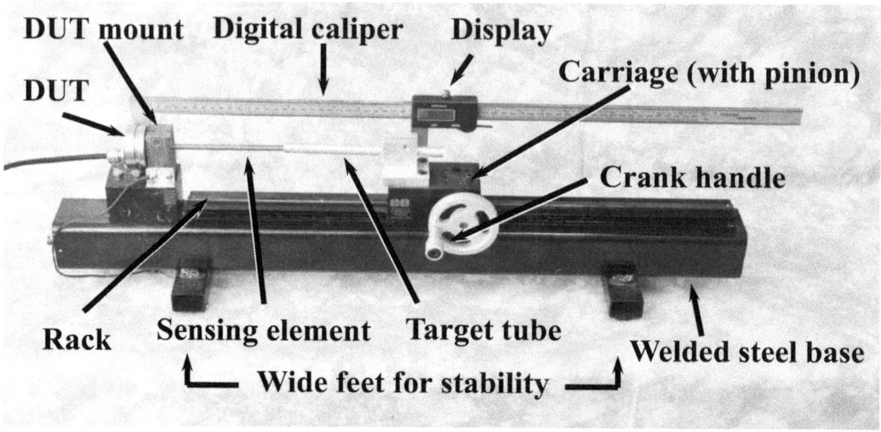

FIGURE 9.13 A test fixture for calibration of a distributed impedance linear position sensor.

9.12 CALIBRATION

Calibration of a distributed impedance sensor takes the same form as with an LVDT/RVDT or an inductive position sensor, except there is no null position, as with an LDVT/RVDT. The distributed impedance sensor is mounted into a test fixture, such as one designed by the author and shown in Figure 9.13.

The test fixture includes means for accurately adjusting the target tube position, while indicating the actual position (reference position) through the use of a reference sensor. The reference sensor in the figure is a set of digital calipers having a display selectable for units of inches or mm. A pushbutton on the display sets zero when the target tube is moved to the zero position in the range of the *device under test* (DUT). A rack is mounted onto a rigid base, the rack having a dovetail milled into its sides to form a rail. A carriage rides along the rail, positioned above the rack. A pinion gear is mounted on a shaft within the carriage so that the pinion gear teeth engage the gear teeth of the rack. A crank is attached to the pinion gear shaft. When a handle on the crank is turned, the target tube is moved precisely, and its position is indicated on the digital display. The error level of the reference sensor should be at least five times lower than the specification that one is intending to meet for the DUT. For an automated version, the movement of the carriage can be motorized.

Distributed impedance sensors inherently exhibit extreme stability, and also very fine repeatability of the curve relating the output signal level versus the input measurand. Plus, this curve has a very smooth and uniform shape, with no abrupt changes. Because of these qualities, it is easy and practical to apply nonlinearity correction. To implement such a correction of nonlinearity, several measurements are taken along the measuring length or rotation angle of the sensor, noting the reference measurement and the corresponding sensor output measurement. These measurements are installed into a correction table of actual (reference) position versus indicated position. Then, in use, a position measurement is first sent through a nonlinearity correction algorithm before being used to generate the signal output. In the algorithm, a position measurement is compared with the correction table. If the present measurement is at the same position for which a reference measurement was recorded in the correction table, that corresponding scaling correction is made. If the present measurement falls between reference points of the correction table, then linear interpolation between the two closest reference points is made, and then the resulting interpolated scaling correction is applied. This procedure is explained further in the Microchip application note AN 942, and can be found here: http://ww1.microchip.com/downloads/en/AppNotes/00942A.pdf.

In addition to nonlinearity correction, temperature sensitivity may also be corrected. The temperature sensitivity of a distributed impedance sensor is very repeatable because it is mainly dependent upon the thermal expansion rate and dielectric constant of the materials forming the sensing element. So, the temperature performance is an inherent property of the design and materials. This means that a correction table can be the same for all units of the same model. The temperature correction table does not need as many reference points as with the nonlinearity correction table. An example would be to use these reference temperatures: 125, 100, 75.50, 25, 0, −25, −40°C, and the corresponding correction percentages required (and whether they are + or − adjustments). As with the nonlinearity correction, an interpolation algorithm is then implemented to make the corrections at in-between temperatures. In order to ensure that the temperature correction does not affect the sensor apparent resolution, the temperature table should have relatively fine resolution, such as in increments of 0.25°C.

Once any nonlinearity and thermal correction algorithms have been implemented in software, it is very simple to calibrate the sensor, and may easily be accomplished on an automated system. Position measurements are obtained in a range spanning the intended measurement range (usually anywhere between 10 points and 100 points), while the reference and measured positions are entered into the nonlinearity correction table. Output after applying the temperature correction algorithm is then run through the nonlinearity correction algorithm, and then usually filtered with a low-pass filter. The low-pass filter may have adjustable frequency response, such as having a corner frequency selectable among 1 kHz, 500 Hz, 250 Hz, 100 Hz, 50 Hz, 20 Hz, 10 Hz, and 1 Hz. This may help to fit the sensor response time to the needs of the application. The low-pass filter may be implemented in hardware or software, or a combination of both, such as having a hardware filter with a corner frequency of 1 kHz, and an adjustable software filter for any remaining lower corner frequencies that may be selected.

9.13 MANUFACTURERS

Some manufacturers of distributed impedance linear and angular/rotary position sensors include the following:

Alliance Sensors	www.alliancesensors.com
LRT Sensors	www.lrtsensors.com
Revolution Sensor Company (design)	www.rev.bz

9.14 QUESTIONS FOR REVIEW

1. **The electrodynamic element of a distributed impedance position sensor, such as a dual helix, is a:**

    a. Periodic structure
    b. Circular target
    c. Metal tube
    d. Semiconductor
    e. Insulator

2. **Closely spaced lines of flux indicates a greater:**

    a. Resistance
    b. Capacitance
    c. Magnetic field strength
    d. Frequency
    e. Length

3. **The velocity of electromagnetic wave propagation in a transmission line is equal to:**

    a. $\sqrt{L_0 C_0 / \varepsilon_0 \mu_0}$
    b. $(\beta + 2\pi m/d)$
    c. 16 bits
    d. $1/\sqrt{L_0 C_0}$
    e. Resolution

4. **Capacitance of an LVDT or LVIT coil can be considered as a:**

    a. High value
    b. Voltage divider
    c. Frequency
    d. Conductor
    e. Lump sum

5. **Pierce is a term specifying this type of circuit:**

    a. Amplifier
    b. Demodulator
    c. Filter
    d. Oscillator
    e. Bridge

6. A crystal is often used in a circuit to provide precise control of:

a. Voltage potential
b. Frequency
c. RFI
d. Conductivity
e. Inductance

7. When calibrating a sensor, the sensor being calibrated is called the:

a. Dummy
b. DUT
c. Calibrator
d. FET
e. Field device

8. When a microcontroller uses a nonlinearity correction table, measurements falling in between the table data points are handled by using:

a. Interpolation
b. Regeneration
c. Heterodyne
d. Modulation
e. Tweezers

REFERENCE

[1] K. Zhang, *Electromagnetic Theory for Microwaves and Optoelectronics*. New York: Springer Science, 2007.

10 The Hall Effect

10.1 HALL EFFECT SENSORS

Position sensors based on the Hall effect are often used in automotive and industrial products because they can provide long life at a relatively low cost. Since the sensitivity of a Hall effect element is based on measuring the magnetic field at a specific point within the device package, a single element provides for a relatively short stroke linear position sensor (less than 25 mm stroke). Longer stroke length sensors can be made by using mechanical advantage or by incorporating an array of sensing elements, but the benefit of lower cost is then reduced.

Hall effect sensors measure the strength and polarity of a magnetic field. A Hall effect linear or rotary position sensor includes at least a Hall device, a position magnet, and associated electronic circuits. The position magnet is attached to the element to be measured. As the magnet approaches the Hall device, the strength of the magnetic field increases, and the output of the Hall device increases. Since the change in magnetic field strength is due to a change in the magnet position, the Hall device produces an electrical output that varies with changes in the location of the position magnet. The polarity of the magnetic field (north or south) is indicated by the polarity of the electrical output (positive or negative).

Hall effect sensing elements have a relatively low output voltage (tens of millivolts), so fabricating a position sensor requires the addition of an amplifier to increase the signal voltage level. A constant voltage or constant current is applied to drive the sensing element, and a differential output voltage is then produced. The output voltage amplitude is proportional to the magnetic field strength and thus related to the distance between the sensing element and a position magnet. As mentioned earlier, the polarity of the output voltage is dependent upon the polarity of the field produced by the position magnet. That is, with lines of flux extending to the Hall device from the north pole or the south pole of the position magnet. So, depending on the connection configuration, as a magnetic *north* pole approaches a given Hall effect element from one side, the resulting output voltage may be positive. Whereas, a negative output would then be produced from a magnetic *south* pole approaching that same Hall effect element from the same side.

A permanent magnet has two poles, north and south. The north pole is the one that seeks the magnetic north of the earth. Since opposite poles attract, this means that the magnetic north of the earth is analogous to what we call the south pole of a permanent magnet. This is important to know when checking the polarity of a permanent magnet by using it as a compass needle. The north pole of a magnet is also called the north-seeking pole.

10.2 THE HALL EFFECT

In a Hall effect sensing element, also called a *Hall device*, an electric current, I, is passed through a conductive material while the material is subjected to a magnetic field, B, at 90° to the direction of the current. Electrical contacts to the conductive

DOI: 10.1201/9781003368991-10

material provide the means for applying the current and for measuring the output voltage. The output from a Hall effect sensing element is called the Hall Voltage, V_H. The direction of V_H is perpendicular to both the electric current and the magnetic field. The magnitude of V_H is proportional to the density of the magnetic flux, B, and to the amount of current, I, flowing through the conductor. The conductor with contacts for applying the current and for measuring the Hall voltage comprises a sensing element, and is sensitive to the magnitude of the magnetic field applied. See Figure 10.1.

In the figure, the current is flowing from left to right through the conductor. The magnetic field is going into the page. The Hall voltage is detected from top to bottom. So, the current, magnetic field, and the Hall voltage detection are all mutually perpendicular to one another.

Magnetic flux density, B, is measured in webers per square meter (Wb/m^2), or tesla in SI units. The magnetic field causes a gradient of carrier concentration to occur across the conductor. The carrier concentration is greater on one side and less on the other, due to the magnetic field. The larger number of carriers on one side of the conductor, compared to the other side, causes the voltage potential, V_H, to develop. In Figure 10.2, with electrons being shown as the charge carriers in an n-type semiconductor, the concentration of electrons is greater toward the upper edge of the conductor. The electrons are supplied by the current flow, and they are directed toward the upper edge of the conductor by the force from the applied magnetic field.

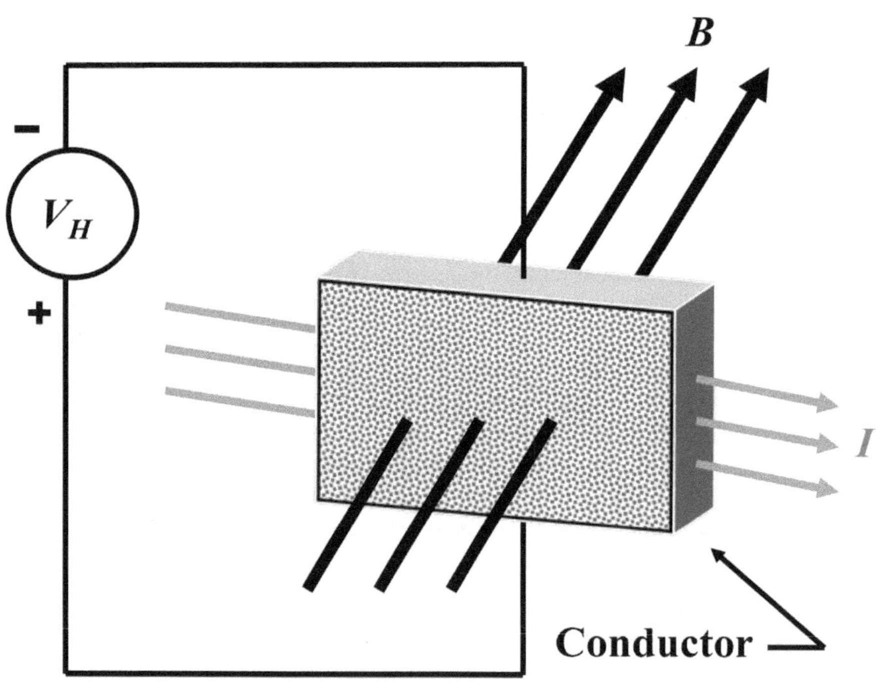

FIGURE 10.1 The Hall effect: mutually perpendicular current, I, and magnetic flux, B, in a conductor, result in generation of the Hall voltage, V_H.

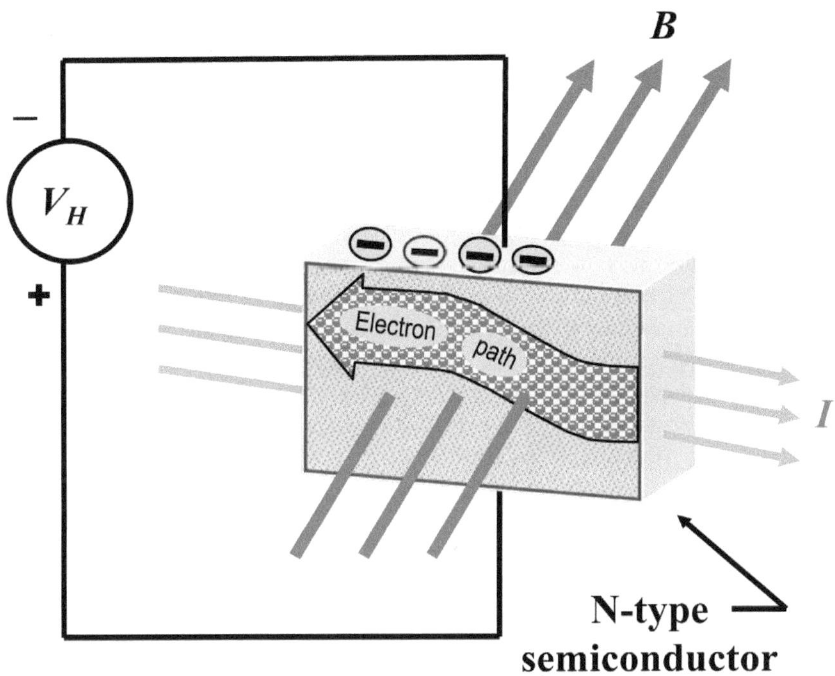

FIGURE 10.2 Concentration of carriers is forced to one side of the Hall device by magnetic force and produces the voltage differential, V_H.

(Note that the direction of electron flow is opposite to that of current flow, according to the standard convention.) A voltage potential difference is thus developed across the conductor (that is, the semiconductor material) of Figure 10.2, with the upper surface at a negative potential with respect to the lower surface.

The top surface of Figure 10.2 is negative because of the negative charge of the greater number of electrons there (note the electron path). This differential voltage, between the top and bottom of the n-type semiconductor of Figure 10.2, is the Hall voltage, V_H.

The amplitude of V_H varies with the amount of current flow and the magnetic flux density [1, p 125] according to:

$$V_H = K_H BI / z \tag{10.1}$$

where V_H is the Hall voltage, K_H is the Hall constant, B is the magnetic flux density, I is the current flowing through the semiconductor, and z is the thickness of the conductor.

With commercially available Hall devices, the conductor (semiconductor) dimension is already known by the manufacturer, and the sensitivity is specified for the particular model. In this case, formula 10.1 can be simplified to:

$$V = KBI \tag{10.2}$$

where K is the sensitivity factor, usually in volts per gauss or volts per milli tesla (milli tesla is abbreviated as mT), that is specified by the manufacturer. 1 tesla = 10,000 gauss.

10.3 HISTORY OF THE HALL EFFECT

In 1879 at Johns Hopkins University, physicist Dr. Edwin H. Hall discovered the effect that now bears his name [2, p 473; 3, p 1]. He found that a voltage potential appears across a conductor when a magnetic field is applied at right angles to the flow of an electric current in the conductor. This voltage potential is called the Hall voltage (refer back to Figure 10.1). The Hall effect has been used to develop sensing elements that measure the strength and polarity of a magnetic field. These magnetic field sensors have been further developed into position sensors, where change in the position of a magnet attached to a target results in change in the magnetic flux density at the location of the Hall device. (A Hall effect-based sensing element is often called a Hall device.)

Although discovered in the late nineteenth century, the Hall effect had little commercial use until the 1950s when development of semiconductor compounds led to the first useful Hall effect laboratory instruments. The use of a Hall effect sensor allowed the measurement of a static magnetic field, which was not possible with the coil assemblies previously used for magnetic field measurement (coils can sense only alternating fields). Once the Hall effect sensor and some signal conditioning electronics were incorporated into a single integrated circuit in the 1960s, many applications became practical. Most of the early sensing applications were for switch-type sensors. These included keyboards, joysticks, and other industrial and commercial products.

The presence of an offset voltage and of temperature sensitivity noticeably limited the performance of single-element Hall effect sensors. To obtain a lower offset voltage and lower temperature sensitivity of the offset voltage, dual elements were used as a partial solution. This was followed by the development of quad elements in a bridge configuration. Most present-day Hall devices are of a bridge configuration. The resulting capability for operating over a wider temperature range made automotive sensors and control systems possible. Further integration of the circuitry and use of the bridge configuration improved the performance of the switch-type devices and also made linear sensors practical. Since the output voltage from the Hall device is very low, chopper-stabilized amplifiers were integrated in order to remove any offset voltage error that originated in the earlier DC amplifiers. A later method, also used to eliminate offset errors, involves constantly switching the polarity of the input current to the Hall device, thereby producing an AC signal and taking away the drift-prone DC bias by using an AC amplifier.

After the introduction of the bridge configuration of the Hall device and improved product uniformity achieved through high volume manufacturing, little of the performance increases developed thereafter were due to improvements in the Hall device itself. Instead, the integration and increasing complexity of the signal conditioning circuitry brought most of the further advancement in performance.

10.4 HALL EFFECT POSITION SENSOR DESIGN

A simple position sensor can be devised by placing a permanent magnet and a Hall device in close proximity to each other as shown in Figure 10.3.

The output voltage across the Hall device will vary as the distance between it and the position magnet varies, due to the change in magnetic field strength at the Hall device. In the figure, the south pole of the magnet is toward the Hall device. As depicted in the output vs. position curve, the relationship is nonlinear. As the position magnet is relatively far away from the Hall device, the sensitivity is low. As the magnet gets closer to the Hall device, the sensitivity steadily increases. This is due to the fact that the magnetic field strength at the Hall device varies approximately as the square of the separation distance between it and the position magnet. The response of the Hall device itself is essentially linear with respect to the magnetic field strength.

An alternate linear sensor configuration is shown in Figure 10.4. A longer sensing stroke is possible because of the different orientation of the position magnet,

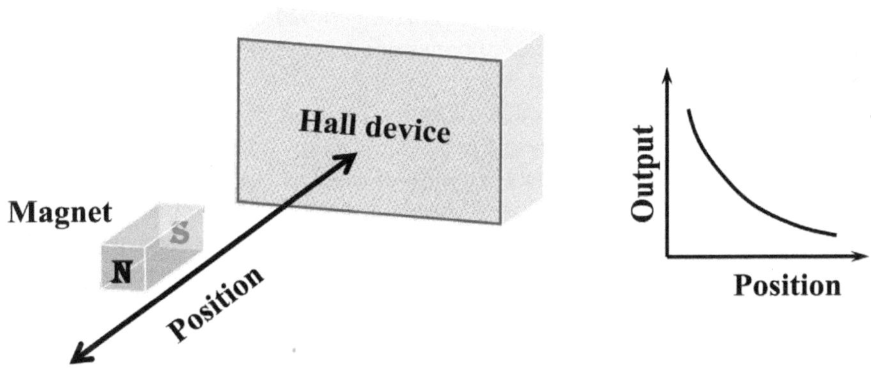

FIGURE 10.3 A simple position sensor based on a Hall device, together with a typical curve of the output voltage vs. position.

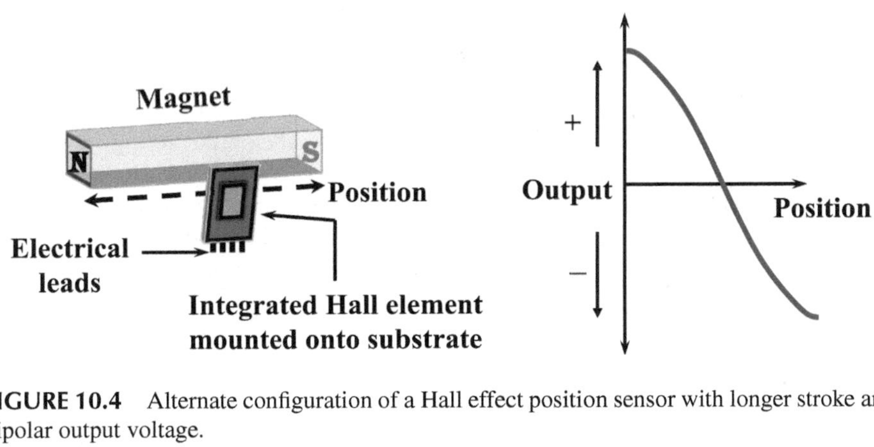

FIGURE 10.4 Alternate configuration of a Hall effect position sensor with longer stroke and bipolar output voltage.

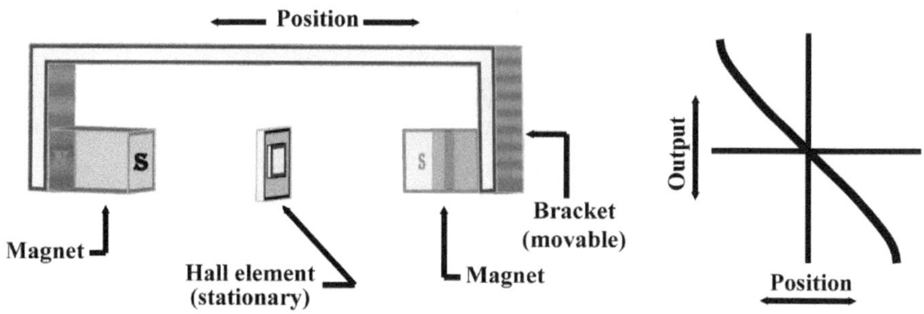

FIGURE 10.5 Hall effect position sensor with improved configuration for greater stability.

as compared with that of the previous figure. As the position magnet moves along its stroke, the stationary Hall device is affected by the north pole of the magnet at one end of the stroke, and then by the south pole at the other end of the stroke. This produces a bipolar output voltage characteristic as shown to the right in Figure 10.4.

An improved linear position sensor configuration is shown in Figure 10.5. Two magnets are held in place by a bracket. Like poles of the two magnets (south, in this case) face the Hall device, which is fixed in a location between the magnets and along their motion axis. The complete bracket and magnet assembly is moved with the measurand. With the Hall device positioned between the two magnets, the field strength in the center is near zero even if the magnetic field strength changes with temperature (assuming that the thermal effect is the same on both magnets). As one of the magnets approaches the Hall device, the magnetic field strength and the output voltage increase.

As the other magnet approaches the Hall device from the opposite side, the polarity of the output voltage is reversed. The bipolar output voltage characteristic shown in the curve is fairly linear if the total separation between the two magnets is not too great. This sensor configuration is suitable for short position sensors in the range of ±1 cm or less because the field strength becomes too low at longer distances, increasing the nonlinearity. At one end of the measuring range (the full measuring range is also called the "stroke"), the magnetic field from the south pole of one position magnet penetrates the Hall device from that side. At the other end of the stroke, the magnetic field from the south pole of the other magnet penetrates the Hall device from the other side. These two directions of the magnetic field are what results in the bipolar output signal from the Hall device. That is, the output voltage goes from negative to positive, as shown in the figure. So, one way to obtain a bipolar output is to subject one side of a Hall device to the field from a magnetic *north* pole and then to a *south* pole (Figure 10.4). Another way to obtain a bipolar output is to first subject one side of a Hall device to the field from a magnetic south pole, and then to subject the other side of the Hall device to a south pole (Figure 10.5).

Figures 10.3 through 10.5 depicted linear position sensors using a single Hall device for simplicity. But as mentioned, most Hall-type sensors on the market implement a bridge configuration of four Hall devices (or at least a half-bridge of two devices) on one semiconductor substrate (along with some of the associated electronics). This allows the thermal and nonlinear response properties of one Hall element

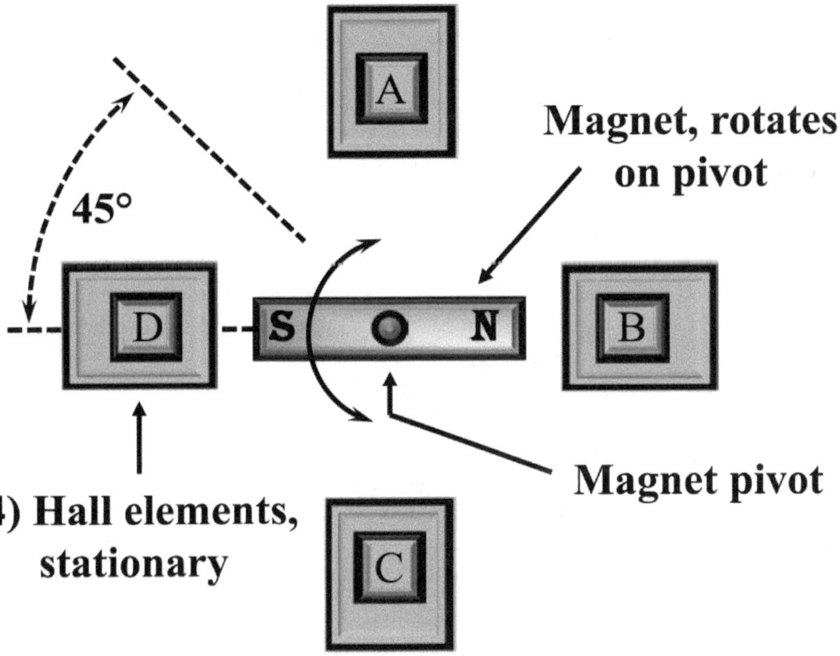

FIGURE 10.6 Angular or rotary Hall sensor.

to be compensated by the other(s). And with multiple Hall devices comes the possibility of realizing an angular sensor as depicted in Figure 10.6.

In the figure, a bar magnet rotates above or near four stationary Hall elements, arranged at intervals of 90°. Signals from the four Hall devices are sent to a microcontroller. The microcontroller applies an algorithm to decode the signals into a measurement of the bar magnet angular position anywhere within its 360° of rotation. For example, in the bar magnet position shown, Hall sensors A and C will have near-zero signal levels (being equally exposed to north and south poles of the magnet), while D responds to a south magnet pole and B responds to a north pole. If the magnet is rotated clockwise by 45°, Hall sensors A and D will have equal response to a south pole, while B and C have equal response to a north pole. At intermediate steps between the position shown and the 45° of clockwise rotation, D will have varying amounts of greater alignment with the south pole, while A has varying degrees of less alignment with the south pole, and the same for B and C, respectively, with respect to the north pole. Proportionate results among the several Hall devices will result from other magnet positions around the 360° of rotation.

10.5 THE HALL EFFECT ELEMENT

The Hall device, or *Hall element*, is made from a thin sheet of electrically conductive material. Input connections, for the introduction of current flow, are positioned perpendicular to the output connections, which are for detecting the Hall voltage. When in the presence of a magnetic field, the Hall voltage is proportional to the strength of the applied field. The polarity of the Hall voltage (+ or −) is determined by the polarity of the applied magnetic field (north or south).

The Hall constant, K_H, is larger in semiconductors than in metals [4, p 79]. Because a larger Hall constant results in greater sensitivity, Hall devices normally utilize semiconductor materials in their construction. The semiconductor material may be either *p* or *n* type, depending on the polarity of the charge carriers (holes or electrons, respectively). A large Hall constant requires high carrier mobility. Semiconductor materials provide high carrier mobility (and lower resistivity) after doping with an impurity selected to provide carriers of electrons or holes. A low resistance value will help to limit thermal noise and allows a higher signal/noise ratio. Since electrons tend to move faster than holes under a given set of conditions, greater sensitivity can be obtained by using n-type semiconductor material [3, p 8].

When a Hall device is manufactured, it is important to accurately space the two contacts for the Hall voltage pickup. If the two contacts are not perfectly aligned, there will be a non-zero output when there is a zero magnetic field strength. This is due to the differential in the voltage drop resulting from the current flow in unequal resistances over the two sides of the element. Since perfect alignment is not likely, a small voltage at zero magnetic field strength will usually be found. This is the zero offset voltage. An adjustment can be included in the signal conditioning electronics to correct the offset voltage to zero.

10.6 ELECTRONICS

When using a single Hall device (containing either a single element or a bridge connection), a circuit similar to the block diagram shown in Figure 10.7 can be used.

Since the Hall voltage is proportional to the current through the Hall device, a constant current source is shown as supplying the drive current. Sometimes a constant voltage source is used instead. Alternatively, if the Hall device is powered directly from the power source with no current or voltage regulator, the output will be ratiometric. Ratiometric operation can allow lower cost by eliminating a voltage reference, while improving performance by the amount of error that would have been added by a voltage reference (see Section 3.1—*Ratiometric output*).

The Hall voltage is a small differential voltage that develops across the Hall device and is then amplified by a differential amplifier. Usually, this amplifier also converts the differential voltage to a single-ended voltage, meaning that the output is referenced to zero volts. Next, a filter passes on the expected range of signal frequencies and rejects (does not pass) other frequencies. The filter frequency response can be a low-pass or a band-pass type, although the low-pass type is normally implemented. A low-pass filter attenuates frequencies above its corner frequency and passes without significant attenuation those signal frequencies below the corner frequency [5, p 6]. The rejected frequencies are not wanted because they are likely due to electrical or electromagnetic noise. If an analog output from the sensor is desired, a buffer amplifier conditions the signal to drive the output. If a digital output is needed, then an A/D converter would be used instead of the buffer amplifier. A digital implementation is shown in Figure 10.8.

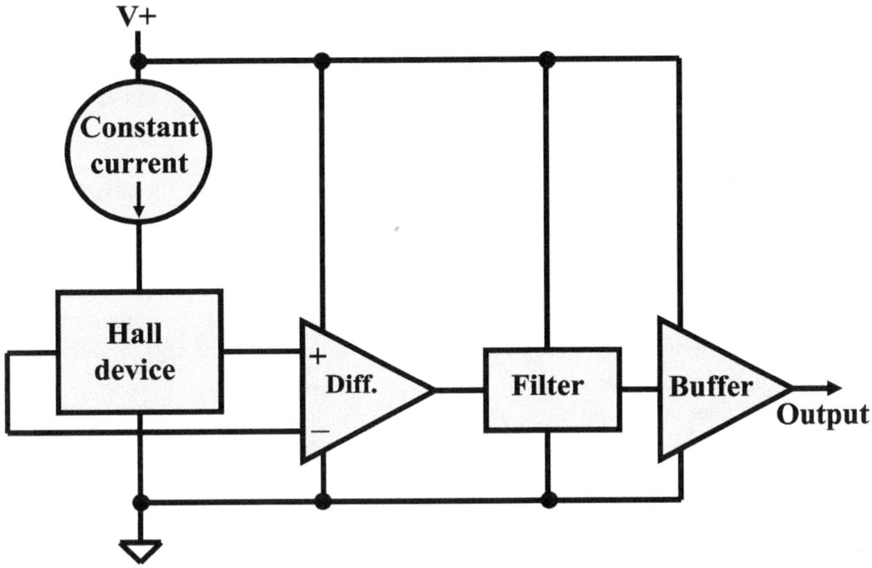

FIGURE 10.7 Block diagram of an analog circuit for a single Hall device, or for a bridge configuration of Hall devices.

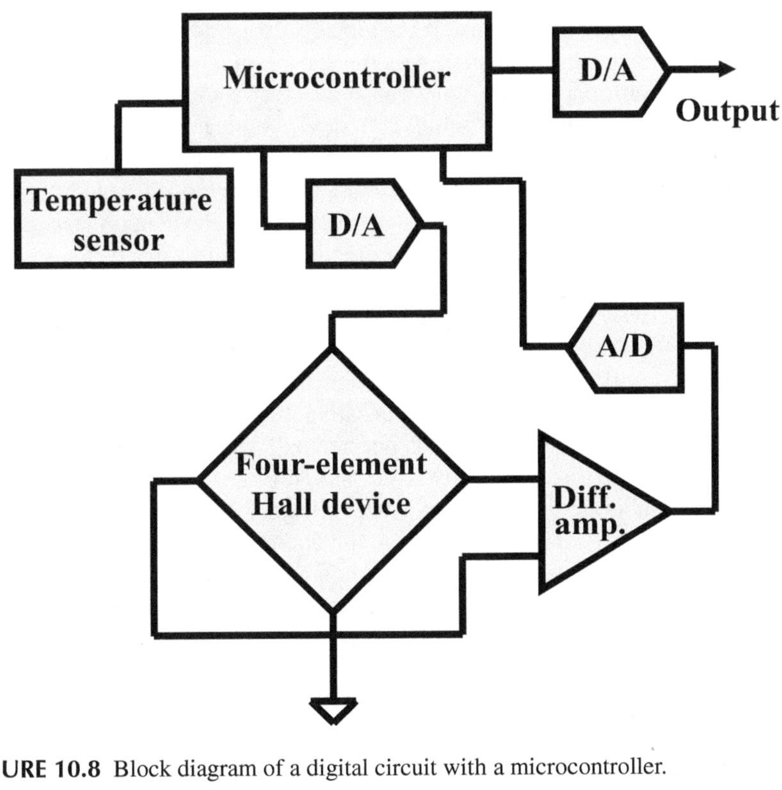

FIGURE 10.8 Block diagram of a digital circuit with a microcontroller.

Note that, even in a digital circuit with a microcontroller, some analog circuitry is still required, including an amplifier, D/A and A/D. A filter is not shown in the digital version, which uses a microcontroller (μC), because a digital filter algorithm can be included in the μC code; although an anti-aliasing analog filter in hardware may still be needed in some cases.

In Figure 10.8, the microcontroller powers the Hall device through a digital-to-analog (D/A) converter. The D/A converter can be a voltage or a current output type. After the differential amplifier, an analog-to-digital (A/D) converter is used to bring the Hall device signal into the μC. (These additional conversion circuits are typically required in order to use a μC with a purely analog sensing element, like a Hall device. A much easier interface to a μC can be had with a frequency-output sensor such as a distributed impedance sensor, or a time variant output sensor such as a magnetostrictive linear position sensor.) After the μC code conditions the Hall device digitized signal properly, an analog output (voltage or current) is produced by a second D/A converter. Alternatively, if a digital output is desired, the μC can provide a parallel or serial output without needing the second D/A converter. If temperature compensation is desired, the μC should be provided with a temperature input as shown in the figure. The temperature sensor can be implemented with a pulse width modulated (PWM) output that can be measured by the μC with a timer. Alternatively, temperature sensors are available that have a direct serial digital output (such as I^2C or SPI, among others). The temperature sensor should be mounted very close to the Hall device, or integrated into the Hall device chip, in order to provide an accurate reading of the Hall device temperature. If the temperature sensor is not close enough to the Hall element, thermal gradients across the sensor (due to power dissipation or to variation in ambient temperature) may cause an error such that the temperature measurement is not indicative of the actual Hall device temperature.

The output from the Hall device will have errors. Some of these include zero offset, gain error, offset voltage change with temperature, gain factor change with temperature, nonlinearity, and hysteresis. Potentiometers (such as digital potentiometers, adjustable by the μC) may be added to the amplifier circuit to provide a means for adjusting the room temperature zero and gain errors. These controls are called zero and span adjustments. Alternatively, this can be done in software if a microcontroller is used.

Compensation for temperature-induced errors is a little more complex. The change in gain error with temperature is often similar for Hall devices made on the same production line. Therefore, one compensation can often be used for all Hall devices if highly accurate compensation is not needed. The temperature sensitivity of zero, however, can vary substantially from one Hall device to the next. For the highest accuracy in laboratory equipment, each Hall device, or the complete sensor, must be characterized by running it throughout the operating temperature range in an environmental chamber. The data are recorded and used to calculate the correction. The correction can be accomplished in analog circuitry by using temperature compensating resistors, or in software if using a μC.

Using four Hall elements together in a Wheatstone bridge configuration helps to compensate for thermal errors and increases the measurement sensitivity. The Wheatstone bridge configuration is usually incorporated into one substrate and used as if it was one single Hall device.

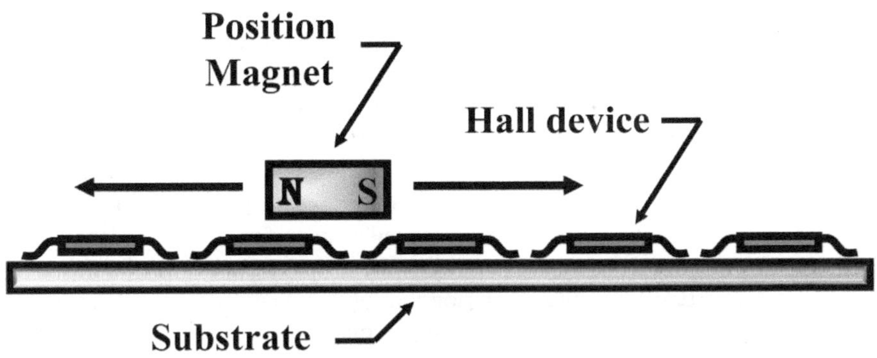

FIGURE 10.9 Expanded range linear position sensor employing many Hall devices in a linear array.

10.7 LINEAR ARRAYS

It was mentioned earlier that a single Hall device, or Wheatstone bridge connection of four devices, is only operable over a range of up to a maximum of 25 mm. Alternatively, multiple Hall devices can be aligned along a linear axis in order to fabricate a sensing element for a longer-stroke linear position sensor, as shown in Figure 10.9.

In this configuration, an array of several or tens of Hall devices is mounted onto a substrate and positioned along a stationary measurement scale. A position magnet (permanent magnet) is moved along the array and represents the position to be measured. The one or two Hall devices in closest proximity to the magnet will indicate their respective magnetic field strength measurements by the voltages of their outputs. Output connections from all of the Hall devices are brought back to a signal processor, usually located at one end of the Hall device array. Multiplexing the Hall device connection lines and scanning the signal amplitudes will indicate which Hall devices are close to the position magnet. The Hall device that is closest to the position magnet is first detected, for example, by way of determining which Hall device has the highest output. The known location of that particular device represents a *coarse position*. The actual readings from that device and the next closest one are combined by interpolation to form a measurement that represents a *fine position*. The coarse and fine position data are next combined in the μC to form a high resolution output signal. This provides a high resolution measurement over a wider range of position, as compared with using only one Hall device or bridge. Position sensors with full-scale ranges of up to several meters can be made using this technique.

10.8 ADVANTAGES

Most commercially available Hall devices are made from semiconductor material. This means they have the advantage to be constructed so that some or all of the required signal conditioning circuitry can be incorporated onto the same semiconductor substrate that comprises the sensor element. This can reduce size and cost.

Hall effect position sensors have an advantage over contact-type position sensors, such as a potentiometer. Since the Hall device is non-contact, there are no wearing parts to limit the life of the sensor. Hall effect-based position sensors also are normally absolute reading: there is no need for re-zeroing after a glitch or power cycling, as would be required with an incremental encoder.

Hall devices can operate at zero speed, this being an advantage over some encoders that have an uncertain output when approaching the edge of a transition between two adjacent readings when the approach speed is near zero. An encoder with this characteristic can have a constant dithering of the output signal. A Hall effect device however, has a virtually infinite resolution and will indicate even small changes in position with low nonlinearity and having a monotonic characteristic. The output having a monotonic characteristic means that as the measurand increases, for example, the output always increases. A non-monotonic sensor may have a characteristic that its output generally increases with an increasing measurand, but there may be a few points where a slight increase in measurand can cause a slight decrease in output at those specific points, but then the output continues to increase as the measurand further increases.

Besides having zero-speed capability, some Hall devices can also operate at frequencies of over 100 kHz. Hall devices have a repeatable response and can operate over a broad temperature range (some as wide as −40 to 150°C).

Nonlinearity and other accuracy limitations of Hall effect position sensors are about average within the range of available position sensors, being about the same as LVDTs (Linear Variable Differential Transformers) but not as accurate as some distributed impedance or magnetostrictive position sensors.

Several position-sensing technologies, such as an LVDT, distributed impedance, or some magnetostrictive position sensors, allow the possibility of design such that the basic sensing element can be separated from the electronics module by a length of cable. This allows the size of the sensing element to be minimized, as well as to be located in a higher temperature environment, where it would be difficult to use low-cost versions of electronics. This is not a possibility with Hall devices because the Hall device itself is made from semiconductor material, thus limiting the maximum temperature. Also, the signal conditioning circuitry that is integrated onto the monolith including the Hall device (to maximize its performance and keep the cost within limits) has temperature limitations because the semiconductor material is formulated to enhance the Hall device performance. High-temperature electronic circuits require different formulations of semiconductor materials and different fabrication techniques from those best suited for manufacturing a Hall device. Some very helpful information about Hall devices and application is available from Honeywell at: www.honeywellscportal.com//hallbook.pdf.

10.9 TYPICAL PERFORMANCE SPECIFICATIONS AND APPLICATIONS

There are many ways to implement a Hall effect device into linear or angular/rotary position sensors. Next is a specification of a linear Hall effect position sensor that has a ratiometric output voltage:

Full-range stroke:	0 to 10 mm
Nonlinearity	≤1%
Supply voltage:	5 VDC, ±10%
Supply current:	16 mA
Output signal:	10 to 90%, ratiometric to supply voltage
Resolution:	0.025%
Hysteresis:	0.1%
Operating temperature:	–30 to 85°C

Here is a specification of an angular/rotary Hall effect position sensor:

Full-range angle:	0 to 360°
Nonlinearity	≤0.3%
Supply voltage:	5 V DC, ±10%
Supply current:	5 mA
Output signal:	10 to 90%, ratiometric to supply voltage
Resolution:	0.025%
Repeatability:	0.03%
Operating temperature:	−40 to 85°C

Besides being utilized in a linear or angular position sensor product, an integrated circuit Hall device can be used as a component of another type of sensor. An example of this is the pressure sensor of Figure 10.10, where the Hall device acts as a transducer of linear motion into a voltage.

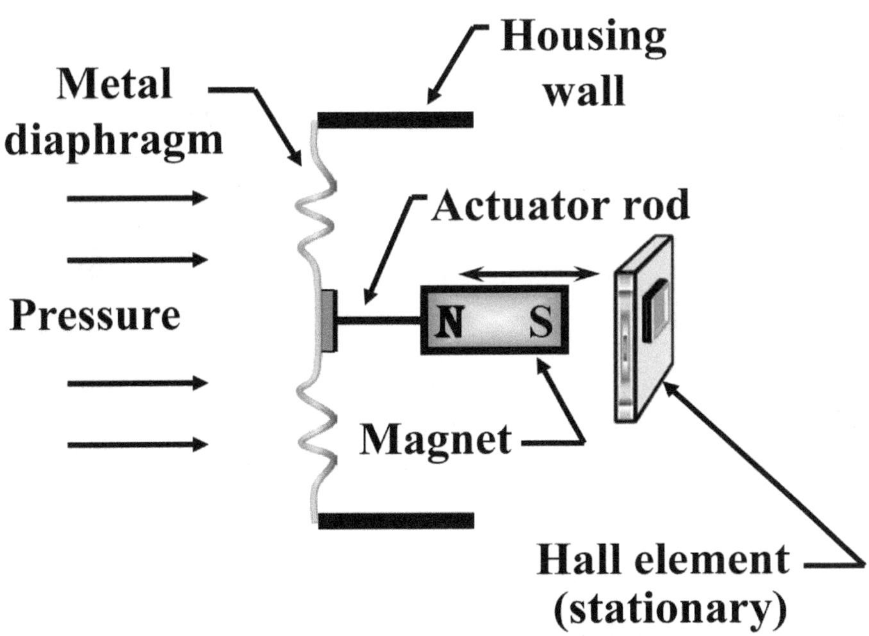

FIGURE 10.10 A pressure sensor configuration utilizing a Hall device to measure the diaphragm position as a function of an applied pressure (sectional view of the diaphragm).

In the sensor of Figure 10.10, a metal diaphragm flexes in response to changes in fluid pressure (gas or liquid). The diaphragm is a circular disk with convolutions pressed into it (convolutions reduce nonlinearity over a longer range of motion), but is shown in a sectional view, cut through its center. A Hall device measures the deflection of the diaphragm, and provides an electrical signal indicative of the diaphragm position. After the signal from the Hall device is amplified, filtered, scaled, and so on, the sensor provides an electrical output indicating the amount of measured pressure.

The Hall device in this case, together with the position magnet and diaphragm, comprise a transducer or sensing element, along with the metal diaphragm, together transducing a pressure change into an electrical signal. A sensor is formed by the addition of the signal conditioning electronics, power supply, housing, and other electronic and mechanical features, to provide a scaled output representative of the measured pressure.

Some additional applications of Hall effect linear position sensors include incorporation into the sending unit for measuring the fuel level in a fuel tank, measuring the accelerator pedal or throttle body position in a car, converting a potentiometric joystick to a non-contact type, measuring liquid level over several meters in a large storage tank by using a linear array of Hall devices, and many others.

10.10 MANUFACTURERS

Some manufacturers of Hall effect linear and angular/rotary position sensors include the following:

Allegro Microsystems	www.allegromicro.com
F W Bell	www.fwbell.com
Honeywell	http://honeywell.com/Pages/Search.aspx?k=hall+effect+position+sensor
Lakeshore	www.lakeshore.com
Melexis	www.melexis.com
Midori	www.midoriamerica.com
Revolution Sensor Company (design)	www.rev.bz
Texas Instruments	www.ti.com
Vishay	www.vishay.com

10.11 QUESTIONS FOR REVIEW

1. A Hall device responds to the magnetic field's:

a. Rotation
b. Duration
c. Convolution
d. Strength and polarity
e. Inductance

2. The output of a Hall device is called the:

a. Hall current
b. Hall resistance
c. Hall voltage
d. Hall reactance
e. Hall of fame

3. DC offset errors can be removed by using a:

a. Balun
b. Filter
c. Chopper-stabilized amplifier
d. Thermistor
e. Toothpick

4. A magnetic field strength of 0.1 tesla is equal to:

a. 10 webers
b. 70 hp
c. 10 milligauss
d. 1,000 gauss
e. 10,000 gauss

5. If the output always shows an increase for any measurand increase, the sensor is:

a. Monotonic
b. Linear
c. Stable
d. Accurate
e. Repeatable

6. Most commercial Hall devices are made from this type of material:

a. Metal
b. Semiconductor
c. Porous
d. Diatomic
e. Plastic

7. A linear position sensor can have extended range with Hall devices aligned in a:

a. Hall way
b. Circle

c. Zigzag
d. Meander line
e. Linear array

8. In an n-type semiconductor material, the charge carriers are:

a. Slow
b. Electrons
c. Crystals
d. Holes
e. Neutrons

REFERENCES

[1] J. R. Carstens, *Electrical Sensors and Transducers*. Upper Saddle River: Regents/Prentice Hall, 1992.
[2] R. Lerner and G. Trigg, *Encyclopedia of Physics*. New York: VCH Publishers, 1990.
[3] E. Ramsden, *Hall Effect Sensors*. Cleveland: Advanstar Communications, 2001.
[4] R. Philippe, *Electrical and Magnetic Properties of Materials*. Norwood: Artech House, 1988.
[5] D. E. Johnson and J. L. Hilburn, *Rapid Practical Designs of Active Filters*. New York: Wiley, 1975.

11 Magnetoresistive Sensing

11.1 MAGNETORESISTIVE SENSORS

A magnetoresistive position sensor includes a transduction element known as a magnetoresistor. As in Hall effect sensors, the sensing element may include a bridge connection of magnetoresistors. Also as in Hall effect sensors, magnetoresistive linear position sensors are non-contact, absolute reading, and have essentially infinite resolution. The measuring range of an individual element or bridge is limited to a maximum of about 25 mm, but practical linear sensors have been designed by using multiple sensing elements in a linear array to obtain sensors with a full-scale range (FSR) of over 2 meters. These linear array sensors can have an overall nonlinearity of less than 0.05%. Popular applications include positioning of industrial machinery, and measuring of small displacements in commercial products and industrial control systems.

As with a Hall device, a single magnetoresistive sensing element transduces a change in magnetic field strength into a change in electrical output. A complete sensor is formed by adding electronic and mechanical components as needed to provide a desired electrical output that is representative of the measured position. An electronics module normally includes an amplifier, filter, scaling capability, and may include a microcontroller with nonlinearity and/or temperature correction.

The coupling between the moving and stationary parts of the sensing element is achieved by means of a magnetic field. Therefore, any number of position changes can be made without incurring wear to the sensing parts. In a single-element magnetoresistive linear position sensor, a position magnet moves in relation to the magnetoresistor. The position magnet is a permanent magnet, and provides a magnetic field strength at the magnetoresistor that changes with the distance between the magnet and the magnetoresistor. As the magnetic field strength increases at the magnetoresistor, the resistance of the magnetoresistor changes. A simple arrangement of a magnetoresistor with position magnet is shown in Figure 11.1.

The resistance of the sensing element responds to the magnitude of the magnetic field, not to its polarity. The magnetic field strength is represented in the figure by lines of flux.

11.2 MAGNETORESISTANCE

Magnetoresistance (MR) is a property that occurs in all metals, in which electrical resistance changes due to the application of an external magnetic field. The property appears in non-magnetic as well as magnetic materials, but is greater in magnetic materials (including magnetic alloys and composites). Therefore, most

DOI: 10.1201/9781003368991-11

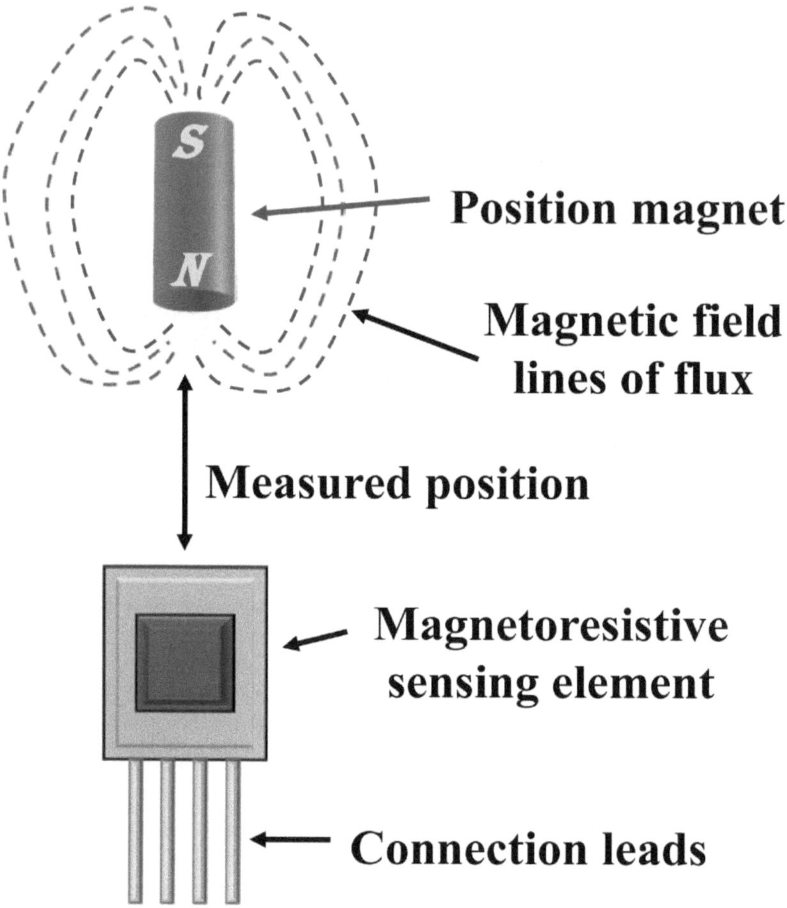

FIGURE 11.1 A magnetoresistive sensing element with position magnet.

magnetoresistive sensors utilize magnetic materials in order to optimize their greater response.

There are several classifications of magnetoresistance, including ordinary, anisotropic, giant, and others. In non-ferromagnetic metals, resistance increases with the application of a magnetic field of any orientation, but the response is somewhat nonlinear [1, p 56]. This is called ordinary magnetoresistance, and the change in resistance versus magnetic field is depicted in Figure 11.2. The resistance change is in response to the magnitude (magnetic flux density, or strength) of the magnetic field, but not to its polarity.

In the figure, as the north pole magnetic field strength increases to the right of center (center = zero magnetic field strength), the resistance of the magnetoresistive element increases. The same effect is evident as the south magnetic field strength increases to the left of center. Ordinary magnetoresistance in non-ferromagnetic metals provides only a small change in resistance of much less than 1% at room temperature.

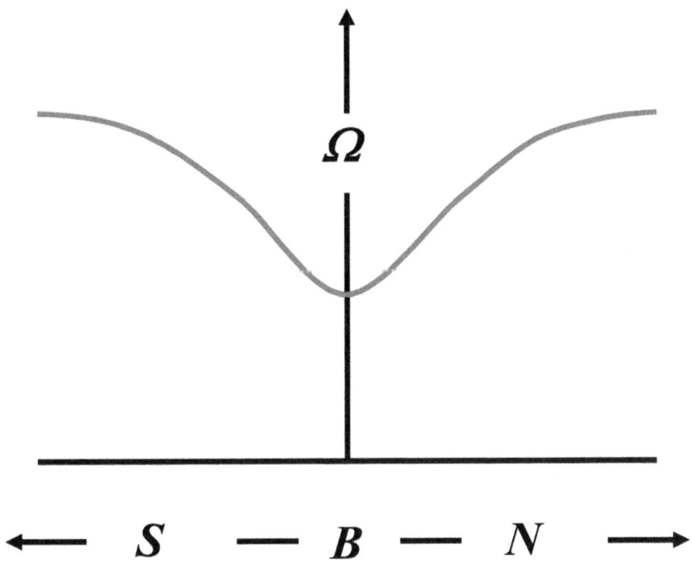

FIGURE 11.2 The resistance of a non-magnetic conductor, in ohms (Ω), varies in response to the magnetic flux density, *B*, but not polarity (north or south).

In ferromagnetic metals (such as nickel, iron, and cobalt), resistance *increases* with the application of an external magnetic field that is parallel to the direction of current flow (similar to ordinary magnetoresistance as shown in Figure 11.2). But conversely, the resistance *decreases* with the application of an external magnetic field that is perpendicular to the direction of current flow, as depicted in Figure 11.3. This characteristic of the resistance being sensitive to the orientation of magnetization with respect to the direction of current flow is called anisotropy. Such magnetoresistors are said to exhibit anisotropic magnetoresistance (AMR). The resistance of an AMR magnetoresistor continues to decrease with application of a stronger perpendicular magnetic field until the effect starts to decline as the material nears magnetic saturation.

In the figure, resistance of the AMR material decreases as the north pole magnetic field strength increases to the right of center (center = zero magnetic field strength), and also as the south pole magnetic field strength increases to the left of center. Anisotropic magnetoresistance in ferromagnetic metals provides a larger change in resistance (as compared with ordinary magnetoresistance) in the range of 1 to 2% (such as 2% for nickel), and up to 5% for some nickel-iron alloys.

A two-terminal device having a single magnetoresistive element is called a magnetoresistor. In an anisotropic magnetic material (such as a nickel-iron alloy), as mentioned, the electrical resistance decreases with the increase of magnetic flux density when the direction of the applied magnetic field is perpendicular to the current flow through the magnetoresistor. See Figure 11.4.

Almost all conductors exhibit some degree of magnetoresistance (but magnetically permeable materials are preferred for magnetoresistors, in order to obtain a

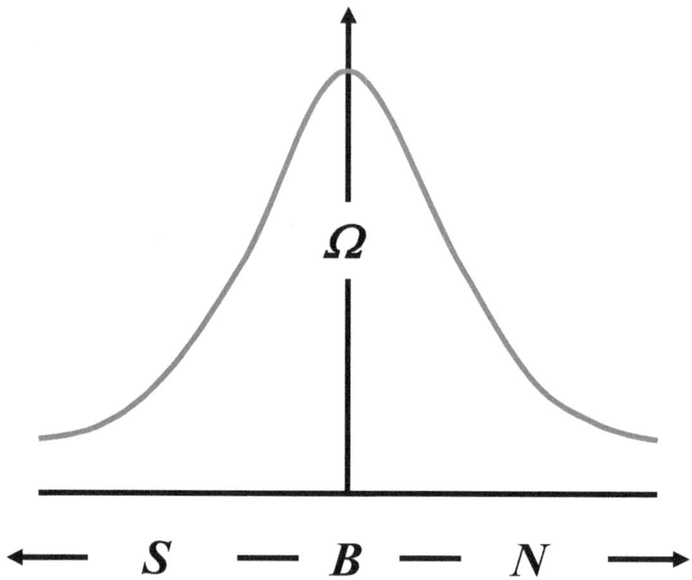

FIGURE 11.3 Resistance of an AMR magnetoresistor decreases with application of a perpendicular magnetic field.

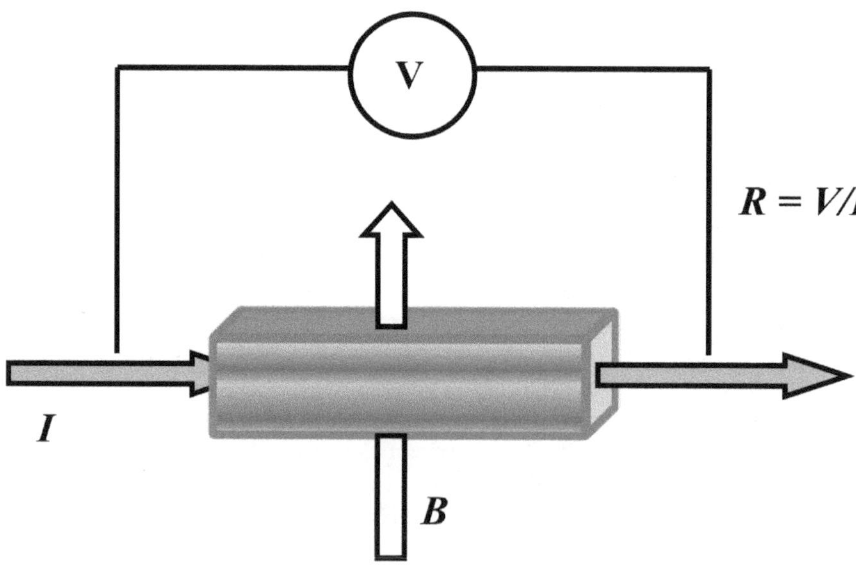

FIGURE 11.4 A single AMR element, with a magnetic field *B*, applied at a right angle to the current flow, *I*.

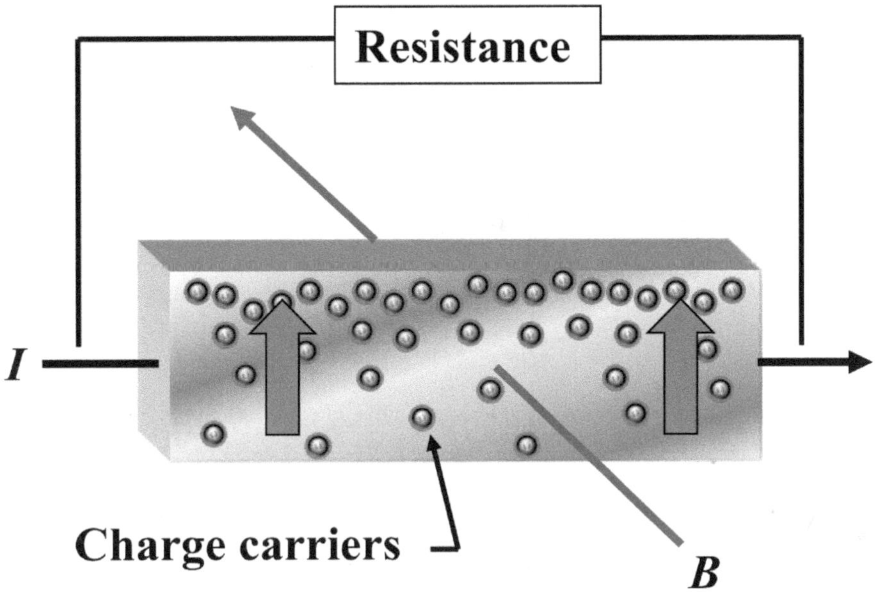

FIGURE 11.5 In a non-ferromagnetic conductor, charge carriers are moved to one side by a magnetic field.

greater response). The physical basis underlying the property of magnetoresistivity in a non-ferromagnetic conductor is the *Lorentz force*. This force causes the charge carriers, electrons that are carrying the current, to move in curved paths [2, p 449]. A Lorentz force is a force that acts upon a moving charge in the presence of a magnetic field. The force is at right angles to the magnetic field vector as well as to the velocity vector of the moving charge. When a magnetic field is applied as shown in Figure 11.5, the charge carriers are aligned with the magnetic field.

More of the charge carriers are forced to one side (the top of the figure), leaving fewer carriers on the other side (the bottom). The magnetoresistor then undergoes an increase in resistance due to the availability of fewer charge carriers when the magnetic field is applied. This is called galvanomagnetoresistance. The change in resistance comes from the combination of two component parts. The two component parts are:

1. A reduction in forward carrier velocity as a result of the carriers being forced to move sideways as well as forward, as described previously.
2. A reduction in the effective cross-sectional area of the conductor as a result of the carriers being crowded to one side [3, p 293].

The resistance changes according to:

$$\text{Resistivity} = \text{Voltage}/(\text{carrier density}\ x\ \text{carrier velocity}) \tag{11.1}$$

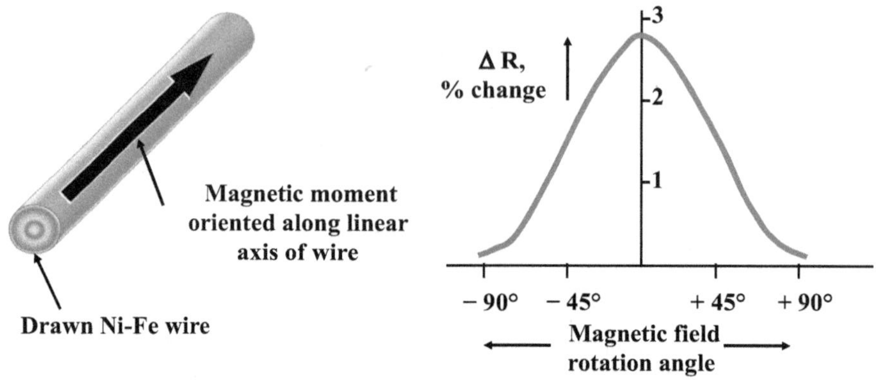

FIGURE 11.6 Magnetic domains tend to align in one direction when a ferromagnetic wire is drawn.

The aforementioned characteristic applies to non-ferromagnetic conductors, but magnetoresistance is greater in ferromagnetic materials, particularly in nickel-iron alloys known as Permalloys, due to the magnetic properties of the material and spin orbit coupling. Before film technologies were used, it was found that a drawn nickel-iron (Ni-Fe) wire had a useful property of magnetoresistance. When the wire is drawn, the magnetic domains tend to align parallel to the linear axis of the wire, shown in Figure 11.6.

This gives the wire an orientation of magnetization along this axis, called the "easy axis". Changing the magnetic axis to be perpendicular to the current flow (the hard axis), by bringing a permanent magnet in close proximity, causes a change in the electrical resistance. As mentioned earlier, this is called the anisotropic magnetoresistive (AMR) effect, anisotropy meaning that the effect is directionally dependent. See Figure 11.7.

A material is *anisotropic* if it has a predictable variation of a property (sensitivity to a magnetic field, in this case) depending on the direction in which the property is measured. The opposite condition is called *isotropic*, which is *directional uniformity*. The charge carriers in a metal are electrons that are free to move within the conductor (charge carriers in a semiconductor may be electrons or holes). The resistance between the two ends of the conductor is relative to the bulk resistance of the material per cross-section times the length. This is assuming that nearly all of the material is participating in the movement of electrons.

The amount of change in the resistance in magnetoresistors varies approximately in proportion to the square of the magnetic flux density. The degree of magnetoresistance of a material for a given change in magnetic flux density is called the magnetoresistive ratio ($R_{Magnetoresistive}$). It is the ratio of the change in resistance (due to the magnetic field) to the original resistance without the magnetic field; see Equation 11.2.

$$R_{Magnetoresistive} = \frac{R_{Max} - R_{Min}}{R_{Min}} \qquad (11.2)$$

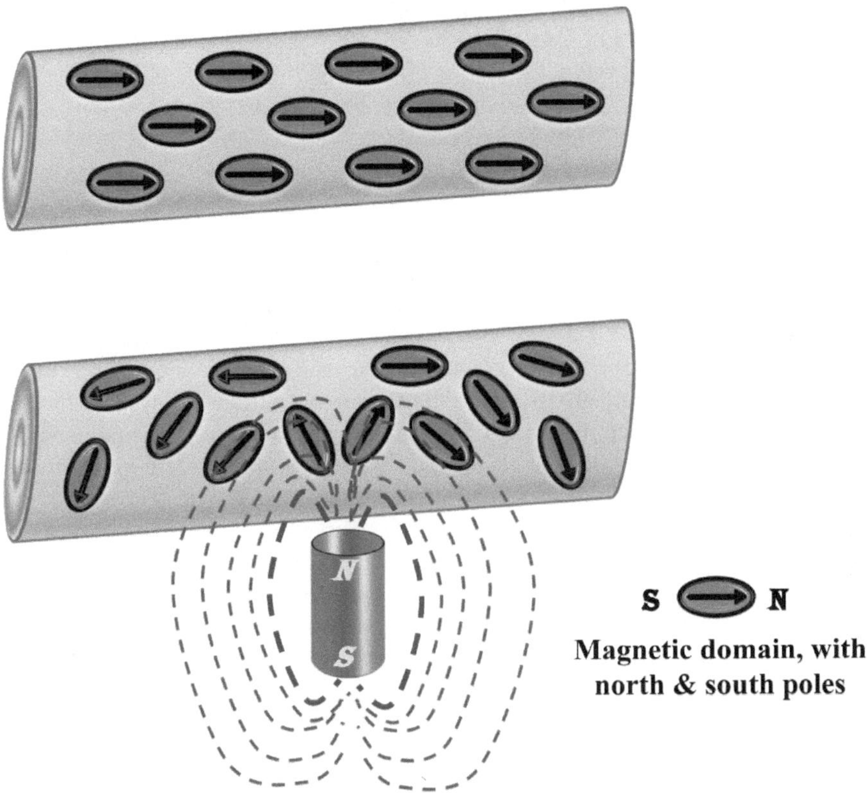

FIGURE 11.7 Anisotropic magnetoresistive effect in a drawn ferromagnetic wire: magnetic domains align along the linear axis of a drawn Ni-Fe wire. When a perpendicular magnetic field is applied, this alignment is changed, and this changes the electrical resistance.

In the case where applying a magnetic field increases the resistance, R_{Min} is the resistance before application of the magnetic field, and R_{Max} is the resistance with the full magnetic field strength applied.

As can be seen from Equation 11.2, if a material is able to maintain a given magnetoresistive sensitivity over a wider range of magnetic field strength, then the magnetoresistive ratio that is possible will be higher. A magnetoresistive material of lesser capability may saturate before reaching the highest magnetic field intensities, and thus limit the operating range of magnetic field strength that can be used.

A drawn Permalloy wire was mentioned earlier, but a more practical and similarly effective magnetoresistor can be made by depositing a thin strip of Permalloy onto a substrate while in the presence of a magnetic field to align the magnetic domains. Most magnetoresistive sensing elements are fabricated as a Permalloy thin film because the anisotropy can be made essentially uniaxial: the magnetic field aligns the magnetic domains in the Permalloy along the desired axis. In this state of aligned magnetic domains, the resistance of the anisotropic magnetoresistance

(AMR) material is at its lowest when the current and the magnetic moment are parallel (the current is parallel to the easy axis). The ratio of the change in resistance ΔR, to the total resistance, R, is approximately proportional to two times the cosine of the angle, a, between them.

$$\frac{\Delta R}{R} \propto 2\cos a \tag{11.3}$$

The angle is changed by bringing a permanent magnet close to the AMR material. The resistance is then highest when the angle between the current and the magnetic moment is 90°.

Because of this relationship, the best linear region is in the area of 45° of rotation of the magnetic moment. To extend the linear range to lower levels of magnetic field strength, a special configuration of conductors has been used. Small strips of a non-magnetic conductor, such as aluminum, have been placed at an angle of 45° to the long axis of an AMR material (see Figure 11.8).

This is called a *barber pole* configuration. The conductive strips have a lower electrical resistance than the base AMR material and cause the current flow to approximate a 45° angle to the magnetization easy axis when no external magnetic field is applied. In addition to improving the linear response to low-strength magnetic fields, this also enables determination of the polarity of the sensed magnetic field. The polarity is indicated by an increase or decrease in resistance, corresponding to an increase or decrease respectively in the 45° angle. One magnet pole causes an increase in the angle, the other causes a decrease. A consequence is that the conductors reduce the active part of the element by shielding it and reduce the magnitude of the sensor signal. As described earlier, the sensitivity of an AMR material is based on a preferred orientation of the magnetic moment. This orientation can be induced by mechanical means, such as drawing a wire, or by applying a magnetic field to orient the magnetic domains during fabrication. One drawback of magnetoresistive materials with oriented domains, is that it is possible to change this orientation accidentally.

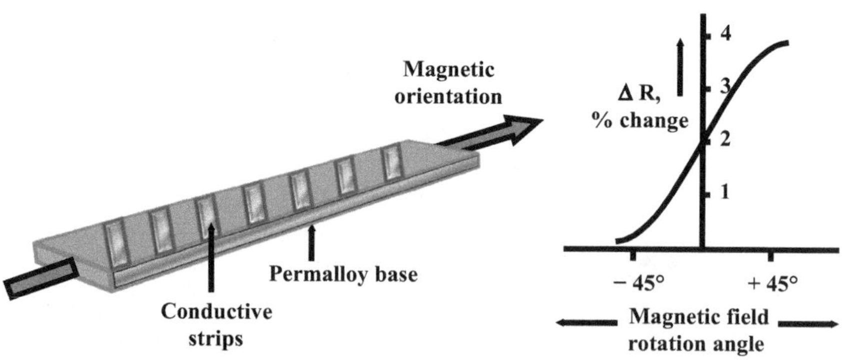

FIGURE 11.8 Placing conductive strips at 45° to improve linear response when sensing low magnitude magnetic fields.

If the MR sensor is exposed to a very strong magnetic field of an orientation different from the original, the MR material may be magnetized in the new direction, drastically changing the performance of the sensor. For this reason, it is important to prevent exposure of this type of MR element to excessively strong magnetic fields.

Simple AMR construction has been used in the fabrication of many magnetically based sensors. The use of magnetoresistance sensing elements accelerated, however, with the development of special materials with a much greater percentage of resistance change for a given level of magnetic field strength. The property of these more sensitive materials is called *giant magnetoresistance* (GMR). Alternately layering ferromagnetic alloys with other conductors into one structure increases the sensitivity (see Figure 11.9).

With such a GMR structure, electrical resistance decreases with an increasing parallel magnetic field strength. Typical GMR structures have a magnetoresistance ratio of 5 to 20% at room temperature, depending on the combination of layered materials and their construction. Further specialized materials designed for magnetoresistive properties at low temperatures can have a magnetoresistance ratio of 70% or more while still in a usable proportional range of resistance vs. magnetic field strength. They are made by using a combination of metal oxides and semiconductor materials. This substantially greater sensitivity is called *colossal magnetoresistance* (CMR). CMR materials have mostly been manganese-based oxides, such as $Ti_2Mn_2O_7$, having a *perovskite*-type of structure. Perovskite is the mineral calcium titanite, $CaTiO_3$, but other minerals having a similar structure may be described as having a perovskite structure, or may be called perovskites. New CMR formulations are still being discovered, including lanthanum manganite, $LaMnO_3$, and sodium chromium oxide, $NaCr_2O_4$. Some formulations have limited temperature range, some require extreme temperature and/or pressure conditions to form, and some operate only under high pressure or in low-temperature environments. Accordingly, CMR materials have not been widely implemented in practical applications as yet.

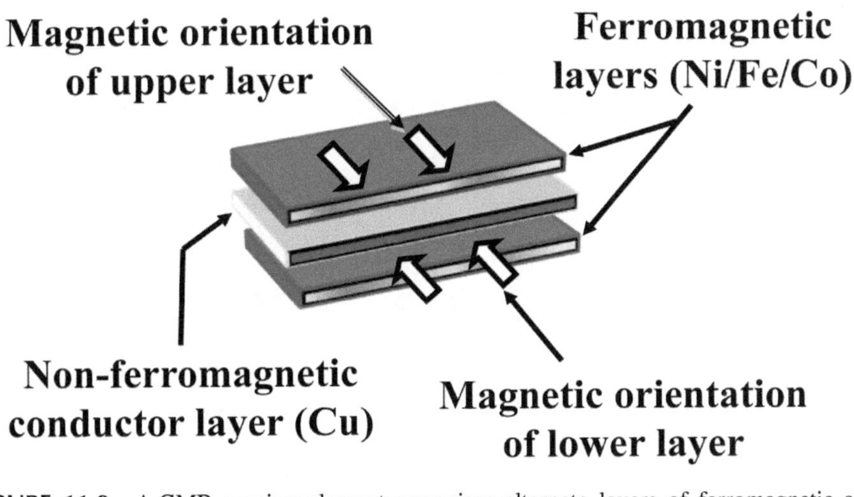

FIGURE 11.9 A GMR sensing element comprises alternate layers of ferromagnetic and non-ferromagnetic conductors sandwiched together.

11.3 HISTORY OF MAGNETORESISTIVE SENSORS

Lord Kelvin discovered magnetoresistance in iron and nickel in 1856. He observed the resistance change of a wire while passing a current through the wire and applying a magnetic field with a permanent magnet. He found that when an iron conductor was subjected to a magnetic field in parallel with the current flow, the resistance of the iron conductor increased. But the resistance decreased when the magnetic field was perpendicular to current flow. He also found that a nickel conductor had the same type of response, but to a greater degree [4, p 745; 5, p 109]. This is called *anisotropic magnetoresistance* (AMR), the property of a material in which electrical resistance depends upon the angle between electrical current flow and the direction of externally applied magnetization. An example is a difference in resistance between application of a parallel or perpendicular magnetic field. Since the change in resistance measured by Lord Kelvin in the iron or nickel conductors was relatively small, few commercial applications were developed using simple magnetoresistance.

Many more commercial applications started to form after discovery of the property of giant magnetoresistance, and many magnetoresistive transducers and sensors have been developed using GMR due to its greater sensitivity as compared with the earlier anisotropic ferromagnetic materials. GMR was first observed in multilayered structures of thin sheets of magnetic and non-magnetic conductors sandwiched together (as was shown in Figure 11.9).

Early GMR transducers comprised a section of this layered material with electrodes attached and packaging added (later versions incorporated integrated circuit techniques). A position magnet was allowed to pass in proximity to the GMR element and cause changes in the resistance of one or more legs of a *Wheatstone bridge* circuit, a Wheatstone bridge being a connection of four elements as shown in Figure 11.10.

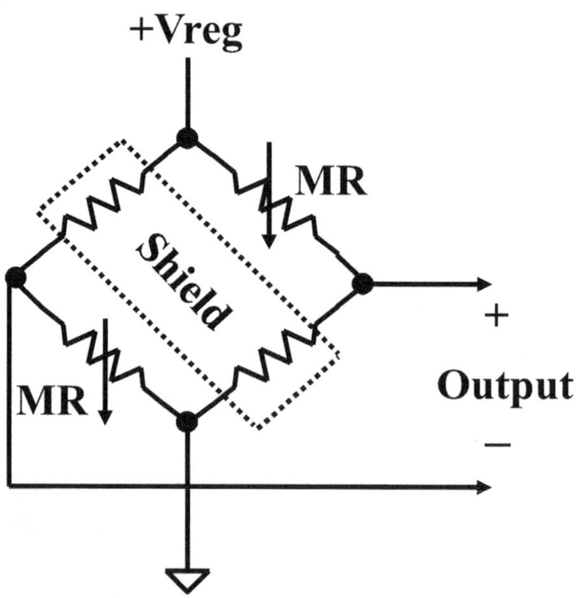

FIGURE 11.10 A set of magnetoresistors connected in a Wheatstone bridge configuration.

When the resistance of one leg, or two diagonal legs, increases (or decreases), the differential output voltage changes to a degree proportional to the resistance change. For example, if the upper right MR value decreases, the voltage + output line will become more positive. If the lower left MR value decreases, the voltage of the – output line will become more negative. Each of these would change the differential voltage between the + and – output lines to make the + output line more positive than it was with respect to the – output line. Two diagonal legs are allowed to respond to the sensed magnetic field. The other two legs are shielded from the magnetic field. This arrangement reduces temperature sensitivity because, although the individual magnetoresistors have a strong temperature dependence, they all respond similarly. The net result is that temperature changes have a much reduced affect on both zero and span.

As with Hall devices of the previous chapter, later GMR sensors incorporated the sensing element into integrated circuits. This has the advantage of allowing for some signal conditioning as well as providing a means for addressing multiple elements. One configuration for their use comprises a linear array of sensing elements, as depicted in Figure 11.11.

In the linear sensing configuration of Figure 11.11, multiple magnetoresistive devices are arranged in a line, to support a longer range of magnet motion. Additional electronics are needed to interface with the array, and to combine the several signals into one output that is representative of the magnet position.

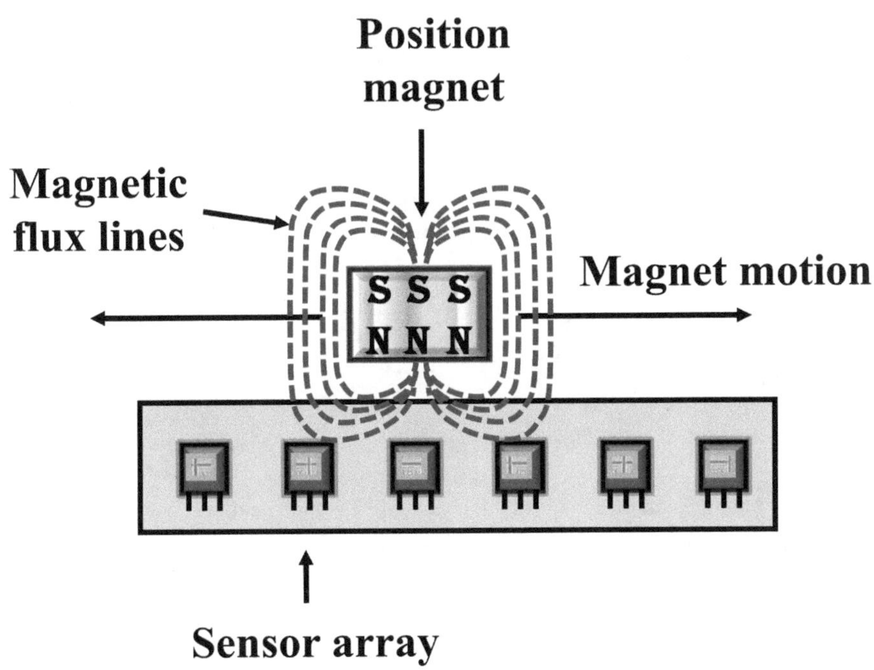

FIGURE 11.11 A linear array of magnetoresistors may be used to form a linear position-sensing element that is longer than is possible with a single magnetoresistor, and having a position magnet movable along the sensing axis.

The basic magnetoresistive sensing elements, packaged and characterized, are available from several sources worldwide. Some of these include AKM, Diodes Inc./Zetex, Fujitsu-Ten, Murata, NVE Corporation, Panasonic, Philips Semiconductor, Siemens, and Sony.

It was noted that Lord Kelvin discovered magnetoresistance in 1856. Giant magnetoresistance was discovered in 1988 independently by two researchers: Albert Fert of the University of Paris-Sud, and Peter Grünberg of Forschungszentrum Jülich in Germany. More recently, the property of extraordinary magnetoresistance (EMR) was discovered by Solin et al. in 2000 at the NEC Research Institute in Princeton, NJ. EMR can provide a change in resistance at room temperature that is several orders of magnitude greater than with AMR or GMR, when a strong magnetic field is applied. EMR elements have been fabricated as hybrids of semiconductor and metal while a magnetic field is applied during the fabrication. The completed EMR element, without external application of a magnetic field, exhibits a low resistance. Upon application of a strong magnetic field, the resistance increases by a much larger factor than with GMR or CMR. Early applications are likely to be in the area of data storage and retrieval. Position sensors based on EMR may be possible at some time after further development.

11.4 MAGNETORESISTIVE POSITION SENSOR DESIGN

A magnetoresistive position sensor comprises a sensing element, together with a housing, electronic circuit, and a permanent magnet. The permanent magnet is coupled with the movable workpiece member of which the position is to be determined. The housing contains the sensing element and electronic circuits, as well as providing the means for mechanical mounting of the sensor to the stationary part of the workpiece. See Figure 11.12 for a pictorial representation of a linear version.

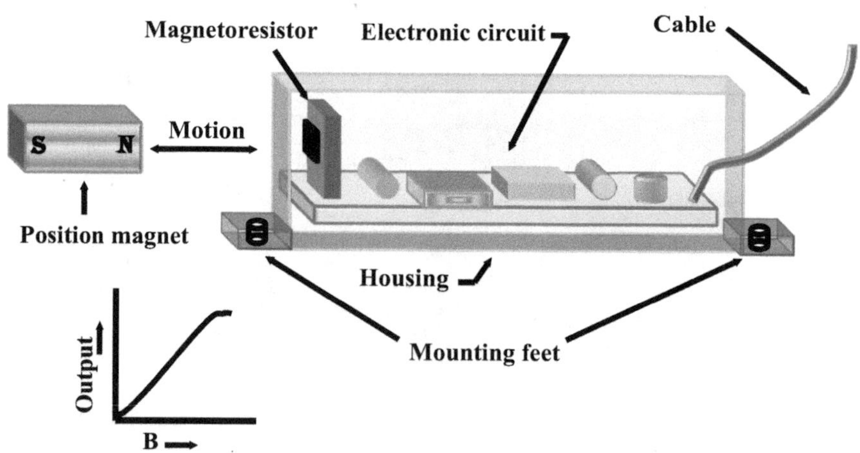

FIGURE 11.12 A magnetoresistive linear position sensor, with sensing element, housing, mounting feet, cable, and position magnet.

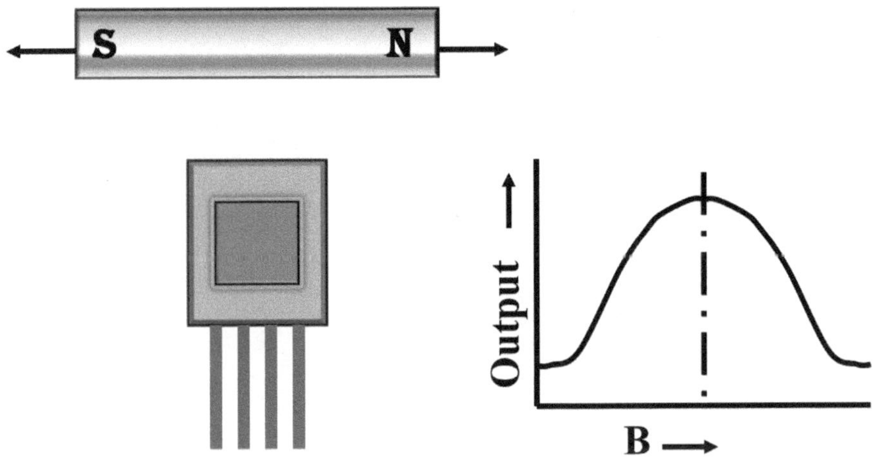

FIGURE 11.13 Configuration with a longer position magnet provides a longer measuring range.

The electronic circuit provides the regulated power supply, MR sensing element measuring circuit, and signal conditioning necessary to provide the desired output type. The housing is typically made from aluminum so that it does not affect the magnetic field experienced by the sensing element. The position magnet itself is sometimes enclosed within an aluminum housing to provide shielding as well as to provide means for mounting it to the movable part of the workpiece.

The position magnet can be configured to approach the sensing element along the axis of the poles, as in Figure 11.12, or can be a longer shape which operates with the sensing element positioned between the poles as in Figure 11.13.

Allowing the use of both of the magnet poles increases the total measuring range of the sensing element.

An angular sensor may be implemented by using a magnetoresistive integrated circuit (ic) having two Wheatstone bridge circuits inside. A magnet is positioned above the ic so that it can rotate with the measurand, as shown in Figure 11.14.

The two Wheatstone bridge circuits are shown in Figure 11.14(a). In each bridge, two opposite resistive elements would be magnetoresistive, and the remaining two only normal resistors (or magnetoresistors that are shielded from the magnetic field). The normal resistors have values to provide the desired voltage offset when the magnetic field strength is zero. Normally, the two V_{CC} terminals would be tied together and connected with a positive regulated voltage, such as +3.3 V DC. The two common connections (Com) would be tied together to circuit common. This allows current to flow through both bridges. Differential amplifiers would measure the output from each bridge at the positive and negative voltage outputs (+Vo and –Vo). An A/D converter would present the signals to a microcontroller, and the μC would be used to interpret the signals and provide a desired output. The MR integrated circuit is shown above the ic in Figure 11.14(b). The ring magnet is polarized with one half as the north pole and the other half as the south pole. As the magnet rotates, the two

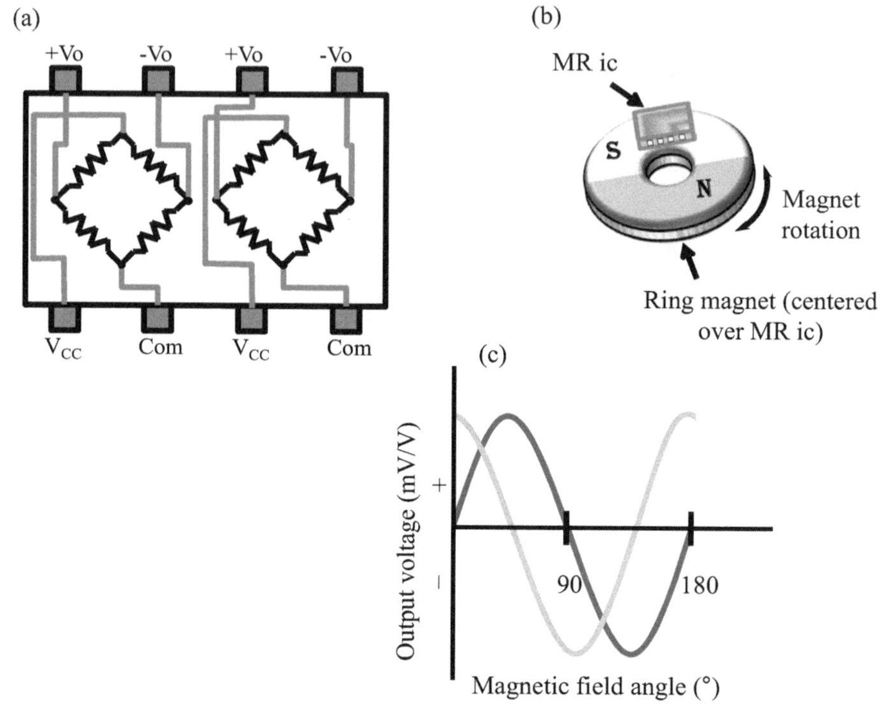

FIGURE 11.14 MR angular position transducer.

MR ics provide output voltages as shown in Figure 11.14(c). The magnet is shown at (b) in approximately the zero position, with a north and south pole equally over one end of the ic, and a south and north pole over the other end of the ic. If the magnet is rotated by 180°, the output would then be the same again. So, this configuration can provide a range of measurement up to 180°. Other configurations can provide a measurement range of up to 360°, including one that uses three bridges instead of two.

11.5 THE MAGNETORESISTIVE ELEMENT

A current-carrying conductor is normally surrounded by a magnetic field (see Section 7.2). The conductor may be coiled around a permeable material, forming an electromagnet having a greater magnetic flux density than a similar coil without a permeable core. Such an electromagnet, or a permanent magnet, can be used as the position magnet with a magnetoresistor in order to measure linear or angular/rotary position. Use of an electromagnet allows the selection of which electromagnet is to be sensed (by energizing the selected electromagnet).

As presented earlier, a magnetoresistor can be designed so that its change in resistance is proportional to the rotation of the surrounding magnetic field by proximity of a position magnet. In a linear position sensor, the field is rotated up to an angle of 90° because the position magnet is arranged to approach the element at a right angle. The change in resistance is approximately proportional to the square of the magnetic

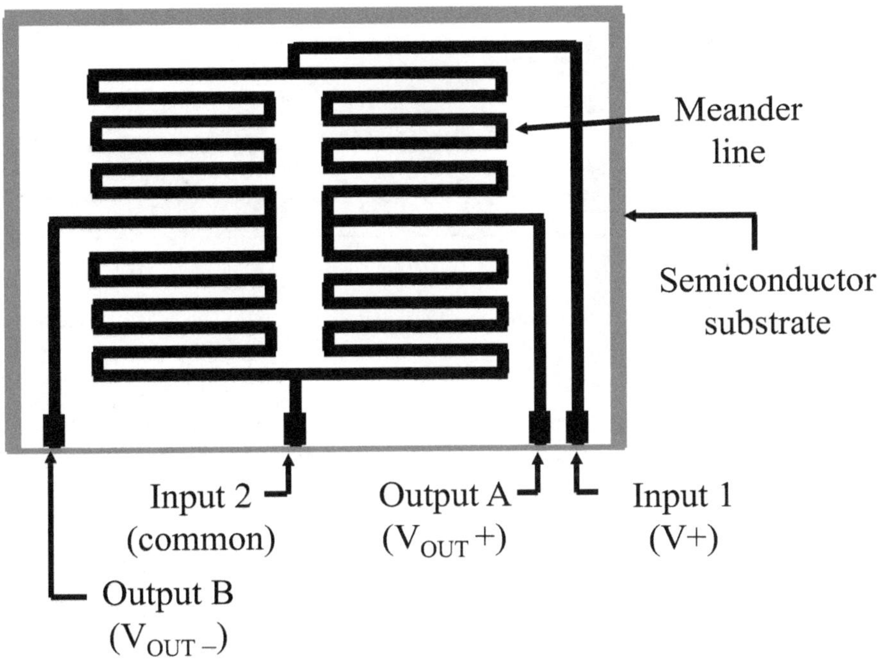

FIGURE 11.15 Wheatstone bridge circuit with meander pattern of four MR legs to form one sensing element.

flux density, until the ferromagnetic material in the sensor approaches magnetic saturation. With an angular/rotary position sensor, the magnetic field may be rotated at various angles, depending on the relative positions of the north and south poles of the position magnet.

The sensing part of a commercially available MR or GMR sensing element or integrated circuit is usually formed by depositing the MR materials or layers of material, onto a semiconductor substrate. This is often done in a *meander* pattern (or meander line) in order to increase the length of the element and thereby increase its sensitivity. Four elements are normally placed on one substrate, and arranged in a Wheatstone bridge connection. See Figure 11.15.

Two of the diagonal elements (or legs) are used for sensing the magnetic field. The remaining two legs are shielded from the effects of a magnetic field as depicted in the schematic of Figure 11.10. This configuration increases the sensitivity to a magnetic field, and also decreases the temperature sensitivity (as compared to using a single MR element).

11.6 LINEAR ARRAYS

Signals from a linear array of individual magnetoresistors (or magnetoresistive bridges) can be digitized, and multiplexed into a microcontroller in order to expand the measuring range to the length of the sensor array. Each of the MR sensing

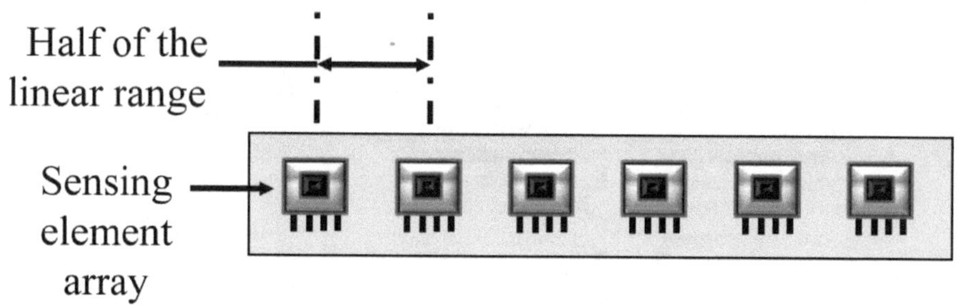

FIGURE 11.16 The MR sensing elements of a linear array should be spaced at one half of their linear range.

elements is usually a Wheatstone bridge connection of four elements, so that temperature sensitivity is minimized and magnetic field sensitivity is maximized. The spacing between the MR elements should be approximately equal to half of their linear range (see Figure 11.16).

The microcontroller (μC) strobes the multiplexer (MUX) to sample the various sensing elements in the array, and retrieve their data. The μC then interprets the result to form a desired sensor output. To accomplish this, the μC analyzes the sensing element signals to derive information that indicates which two MR sensing elements are in the closest proximity to the position magnet, as well as to measure the magnetic field strength indicated by those particular two elements in the array. Linear interpolation between the known locations of these two closest MR sensing elements provides the exact location of the position magnet, and is used to form a desired output signal that is indicative of the location of the position magnet.

11.7 ELECTRONICS

In a single-point position sensor (with a correspondingly limited measuring range) the Wheatstone bridge connection of four MR elements is wired to a regulated voltage source and a differential amplifier as shown in Figure 11.17. This is a simple analog circuit configuration, without a μC.

The amplifier output is fed to a filter to remove noise that is not within the desired frequency response characteristic of the sensor (for example, a 250 Hz low-pass filter). After the filter, the position signal is calibrated (offset for desired zero, and scaled for desired span) and buffered in an output circuit that provides the desired type of output voltage, current, or digital signal to the end user.

In a longer position sensor that uses an array of MR sensors, it is required that a microcontroller be used to effectively derive the required mathematical interpretation of the signals from each of the sensors (see Figure 11.18).

Signals from the sensor array are brought into a multiplexing (MUX) circuit. Switching of the MUX is directed by the microcontroller. The selected signal is then amplified by the differential amplifier, filtered, and converted to a digital representation by an analog-to-digital converter (A/D). With the digitized signals from the A/D converter, the microcontroller can then interpret the correct reading and output

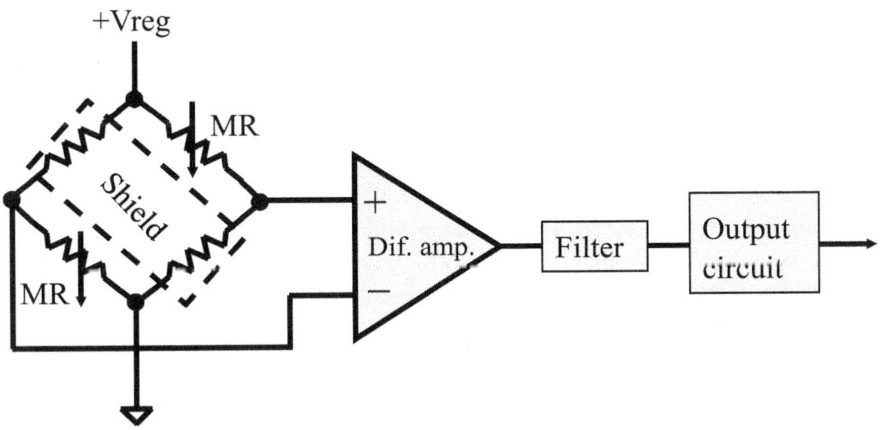

FIGURE 11.17 A single MR element position sensor with analog output.

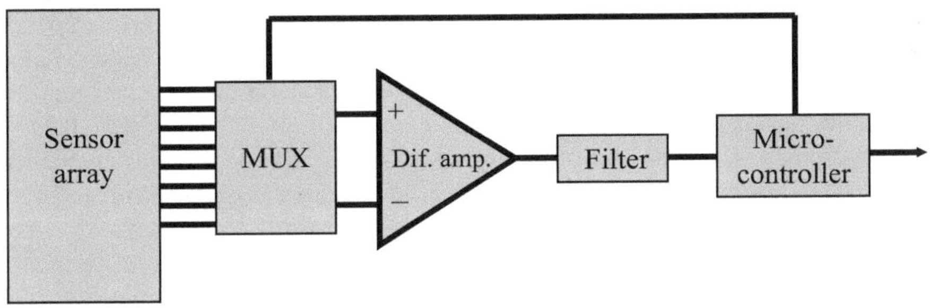

FIGURE 11.18 A linear array with microcontroller and digital or analog output.

a digital position directly, or provide an analog output by means of a digital-to-analog converter (D/A).

11.8 ADVANTAGES OF MAGNETORESISTIVE SENSORS

Magnetoresistors have an advantage over inductive sensors in that they can operate in a linear position sensor without a modulator (oscillator). Compared to a Hall device, a magnetoresistor can generally operate at a higher frequency (or rate of change of position) before attenuation occurs. Magnetoresistors can operate over a wide temperature range (for example, –55 to 200°C). A GMR magnetoresistor generally has greater sensitivity to a magnetic field than does a Hall device, so a less powerful magnet can be used, or a stronger magnet can be used at a greater distance.

The application of a magnetoresistive sensing element is generally similar to that of a Hall device. The primary difference being that a Hall device produces a change in voltage due to a change in magnetic field strength, whereas a magnetoresistor produces a change in resistance under similar conditions. Other differences include the sensitivity of a magnetoresistor to the magnitude of the magnetic field strength, versus the Hall device being also sensitive to the magnetic field polarity. But special construction, such as the barber pole configuration, can enable a magnetoresistor to sense magnetic polarity. A magnetoresistor typically requires a lower amount of current to power it, since a Hall device needs a driving current to produce the Hall voltage. Both technologies are compatible with integrated circuit fabrication processes, and so the sensing elements may be fabricated together with circuitry on a common substrate. Hall effect devices can withstand much stronger magnetic fields without saturation. An excessively strong magnetic field can damage some MR configurations, such as the barber pole. The Hall effect senses a magnetic field that is perpendicular to the element surface, while most commercially available MR elements sense a magnetic field that is applied in the long direction of the sensing element.

11.9 TYPICAL PERFORMANCE SPECIFICATIONS AND APPLICATIONS

When using a magnetoresistive position sensor to monitor the motion of an industrial machine, for example, the position magnet is usually a permanent magnet that is fitted to the moving part of the machine. In a linear position sensor, the distance to the position magnet is read by a magnetoresistive sensing element mounted along and parallel to the motion axis of the machine. The position magnet does not need to touch the sensing element, because the coupling is through the magnetic field. As the position magnet moves along the sensing axis in parallel to the sensing element, the distance between the position magnet and the sensor electronics (head) end of the sensor is read electronically.

Though measuring ranges of longer than 25 mm can be handled by using an array of MR sensors and a microcontroller, even longer ranges are possible by using a coded assembly of position magnets. Overall ranges of hundreds of meters can be outfitted with position magnet assemblies, each assembly comprising a set of position magnets having its own location code based on the spacing among the individual magnets in the assembly, as shown in Figure 11.19.

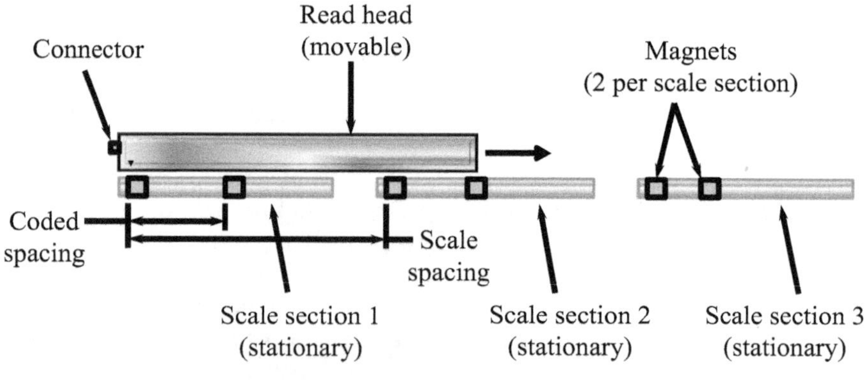

FIGURE 11.19 A long-range linear position sensor with coded magnets.

In this case, the movable *read head* comprises an array of MR sensors, and is long enough to always read at least two of the coded magnet assemblies. Instead of reading the position of only one magnet, the software allows the read head to simultaneously measure the positions of up to four magnets. The coded magnet assemblies are installed within scale sections, as shown in the figure, and each scale section is mounted in a stationary position. Several scale sections are arranged along a straight line, each scale section being of the same length, and each spaced to the adjacent scale sections by the same scale spacing. The spacing between the two magnets of a coded magnet assembly indicates in which scale section it is located. It can be seen in the figure that the spacing between the two magnets is different for each scale section. When four magnets (from two adjacent scale sections) are simultaneously within the measuring range of the read head, software in the read head selects which set of magnets to use in the measurement. As the read head moves out of the range of the magnets in one scale section, it then switches to start using the next set of magnets for the measurement. The scale section number being read represents a coarse reading, and the individual reading from the first magnet in the magnet assembly of that scale section represents the fine reading. The combination of these two readings provides a measurement of the absolute position of the read head with respect to the first magnet in the first scale section. The illustration of Figure 11.19 is an example showing one method of using coded magnets. Long-range MR linear position sensors may also be implemented by using other magnet coding methods.

The specification of this long-range type of sensor depends on the measuring range, but here is a representative example:

Specification of long-range MR linear position sensor array (as in* Figure 11.19*)

Range	10 m (depending on the number of scale sections)
Resolution	100 µm
Nonlinearity	±1000 µm
Operating temperature	–20 to 60°C
Sampling interval	0.8 ms
Supply voltage	10 to 32 V DC
Supply current	0.12 A
Output	24-bit SSI

Applications for the long-range MR linear sensor include elevator shafts, where the elevator car position is monitored relative to the floor designation; inventory control systems in which the location of the inventory retrieval robot is measured with respect to the inventory storage shelf; and directing the movement of mules (guided vehicles) where materials are delivered to work stations on a production floor.

Applications for short range, single sensor element sensors, and up to 1 meter length arrays include LVDT replacements, valve positioners, injection molding and other machine control, and pedal position for onboard automotive use.

Here are some representative specifications for MR position sensors of other configurations:

Specification of single MR bridge element linear position sensor (as in* Figure 11.12*)

Range	15 mm
Resolution	0.05 mm
Nonlinearity	0.4% FRO (full-range output)
Operating temperature	−40 to 125°C
Supply voltage	6 to 24 V DC
Supply current	32 mA
Output	0 to 5 V DC

Specification of MR element array linear position sensor (as in* Figure 11.16*)

Range	225 mm
Resolution	0.14 mm
Nonlinearity	0.4% FRO (full-range output)
Operating temperature	−40 to 125°C
Update time	400 μs
Supply voltage	6 to 24 V DC
Supply current	34 mA
Output	0 to 5 V DC

Specification of angular/rotary MR position sensor (as in* Figure 11.14*)

Range	0 to 100°
Resolution	0.06°
Nonlinearity	0.4% FRO (full-range output)
Operating temperature	−40 to 85°C
Supply voltage	5 V DC
Supply current	30 mA
Output	0.5 to 4.5 VDC (ratiometric)

11.10 MANUFACTURERS

Some manufacturers of magnetoresistive linear and angular/rotary position sensors include the following:

Anasem Semiconductor	www.anasemi.com
Asahi Kasei Microdevices (AKM)	www.akm.com
Diodes Inc./Zetex	www.diodes.com
Farnell	www.farnell.com
Fujitsu-Ten	www.fujitsu-ten.com
Honeywell	www.sensing.honeywell.com
Murata	www.murata.com/en-us/products/sensor/amr

NVE Corporation	www.nve.com
Panasonic	www.industrial.panasonic.com
NXP (Philips Semiconductor)	www.nxp.com
Revolution Sensor Company (design)	www.rev.bz
Sensitec	www.sensitec.com
Sick-Stegmann (Pomux)	www.sick.com

11.11 QUESTIONS FOR REVIEW

1. **Upon application of a magnetic field perpendicular to current flow, the resistance of an iron wire:**

    a. Increases
    b. Decreases
    c. Oscillates
    d. Compensates
    e. Stabilizes

2. **A series connection of four resistors with power applied at two opposite nodes and voltage measured across the two remaining nodes is called a:**

    a. Cornhole
    b. Wheatstone bridge
    c. Quadrilateral
    d. Quadsistor
    e. Thévenin equivalent

3. **Magnetoresistance that depends on the direction of the magnetic field is called:**

    a. AMR
    b. CMR
    c. GMR
    d. EMR
    e. TMR

4. **MR formed of alternating ferromagnetic and non-magnetic conductive layers:**

    a. GMR
    b. Anti-magnetic
    c. Tunneling MR
    d. Positor
    e. Thermistor

5. Permalloy is a ferromagnetic alloy of these metals:

a. Iron, copper
b. Cobalt, aluminum
c. Nickel, iron
d. Iron, cobalt
e. Nickel, manganese

6. In an MR linear array, the sensing elements are typically spaced by:

a. Approximation
b. Interpolation
c. Cobalt
d. Linearization
e. One half their linear range

7. Digitizing an analog voltage or current is accomplished by using a:

a. VCO
b. D/A
c. VFO
d. A/D
e. NRZ

8. The + and – voltage outputs of an MR bridge circuit are usually connected to a:

a. Oscillator
b. Multiplier
c. Differential amplifier
d. Integrator
e. Kelvin-Varley divider

REFERENCES

[1] D. Jiles, *Introduction to Magnetism and Magnetic Materials*. London: Chapman & Hall, 1991.
[2] J. Fraden, *Handbook of Modern Sensors*. New York: Springer-Verlag, 2010.
[3] H. Burke, *Handbook of Magnetic Phenomena*. New York: Van Nostrand Reinhold, 1986.
[4] R. M. Bozorth, *Ferromagnetism*. New York: D. Van Nostrand, 1951.
[5] R. Pallas-Areny and J. G. Webster, *Sensors and Signal Conditioning* (2nd ed). New York: Wiley, 2001.

12 Magnetostrictive Sensing

12.1 MAGNETOSTRICTIVE SENSORS

Magnetostrictive position sensors are non-contact, absolute reading, and have essentially infinitely fine resolution relative to the requirements of most applications (having a waveguide system with inherent resolution of less than 1 μm). Practical linear sensors can be designed with a nonlinearity of less than 0.01% and full-scale ranges (FSR) from less than 10 mm to over 20 meters. But many industrial type magnetostrictive position sensors are designed with nonlinearity of about 0.02 to 0.05% nonlinearity, in order to provide a more economical product. Curved and rotary types of angular sensors are also possible. Popular applications include industrial machinery, such as injection molding machines and hydraulic cylinders, as well as automotive, military, agricultural, woodworking, and medical devices.

The coupling between the moving and stationary parts of the sensor is achieved by means of a magnetic field. Therefore, any number of position changes can be made without incurring wear to the sensor parts. When using a magnetostrictive position sensor to monitor the motion of a machine tool, for example, a permanent magnet is fitted to the tool holder. The permanent magnet is called the position magnet and is read by a sensing element mounted along and parallel to the motion axis of the tool. The position magnet does not touch the sensing element. As the position magnet moves along parallel to the sensing element, the distance between the position magnet and a pickup device near the sensor electronics (head) end of the sensor is read electronically. A magnetostrictive linear position sensor designed by the author is shown in Figure 12.1.

The sensor in the figure has a 316 stainless steel housing that contains the pickup and the electronics (aka the electronics head), and has a screw thread (mounting threads) for mounting the sensor into a hydraulic cylinder. The sensing rod (probe) extends into a hole bored into the piston head and cylinder rod. The position magnet is mounted into the piston, so that stroking the cylinder piston will move the position magnet along the length of the sensing rod. The position magnet must be mounted into non-ferromagnetic material, such as 316 stainless steel, aluminum, or plastic.

12.2 MAGNETOSTRICTION

Magnetostriction is a property of ferromagnetic materials, including nickel, iron, cobalt, and some of their alloys. When placed in a magnetic field, these materials change size and/or shape. James Prescott Joule discovered the magnetostriction of iron in 1842. He established the fact that, when magnetized, iron increases in length in the direction of magnetization, and contracts at right angles to this direction [1, p 630; 2, p 4].

DOI: 10.1201/9781003368991-12

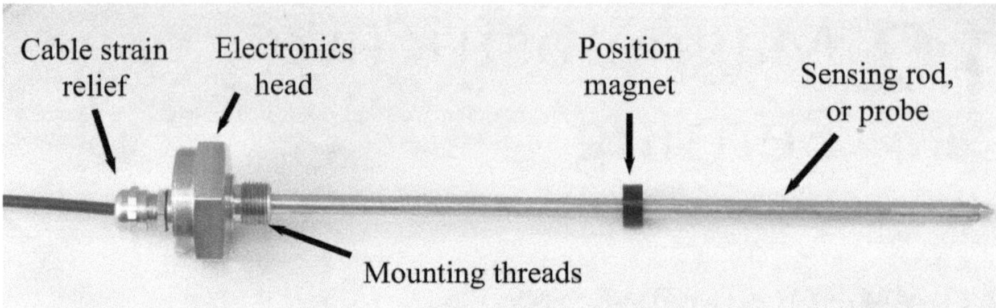

FIGURE 12.1 Magnetostrictive linear position sensor with position magnet.

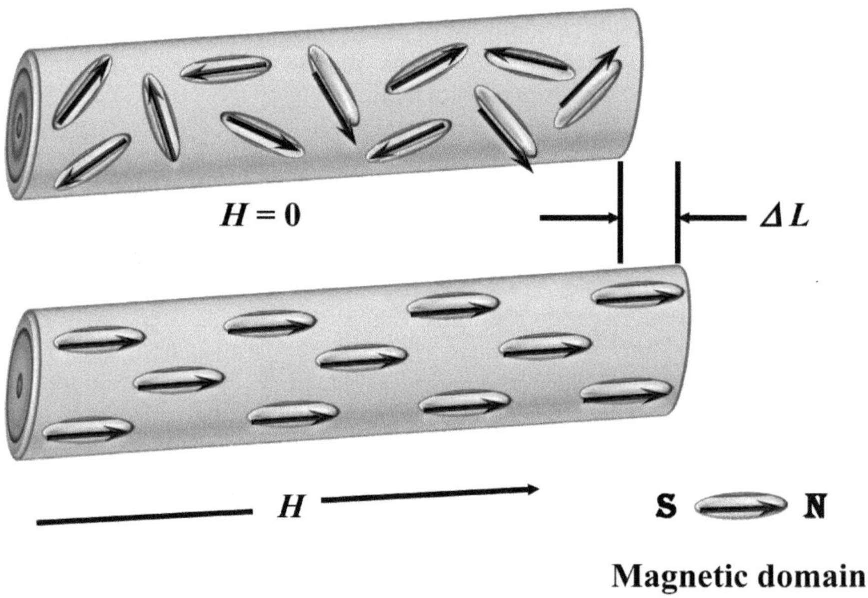

FIGURE 12.2 Positive magnetostriction: magnetic domains align with magnetic field, H, and cause stress, inducing an increase in mechanical dimension, ΔL.

The physical response of a ferromagnetic material is due to the presence of magnetic moments. Generally speaking, most materials have approximately as many electrons spinning in one direction as the other, resulting in a magnetically insensitive structure [3, p 123]. However, in an element with unfilled subvalence shells, more electrons spin in one direction than in the other. Therefore, these elements have a net magnetic moment and can be understood by considering the material as a collection of tiny permanent magnets, called domains, as depicted in Figure 12.2.

Each domain comprises many atoms. When a material is not magnetized, the domains are arranged randomly. However, when the material is magnetized, the domains are oriented with their axes approximately parallel to each other. Interaction of an external magnetic field with the domains causes the magnetostrictive effect.

Controlling the ordering of the domains through alloy selection, thermal annealing, cold working, and magnetic field strength can optimize this effect [4].

The term *magnetostriction* generally refers to any change in dimensions due to magnetization. Magnetostrictive position sensors however, utilize *Joule magnetostriction*, which is the change in length of a ferromagnetic material due to magnetization. When a material has positive magnetostriction, it enlarges when placed in a magnetic field. Conversely, with negative magnetostriction, the material shrinks. The amount of magnetostriction in base elements and simple alloys is on the order of a few parts per million. Some exotic materials, when magnetized, change dimension by a factor of hundreds of times greater than this when under a very strong magnetizing force. One such material developed in the 1980s is Terfenol D, but it has a high cost, and limited practical use in sensors because it is not ductile and therefore cannot be drawn into a wire or ribbon shape.

Just as magnetization stresses the material, causing dimensional changes, the reverse is also true: applying stress to a magnetostrictive material changes its magnetic properties (for example, its magnetic permeability). This stress-induced effect on magnetic properties is called the *Villari effect*, after it was discovered in 1865 by physicist Emilio Villari of Naples, Italy. The Villari effect may sometimes be called the inverse Joule effect, or inverse magnetostriction.

An important characteristic of a wire that is made of a magnetostrictive material is the Wiedemann effect. It is the torsional force produced in a ferromagnetic wire at the location of an axial magnetic field when a current flows in the wire (see Figure 12.3).

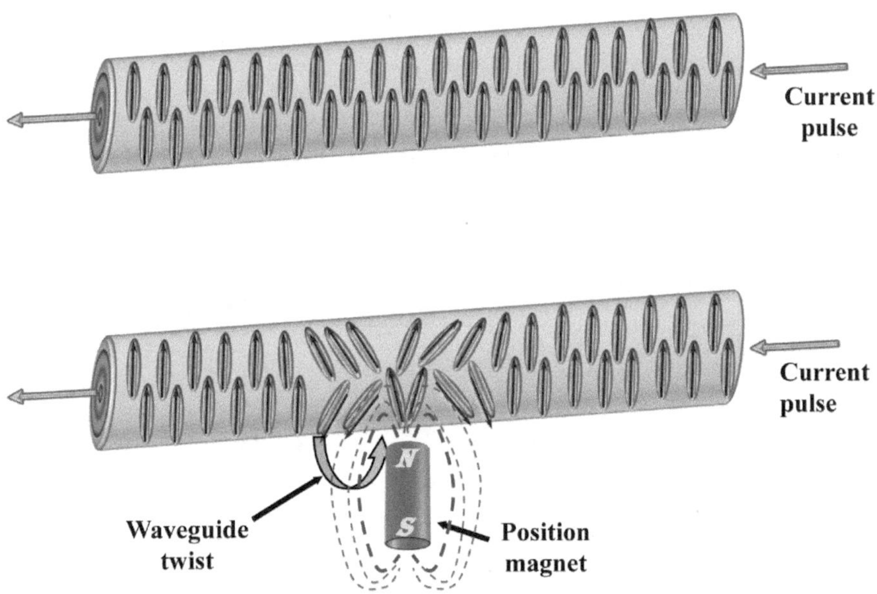

FIGURE 12.3 The Wiedemann effect: a torsional force occurs at the location of an axialmagnetic field (position magnet) when current is applied to a ferromagnetic wire, forming a torsional wave.

The current pulse through the waveguide tends to align the magnetic domains of the waveguide material along its lines of magnetic flux (the domains will not be as uniform or uniformly aligned as shown here, but are shown this way for ease of illustration). The torsional force results from the vector addition of the magnetic field surrounding the waveguide, due to the current pulse, and the magnetic field supplied by the position magnet. In a magnetostrictive position sensor, the current is applied as a short duration pulse, of about 0.5 to 5.0 μs. The minimum current density is along the center of the wire and the maximum is at the wire circumference. This is due to the *skin effect*, and means that the magnetic flux density, due to the current, is also greatest at the wire surface. This aids in developing the waveguide twist.

Since the waveguide twist is applied as a short duration pulse, the resulting mechanical torsional force is sharply applied, and causes a torsional wave to form at the location of the position magnet, and to propagate down the length of the waveguide toward each end of the waveguide. It is similar to a soundwave being conducted down the waveguide, but travels faster than sound (at a little less than 3,000 m/s).

12.3 HISTORY OF MAGNETOSTRICTION

An important precursor of the magnetostrictive position sensor was the magnetostrictive delay line as used in an early memory device of the 1960s. A wire made from a nickel-iron alloy acted as a sonic waveguide, forming the memory core for storing serial data. A current-to-magnetic field transducer (a coil) at the memory input enabled excitation of sonic pulses onto the waveguide. A magnetic field-to-voltage transducer (another coil and a small magnet) at the output end of the wire detected the presence or absence of the sonic pulses (see Figure 12.4).

To store data, a series of pulses (data in) was impressed onto the waveguide at the input (transmitter) end through an input transducer, such as by applying current pulses through a coil of wire around the waveguide. This produced sonic pulses as torsional waves traveling in the waveguide. The presence of a sonic pulse on the waveguide represented a logical 1, and the absence of a pulse at the expected time represented a

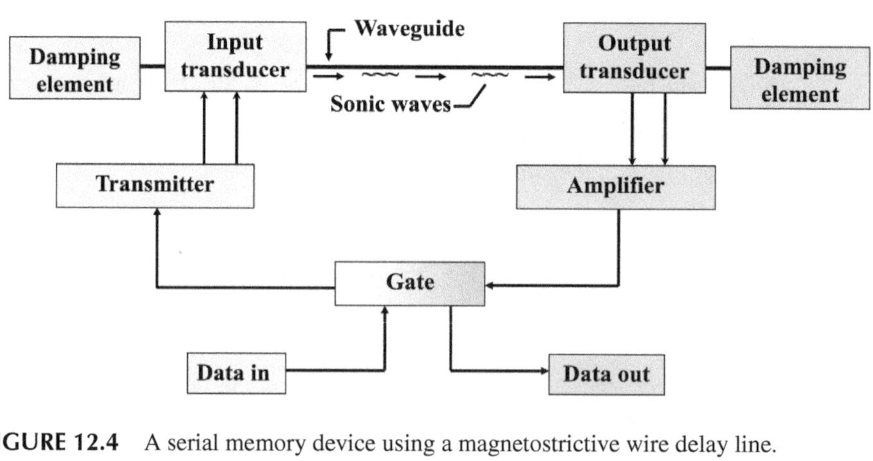

FIGURE 12.4 A serial memory device using a magnetostrictive wire delay line.

logical 0. The sonic waves traveled along the waveguide at the "speed of sound" in that material (approximately 2,800 m/s). Additional data could be fed into the input end until just before the first pulse reached the output (amplifier) end. When the first pulse was just about to reach the output end, the memory was full and no further data could be entered. As a sonic pulse reached the output end, it would be detected by a pickup coil and magnet assembly, amplified (to account for attenuation in the waveguide), and fed again into the input end. The data could be stored in the waveguide continuously. The data could be read serially after a synchronizing space that was imposed between the last data pulse and repeating of the first pulse. Building on this delay line technology, Jacob Tellerman invented the magnetostrictive position sensor by adding a position magnet and introducing a current pulse into the waveguide. The current pulse is called an interrogation pulse, and provides one reading of the magnet position. The first products based on this technology were called Temposonics®. Tempo, because the actual measurement is the time between the current pulse application and receipt of the position magnet pulse. Sonics, because a sonic wave enables the measurement.

Until the early 1990s, the several manufacturers of magnetostrictive position sensors designed their products with three- or four-wire connections for power and analog signal. It was often stated that it was not possible to design a two-wire current loop transmitter due to the need for a high-current interrogation pulse. Nonetheless, in 1990, the author invented the first 4 to 20 mA loop-powered magnetostrictive position sensor, US patent 5,070,485, issued in 1991. The low-power design required that all electronics, including the average of the interrogation pulse current, add up to a total current draw of less than 4 mA. This was accomplished by designing very low-power amplifier, filter, and comparator circuits, and driving the waveguide through a transformer. A few years after the patent issued, some other manufacturers followed with their own designs of loop-powered magnetostrictive position sensors.

Later designs included microcontrollers, and this made it practical to implement adjustable interrogation capability, linearization, easy calibration, self-checking, and a wide selection of digital communication protocols.

12.4 MAGNETOSTRICTIVE POSITION SENSOR DESIGN

To incorporate these ideas into a linear or angular position sensor, the current is introduced as a short duration pulse (called the interrogation pulse). The interrogation pulse duration is usually around 1.0 μs, but can range from about 0.5 to 5.0 μs, depending on the characteristics of the waveguide and pickup device used. This current pulse, together with the field from the position magnet, produces a mechanical torsional pulse at the location of the position magnet. The torsional pulse travels as a sonic wave in the waveguide, at about 2800 m/s. The waveguide is typically a wire made from a nickel-iron (Ni/Fe) alloy and is usually about 0.25 to 1 mm in diameter, with 0.375 mm (0.015 inch) diameter being commonly selected. With a waveguide diameter of less than 0.375 mm, it becomes more delicate, and may be difficult to handle without damage. With greater diameter, larger current levels are needed to magnetize the waveguide sufficiently to overcome hysteresis.

Upon application of the interrogation pulse, the sonic pulse is generated and travels in both directions on the waveguide, starting from the position magnet. The pulse traveling toward a pickup device is detected by the pickup. The pulse traveling in the other direction is eliminated by a damping device. The damping device prevents interference due to sonic waves that may reflect from the waveguide tip (the waveguide tip is the end of the waveguide that is opposite from the pickup location on the waveguide). In operation, a timer is started when the interrogation pulse is applied to the waveguide. An amount of time elapses as the sonic wave travels from the location of the position magnet to the pickup. The timer is stopped when the pickup detects the sonic wave. The elapsed time as measured by the timer indicates the distance between the position magnet and the pickup.

Sometimes the magnetostrictive position-sensing technology is referred to as performing a time-of-flight measurement, but this author refrains from doing so. Referring to magnetostrictive technology as time-of-flight causes many people to confuse the way it works with radar, sonar, ultrasonic distance measurement, and other two-way time-of-flight measurements. As with an ultrasonic distance measurement, a typical time-of-flight measurement involves the sending of a pulse or wave from a transceiver, the pulse or wave travels to a target over a first period of time, is reflected from the target, travels back to the transceiver over a second period of time similar to the first, and the reflected pulse or wave is then detected. Such a time-of-flight system provides a round-trip measurement of time. Then, the speed of the wave in a particular medium (such as air) multiplied by one half the round-trip measured time is then indicative of the distance to the target. But to the contrary, in a magnetostrictive position sensor, the waveguide is energized at once all along its length. The strain pulse originates at the location of the position magnet, and then travels in the waveguide until it is detected. So, there is no reflection, and the pulse or wave only travels one way: from the position magnet to the pickup (that is: there is no round-trip time-of-flight to be measured).

A practical magnetostrictive position sensor comprises five basic components: waveguide assembly, position magnet, pickup, damping element, and the electronic circuit. Design and optimization of each of these components for cost and performance depends partly on the type of application for which the sensor is intended. Industrial applications, like position feedback for control of hydraulic cylinders, require high resolution and fast update time. High-volume applications, like struts for automatic body control in cars, require high reliability and low cost. The following paragraphs provide a closer look at some of these design considerations.

12.5 WAVEGUIDE

A waveguide assembly usually comprises a waveguide, return conductor, waveguide suspension component(s), damping element, a tubular housing, and may include the pickup device.

The waveguide material and manufacturing process are developed to control many performance characteristics. These include the coefficient of magnetostriction, sonic wave attenuation, temperature coefficient of the sonic velocity (also called t/c), variation of sonic velocity from unit to unit, variation of sonic velocity along the length of waveguide within one unit (this is the main source of nonlinearity), and hysteresis. Other less important properties include straightness, solderability, bendability, weldability, electrical resistance, and permeability. Short- or long-term drifts are not a property of the waveguide as long as it is operated at temperatures substantially lower than the *Curie point*. The technical properties of a magnetic material depend on both the intrinsic properties of the material alloy selected, and on its microstructure [5, p 377]. Microstructure is affected by coldworking and by thermal treatments. In some alloys, microstructure may be further affected by application of an electric current or magnetic field during a thermal treatment.

The *coefficient of magnetostriction* (c) is the factor that relates the amount of strain (ε) produced in a magnetostrictive material to a given amount of magnetic flux density (B).

$$\varepsilon = Bc \tag{12.1}$$

One might think that it is important to have a high coefficient of magnetostriction in the waveguide material of a magnetostrictive sensor. But to the contrary, it is more important to optimize parameters like low hysteresis, low attenuation, and low sensitivity to temperature changes, while still retaining a sufficiently high coefficient of magnetostriction for accurate transduction. For that reason, the coefficient of magnetostriction of the waveguide is normally only that of a simple nickel-iron alloy. A popular alloy to use for the waveguide is Ni-Span C, developed by the International Nickel Company. It was developed for use in springs to maintain a constant modulus of elasticity over a wide temperature range. Its magnetostrictive effect was also used in sonar transducers to produce the sonic wave while optimizing the transducer performance over a range of temperature.

The original magnetostrictive linear position sensors developed by Jacob Tellerman utilized a tubular waveguide made of Ni-Span C. A magnet wire ran inside the tubing, and then returned along the outside of the tubing. In order to reduce any dampening effect from the internal magnet wire touching the waveguide, the magnet wire had very small beads of silicone rubber applied along its length at a fairly uniform spacing. This *beaded wire* was carefully threaded into the waveguide tube by hand, and was one of the labor-intensive parts of the manual assembly of the early magnetostrictive sensors. In the early 1990s, the author, as director of technology of the major worldwide magnetostrictive position sensor manufacturer, proposed the use of a solid wire waveguide instead of the tubular construction. An engineering manager disagreed and hired a contractor (a mechanical engineer) to analyze the expected performance differences between using the tubular waveguide versus a solid wire. The report appeared to prove that a solid wire waveguide could not approach the performance of the tubular waveguide. But on reading the first paragraph of the report, this author noticed an error in the primary mathematical relation on which the whole analysis was based, regarding the skin effect. The equation became meaningless with smaller wire diameters. After this error was corrected, the report then became proof that a solid wire waveguide could have similar performance to a tubular waveguide if the wire diameter would be in a certain range. After hiring a materials scientist, a solid wire waveguide was developed that provided the highest performance magnetostrictive sensors available at that time. Since then, most manufacturers of magnetostrictive position sensors utilize a solid wire waveguide material.

In a magnetostrictive waveguide material, *attenuation* is the gradual reduction in amplitude of the sonic wave as it travels over a longer distance in the waveguide. The amount varies with waveguide alloy and with the processing of the waveguide material, and may also vary somewhat for large changes in diameter of the waveguide. The amplitude of the sonic wave is usually measured as a voltage output from the pickup, and attenuation of the signal amplitude versus distance between the position magnet and pickup varies approximately according to the equation:

$$a = Axc^{N} \tag{12.2}$$

where a is the signal amplitude at a given distance between position magnet and pickup, A is the signal amplitude with a minimum distance between the position magnet and pickup (usually, the maximum amplitude), c is a constant derived for a particular system, and N is an exponent derived to represent the given distance between the position magnet and pickup. The value of A is readily determined by measuring the pickup voltage with the position magnet near to the pickup. Values for c and N must be derived from test data, but then may be used to calculate signal amplitudes for any magnet position thereafter for a given waveguide assembly design. For example, with a tubular waveguide material of Ni-Span C having a beaded wire inside, according to inventor Jacob Tellerman in US patent 5,334,933, c could be 0.87, with N being the distance between the position magnet and pickup in feet, divided by ten. Other materials and configurations may provide coefficients that are substantially different.

This attenuation can be optimized (reduced) by proper selection of the waveguide material, the amount of cold work, and the annealing process (as well as reducing any attenuation that may be caused by a beaded wire and/or waveguide suspension components). When the waveguide wire is sized by passing it through progressively smaller dies, it becomes harder and is called *cold work hardening*. Several times during the process of forming the wire down to the desired diameter, the wire must be heated to soften it, called *annealing*. The last stages of working and annealing result in a *percent cold work*. Various manufacturers have their preferred targets for the amount of cold work in the finished waveguide material, but this information is often closely held as a trade secret.

The velocity of the torsion wave is often called the *sonic velocity*. Its value depends on the torsion spring constant and the moment of inertia of the wire, according to the equation:

$$c = \sqrt{K / I} \tag{12.3}$$

where c is the sonic velocity, K is torsional spring constant, and I is moment of inertia. This equation can be more of a help in understanding the property of sonic velocity, rather than for use in performing calculations. It is easier to measure sonic velocity on a sample than it is to find a torsional spring constant and moment of inertia for the waveguide and processing of a given waveguide specimen.

To some degree, sonic velocity may vary from *unit to unit*. This can be minimized by tight control over the alloy composition, cold working, and annealing process.

The remaining variation is adjusted by individual calibration of the sensor, loading a calibration factor into onboard memory that can be downloaded by the user, or by publishing a calibration factor for use with that sensor.

Also, sonic velocity will have some small amount of variation along the length of the waveguide within a single sensor. This is one of the sources of *nonlinearity*. Nonlinearity that originates with the waveguide has two main sources: processing/handling induced, and signal interference. The process/handling-induced nonlinearity errors can be reduced by the same methods as outlined earlier for sonic velocity variation from unit to unit, but with the greatest emphasis on applying the process uniformly along the length of the waveguide. If the percent coldworking or thermal treatment has any nonuniformity along the waveguide length, it will result in a greater amount of nonlinearity in the completed sensor. Signal interference-induced error occurs mainly toward the head end of the sensor, near the pickup: when the waveguide is interrogated, the pickup coil receives some of the magnetic field formed around the waveguide, and *rings* for a short period of time. This means that a voltage pulse will form in the pickup output due to the interrogation pulse. Rather than appearing as one simple square pulse, the voltage pulse at the pickup will be a decaying sinusoid, taking a few to tens of μs to dissipate, as shown in Figure 12.5 (and labeled in the figure as interrogation pulse). If it takes, for example, 14 μs for the interrogation pulse in the pickup to dissipate, and the sonic velocity is 2,800 m/s, then 14 μs represents 39 mm of waveguide length that can't be used. In that case, the position magnet should not come any closer to the pickup than 39 mm, or the signal will be distorted by the pickup ringing. If the damping device (at the other end of the waveguide) is not sufficiently reducing the amplitude of the sonic wave reflected from the waveguide tip, then this reflected wave can also cause nonlinearity.

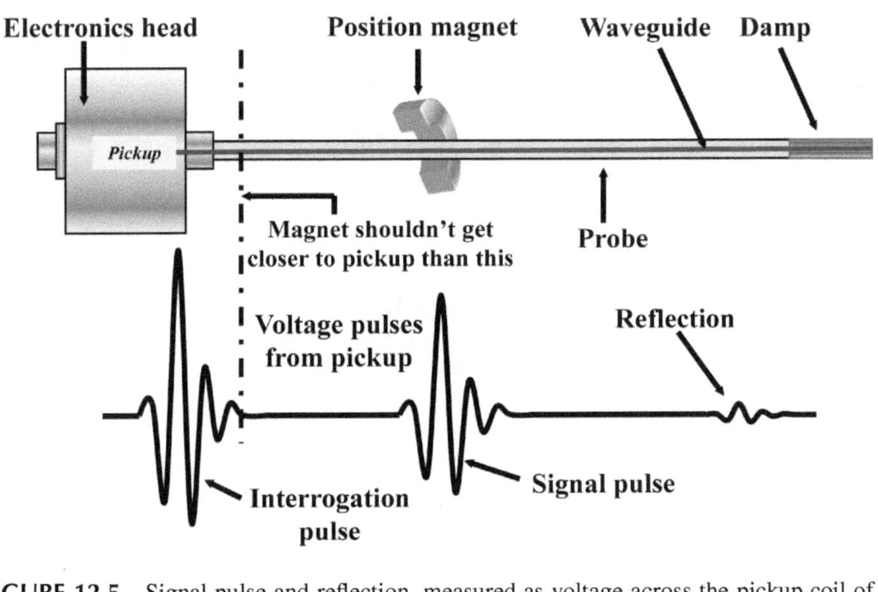

FIGURE 12.5 Signal pulse and reflection, measured as voltage across the pickup coil of a magnetostrictive linear position sensor.

This is because the reflected pulse amplitude will add to that of the signal pulse at any position magnet locations where a signal pulse arrives at the pickup at the same time as the reflected pulse from the previous interrogation. To consider the damping device to be doing a sufficient job, the amplitude of the reflected wave should be no more than 5% of the signal wave (see Figure 12.5).

When measured as a voltage across the pickup, the interrogation pulse will typically have an amplitude greater than that of the signal pulse. The reflection should have a much smaller amplitude than the signal pulse, and it can be seen in Figure 12.5 that the phase of the reflection becomes inverted (as compared with the interrogation and signal pulses) upon reflection from the waveguide tip.

The *temperature coefficient* of sonic velocity affects the thermal sensitivity of both the zero setting and the scale factor of the output. The effect on zero is because all readings are made relative to the distance of the position magnet from the pickup. The position magnet should not come too close to the pickup because it interferes with the pickup signal. So the zero location of the position magnet is a finite distance from the pickup (and indicative of a non-zero amount of time at the zero location of the position magnet). Since the popular alloy Ni-Span C has a low-temperature coefficient of the torsional modulus of elasticity, it will also have a very low-temperature coefficient of torsion spring constant, and thus a low-temperature coefficient of sonic velocity (the relationship between torsion spring constant and sonic velocity was shown in Equation 12.3). With optimum processing, the temperature coefficient of sonic velocity with a Ni-Span C waveguide can be less than 5 ppm/°C.

Since the waveguide is a ferromagnetic material, it has a magnetic *hysteresis* when being magnetized/demagnetized. This manifests itself as a difference in sensor-indicated position when the position magnet approaches a given point from an upscale position, as compared to approaching that same point from a downscale position. Once the waveguide material, processing, and diameter have been selected, increasing the current level of the interrogation pulse can further reduce the hysteresis and increase the signal pulse amplitude. Further increasing the interrogation current past magnetic saturation of the waveguide will not further reduce hysteresis, nor further increase signal pulse amplitude, and will only increase power dissipation.

12.6 POSITION MAGNET

A position magnet can be an electromagnet, if it is desired to have the capability of turning the magnet on and off as needed. But most position magnets are any one of three types of permanent magnets: neodymium-iron-boron (NdFeB), samarium cobalt (SmCo), or ceramic (ferrite). Alnico magnets are not normally used because they can easily be demagnetized. NdFeB magnets have very high coercive force, and are useful when a small rod magnet is needed. They are brittle, and should be nickel plated to avoid corrosion. SmCo magnets have a relatively high coercive force, are oxidation-resistant, can operate to a higher temperature (300°C), but are expensive. Many ring magnets are made of ferrite, through pressing or sintering. They are strong and inexpensive. Ring magnets are often injection molded. Molded magnets are a composite of various magnetic powders and a resin. These can be relatively low

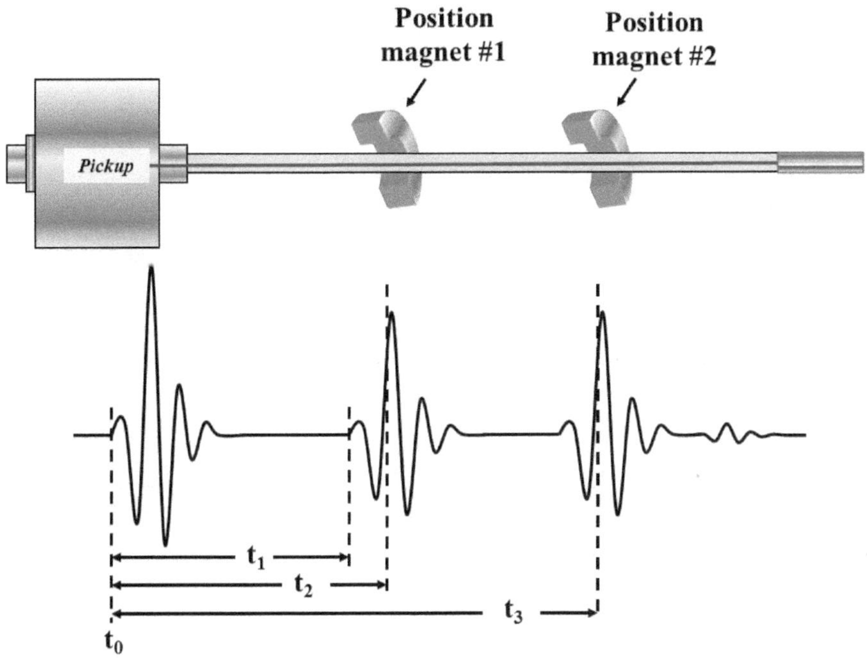

FIGURE 12.6 Two or more position magnets may be implemented on one waveguide.

cost, and can be molded into any shape. Since they are somewhat weaker, molded magnets may not be suitable for larger-diameter ring magnets.

More than one position magnet can be implemented onto one magnetostrictive position sensor. Figure 12.6 shows two position magnets. In operation, two timers are started when the interrogation pulse is applied. The first timer is stopped when the first signal pulse is detected at the pickup, but the second timer continues to run. The second timer is stopped on detection of the second signal pulse (that is, the signal pulse from the magnet farther away from the pickup). The two timers indicate the two positions, respectively.

In the figure, the interrogation pulse (for example, a 2 A current pulse of 2 μs duration) is applied at t_0. The sonic wave from the first magnet starts to appear after time period t_1, but the pulse is normally detected at t_2, closer to its positive peak voltage. Accordingly, t_3 shows the elapsed time until the second magnet is detected. (Note that the term *interrogation pulse* is used to indicate the current pulse applied to the waveguide, as well as to the voltage pulse that it generates in the pickup, the actual meaning depending on context.)

In some cases, more than two position magnets may be used. When applied as a sensor on an industrial slitting machine (or slitter machine, or slitter), there may be ten or more position magnets on one magnetostrictive sensing probe. A slitter has many blades to slit a thin roll or sheet, such as paper, film, tape, and so on. A position magnet may be coupled with each blade to allow precise positioning of the blade.

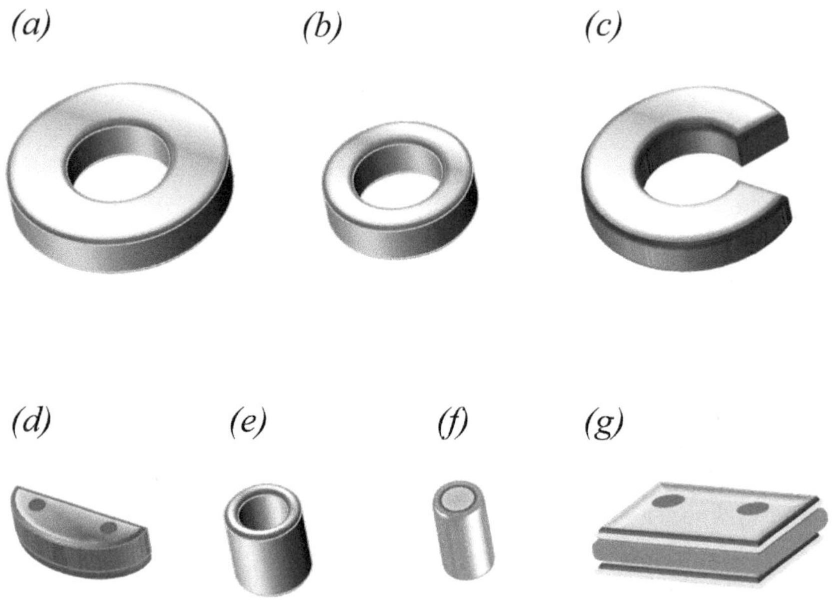

FIGURE 12.7 Position magnet shapes.

There is a minimum separation distance required between any two position magnets to avoid having their signal pulses interfere with one another. That minimum separation distance depends on the time period taken up by the decaying sinusoid of a signal pulse from one magnet, such that the leading edge of the pulse from one magnet must not encroach on the trailing edge of the preceding magnet pulse. The minimum separation requirement is normally between 65 and 100 mm. In Figure 12.6, the signal pulses would appear to start interfering with one another if $t_3 - t_2$, the distance between the two position magnets, was about half as much as shown.

The position magnet may be fabricated according to any one of several shapes in order to fit the application requirement. See Figure 12.7. The most popular shape is the ring magnet of Figure 12.7(a). (The ring may be a molded magnet material, or can be a non-magnetic ring with magnet rods inserted into a plastic housing and epoxied in place.) With the waveguide assembly running through the center of a ring magnet, the sensor output is less sensitive to the radial position of the waveguide within the ring. For example, if the ring magnet maintains the same position with respect to the pickup, but it is moved 1 mm off center, the output to the sensor will typically indicate an error of less than 1/30 of that amount. That's because, with a ring magnet, when the waveguide is getting farther from one side of the magnet, it is also getting closer to the other side, causing minimal fluctuation in the magnetic field strength.

The several shapes of position magnets of the figure are:

(a) standard ring
(b) small ring
(c) C-shaped or open-ring

(d) D-shaped, floating
(e) small multi-pole ring
(f) rod, floating
(g) bar, floating

By convention (initiated by Temposonics/MTS in the 1980s), position magnets are polarized with the north pole facing the waveguide. So, for a ring magnet, the north pole is facing in toward the center bore of the ring magnet, and the south pole is facing out.

The various magnet shapes are made to accommodate requirements of the application. A ring magnet, or a C-shaped one, being the most common; but rod or bar magnets can also be used. The bar magnet shown has a flat magnet in the center, with steel plates above and below it to shape the magnetic field. With a rod or bar magnet, the application must be designed so that the spacing between the magnet and the waveguide remains relatively constant in order to keep the signal amplitude the same as the position magnet moves throughout its range. A ring magnet is preferred, because this provides the most uniform magnetic field when the waveguide spacing to the magnet may vary. A cut into the ring (C-shaped magnet) allows the position magnet assembly to be added or taken away from the sensor probe (containing the waveguide) without having access to the end of the probe. A C-shaped ring can provide the same performance as a ring magnet when designed as shown in Figure 12.8.

In the figure, four small rod magnets are arranged at 90° intervals around a plastic shell into which the rod magnets fit. Then the magnets are potted in place with epoxy resin to form a magnet assembly. It can be seen that the ring magnet assembly at left and the C-shaped magnet assembly at right actually have identical magnetic properties, since the plastic shell has no effect on the magnetic field. It is of interest to note

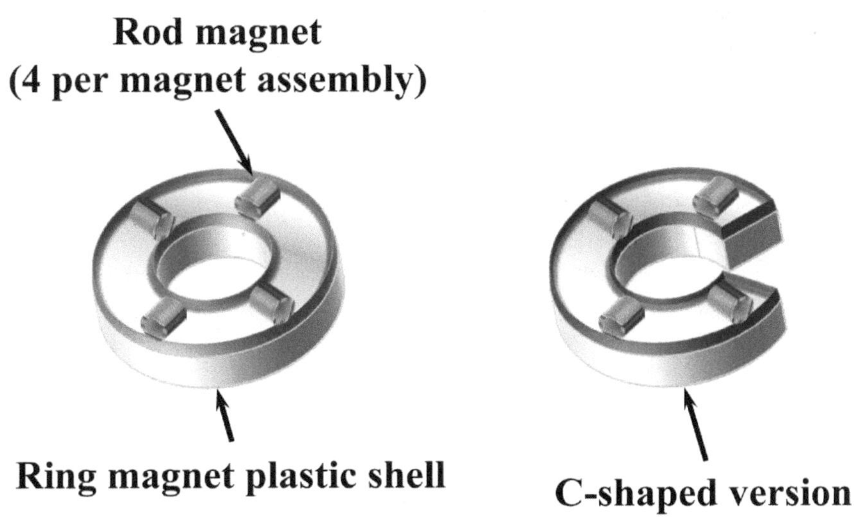

FIGURE 12.8 A C-shaped magnet with the same characteristics as a ring magnet.

(a)

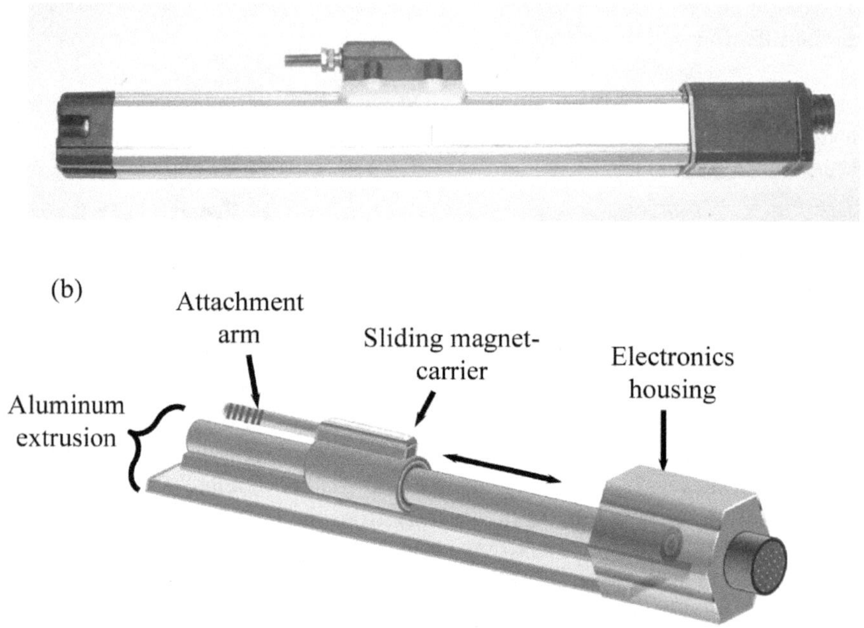

FIGURE 12.9 Sensors with track or rail to guide a sliding position magnet: (a) photo of a magnetostrictive linear position sensor developed by the author with one version of a sliding magnet-carrier; (b) illustration of an alternate version of a sliding magnet-carrier.

that such a magnet assembly having four rod magnets can have the same insensitivity to waveguide position as with a fully molded ring magnet.

A rod is the simplest configuration of position magnet, but must track directly above the waveguide as the magnet moves in order to preserve measurement accuracy. A bar magnet assembly has a small amount of leeway in the cross axis, but must track relatively parallel to the waveguide (keeping distance to the waveguide constant) during its travel to preserve accuracy. When not using a ring magnet, tracking of the position magnet with respect to the waveguide can be controlled by a sliding design, as shown in Figure 12.9.

In the figure, a rod or bar magnet is mounted within a sliding magnet-carrier. The sliding magnet-carrier is made of plastic, and follows a track that is part of an aluminum extrusion. This maintains parallel alignment of the waveguide with the position magnet path, without any effort by the user. The sliding magnet-carrier incorporates an attachment arm for connection with a moving member that is to be measured. The aluminum extrusion has a flat base for mounting. The waveguide assembly protrudes from the electronics housing and slides inside of the aluminum extrusion. The electronics housing is attached at the end of the extrusion. This configuration is popular in some factory automation machinery such as plastic injection molding machines.

Another possible magnet configuration that also requires good alignment is an outside ring or partial outside ring. In this case, the waveguide is located on the outside of the ring. The ring may be allowed to rotate, but will still energize the waveguide. When the waveguide is on the outside of the ring magnet, the ring magnet should be polarized with the north pole facing out.

There are other possibilities for ring magnets. A ring magnet can be made that has a first portion with the north pole facing in, and an adjacent portion with the south pole facing in. This north-south magnet configuration can provide a greater output signal than just a single pole with north facing in. Another possibility builds on that idea, with a north-south-north configuration. But in this last configuration, it is important to space the pole reversals to coincide with the sonic wave shape so that the greatest output signal is obtained.

12.7 PICKUP DEVICES

At least three general types of *pickup* devices are in common use: *radial tape*, *coaxial coil*, and *piezoelectric* pickup element (see Figure 12.10). The radial tape is the most difficult to manufacture, but has the advantage that the tape material can be optimized as needed for a pickup device. This can result in a higher signal-to-noise ratio than in the other types, since the tape can be optimized for maximum output instead of for low thermal coefficient of sonic velocity, or low attenuation versus position. The tape is a ribbon of ferromagnetic alloy that is selected and processed to achieve the highest possible coefficient of magnetostriction (much higher than that of the waveguide), and is welded to the waveguide. This tape material is not suitable for use as a waveguide due to its high rate of attenuation, but may be used for the radial tape pickup because attenuation is unimportant, since only a small length is used. As the torsional sonic wave reaches the intersection of the waveguide and tape, some of the wave energy travels down the tape (and some continues on in the waveguide). In the tape, the sonic wave is converted to a compression wave, rather than remaining a torsional wave.

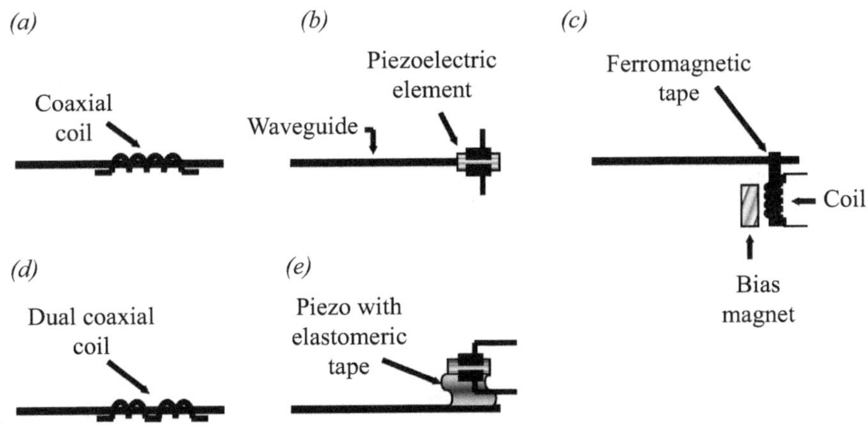

FIGURE 12.10 Types of pickup devices.

Due to the Villari effect (also called *reverse magnetostriction*), the compression wave causes an area of different permeability to exist in the tape as compared with the remaining unaffected area of the tape material. This area of different permeability travels as a wave along the length of the tape. The free end of the tape (the end not welded to the waveguide) protrudes into the center of a coil and is magnetized by a small adjacent *bias magnet.* As the sonic wave traveling on the tape passes through the coil, the associated area of different permeability also passes through the coil. This causes a change in magnetic flux density (by changing the flux density supplied by the bias magnet) in the section of the tape having different permeability. Due to the Faraday effect, a voltage pulse is produced across the coil ends as the wave of variable magnetic flux density passes through the coil. This voltage pulse is used to detect the presence of the sonic pulse.

The coaxial coil pickup of Figure 12.10(a) is the simplest type of pickup. It comprises a coil of wire through which the waveguide passes. This is simpler than the tape pickup because it eliminates the need for the tape and for the bias magnet. The current pulse used to generate the sonic wave also leaves a remanent magnetic field on the waveguide. This is enough to provide the magnetic field needed for the pickup coil to produce a voltage pulse when the sonic wave passes through the coil. As the sonic wave on the waveguide passes through the coil, the associated area of different permeability also passes through the coil. This produces a voltage pulse across the coil leads in the same way as explained for the tape pickup.

The disadvantage of this simpler design is that, since the waveguide must be designed for low attenuation and low temperature sensitivity, the coefficient of magnetostriction that results is relatively low. For a given interrogation pulse current, this produces a lower amplitude of signal from the pickup and an associated lower signal/noise ratio. This can be accommodated by optimizing the number of turns on the pickup coil, using a small-diameter wire to wind the coil, and minimizing the inside diameter of the coil. Using more turns on the coil increases the voltage pulse amplitude for a given mean coil diameter and coil length, because more turns mean a greater inductive capability. A smaller mean coil diameter (by using a smaller diameter wire) allows a given magnetic field change to have a greater effect on a given number of coil turns. But if the inside diameter of the coil is made too small, then there is a risk of the waveguide touching the coil, and strongly dampening the sonic wave energy. So, with a waveguide diameter of 0.375 mm, the coil inside diameter should be at least about 1.5 mm.

A dual-coil pickup has also been used. One coil is wound in the clockwise direction, while the other is wound anti-clockwise. The two coils may be spaced apart by one-half wavelength of the sonic wave. With the coils spaced by one-half wavelength, a greater signal pulse voltage is generated. As the peak of a sonic wave is affecting one coil, the opposite peak of the same wave is affecting the other coil, and the coils are connected together so that the two voltages add. (Or, the dual coil assembly may be spaced from the waveguide end so that a first reflected peak is entering one coil while another peak of the same wave is still in the other coil.) But when an external electromagnetic field affects the coils, their outputs subtract and provide very little output voltage in response to the external field that affects both coils similarly.

A piezoelectric pickup can be used to detect a sonic wave from the waveguide without using a magnetic field. This provides the hope that it would be less sensitive to interference from external magnetic fields. In actual sensors however, it has proven difficult to get as good an energy transfer as can be had with the tape or coaxial coil methods. The associated low signal level results in a low signal/noise ratio, confounding the effort to reduce corruption from external interference. An additional area of concern with the piezoelectric pickups is that they are generally made from brittle ceramic materials. A lot of field problems have been caused by fracture of the piezoelectric element when the sensor was inadvertently dropped onto a hard surface.

A piezoelectric element has also been coupled to the waveguide by an elastomeric tape with the intent to reduce sensitivity to shock and vibration. But limited success is achieved due to the accompanying reduction in signal amplitude, as well as an increase in manufacturing time.

12.8 DAMP

A damping element (also called the *damp*) is incorporated at the tip end of the waveguide, as mentioned earlier, in order to remove sonic pulses that are moving away from the pickup and toward the waveguide tip. If they were not removed, they would reflect from the waveguide tip and travel back toward the pickup, causing interference with the desired signal pulse. The damp must have sufficient damping quality to remove the unwanted sonic pulse, but not be so effective at the front end that it causes a reflection directly from its own front end. Various combinations of shape and hardness of elastomeric materials are used to approximate the desired performance.

In the simple damp of Figure 12.11(a), a silicone rubber sleeve is slid over the waveguide. The fit should be almost tight, but still be able to slide (that is, the inside diameter

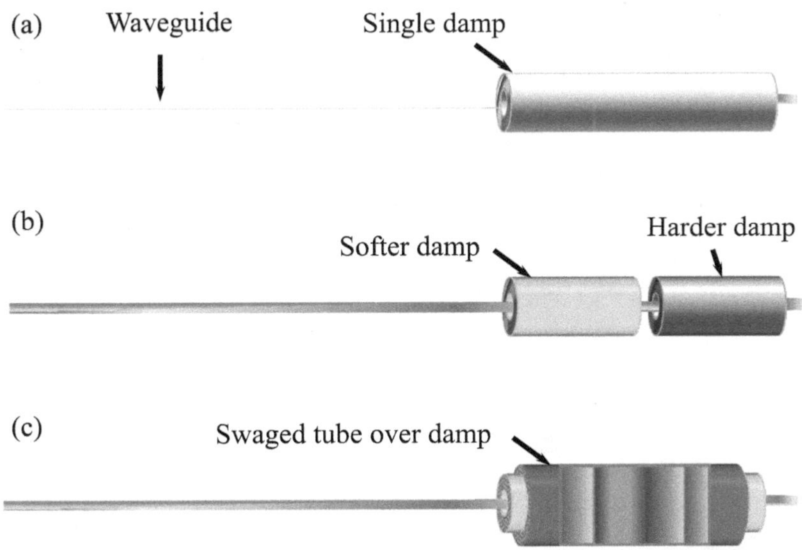

FIGURE 12.11 Some examples of damping elements.

of the damp tubing should be close to the same size as the outside diameter of the waveguide wire). This single damp method requires a softer damping material (such as Shore A 35 durometer) and longer length (such as 65 mm), as compared with the other types.

With a dual damp, the first damp (on the left in Figure 12.11(b)), should be soft, so that there is no reflection from its face. The second damp can be somewhat harder, to provide stronger damping. For example, the first damp might be 35 durometer and 15 mm long, with the second damp of 55 durometer and 25 mm long. Selection of the damping element sizing and hardness requires some trial and error.

Figure 12.10(c) shows a different type of damp. It utilizes an elastomer or fiberglass sleeve, with pressure applied to obtain the desired amount of dampening of the sonic wave. The pressure is applied by sliding a metal tube (such as one made of brass) over the damping sleeve and swaging down on the metal tube to compress it by the desired amount. This configuration of damp may be difficult to fixture for easy manufacture, but can result in a damping element that is shorter than the other two for a given amount of damping.

12.9 WAVEGUIDE SUSPENSION

The interrogation pulse current is usually applied at one end of the waveguide. To form a complete circuit, a *return wire* is attached to the other end of the waveguide (the waveguide tip) and brought back to the first waveguide end, for the current return path. It is important to prevent the return wire from touching the waveguide at any point along the waveguide length so that additional reflection points may be avoided. It is also important to suspend the waveguide within its housing (the housing is usually an outer metal tube of aluminum or non-magnetic stainless steel, such as 316SS). Another related requirement is that the waveguide must be mechanically isolated from external shock and vibration so that externally applied shock and vibrations do not cause waveguide disturbances that might conflict with the sonic wave that is to be detected.

The components that accomplish these three tasks comprise the waveguide suspension system. This function may be accomplished in many ways, and two examples are shown in Figure 12.12.

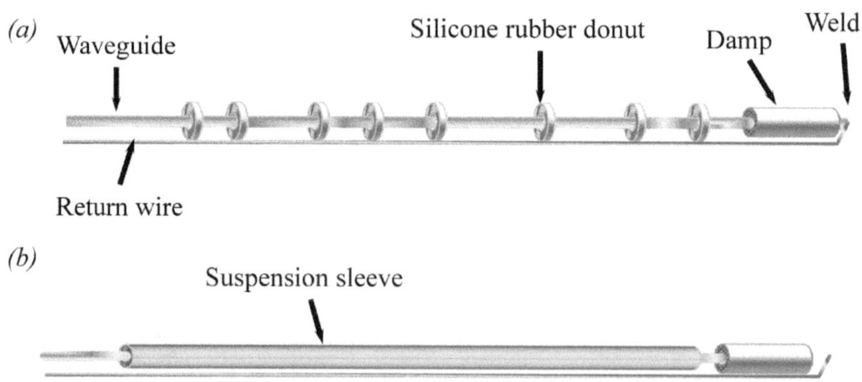

FIGURE 12.12 Two examples of waveguide suspension system components.

The suspension system of Figure 12.12(a) uses thin, soft, silicone rubber donuts spaced along the waveguide length. The spacing of the donuts should be random rather than regular, to avoid the generation of standing waves on the waveguide (due to the excitation from the interrogation pulse current and the mechanical motion that may result). The return wire is adhered to the edges of the silicone rubber donuts by application of small dots of silicone RTV silicone rubber (room temperature vulcanizing). An insulating sleeve should be placed over the assembly shown in order to insulate it electrically and mechanically from the metal tube into which it will slide. The silicone rubber donuts must be soft enough and thin enough to avoid substantially attenuating the sonic wave.

The suspension system of Figure 12.12(b) utilizes a suspension sleeve that is capable of suspending the waveguide without substantially attenuating the sonic wave, while insulating it from the outer housing. Rubber materials have been found to cause too much attenuation, even with very soft silicone rubber. One material that has the required qualities is the standard fiberglass insulating sleeve that is normally used to electrically insulate wiring that doesn't already have an insulating layer. To avoid attenuation, the sleeve should have an inside diameter that is much greater than the waveguide diameter (for example, a fiberglass sleeve having an inside diameter of about 1.5 mm for a waveguide of 0.375 mm outside diameter). When using such a sleeve, it is important to avoid moisture or humidity that can form ice when the waveguide assembly is very cold, because the ice will attenuate the sonic wave.

The waveguide is shown in Figure 12.12 as being free-floating at the damp end. It is supported along its length by the silicone rubber donuts, or by the fiberglass suspension sleeve. Alternatively, some magnetostrictive sensors incorporate means to pull on the waveguide tip with spring force, holding the waveguide centered within its suspension system or within its housing. In this case, care must be taken to avoid excessive pull-force, because this will reduce the signal level from the sonic wave. With a 0.375 mm diameter waveguide, for example, the pull-force should not exceed about 2.0 newtons (0.2 kg), and may need to be less than that, depending on the waveguide material properties. Spring suspension of the waveguide within a housing without using any silicone rubber donuts or suspension sleeve may be a useful method for sensors operating at cryogenic temperatures. That would avoid hardening of the suspension elements at low temperature and avoid problems of ice in the suspension sleeve. But the gas inside the waveguide housing would still need to be dry to avoid ice forming directly onto the waveguide itself.

12.10 ELECTRONICS

The electronic circuit provides the interrogation pulses, receives signal pulses from the pickup, amplifies, filters, and detects the signal pulses, and then provides the desired analog or digital output. See Figure 12.13.

The electronic circuit of Figure 12.13(a) uses an oscillator circuit to provide a pulse train for timing of the interrogation circuit. The interrogation frequency is usually in the range of 10 Hz to 1 kHz. The interrogation pulse width is usually 0.5 to 5.0 μs. So, for example, the oscillator could provide pulses at 250 Hz, with each pulse having a duration of 1.0 μs. The interrogation circuit pulses the waveguide accordingly, but

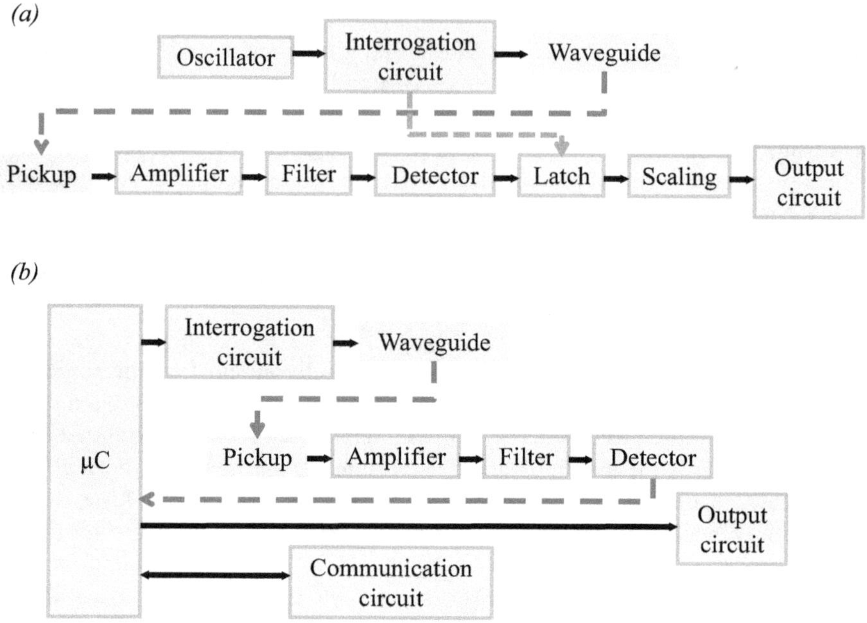

FIGURE 12.13 Electronic functions block diagram without a μC (a), and with a μC (b).

with a current level usually between 0.5 and 4.0 amps. Since it would not be desirable to draw several amps directly from the power supply, a charge is stored in a capacitor. Then the capacitor is discharged into the waveguide as needed. A circuit combining the oscillator (or it can be called a pulse generator) and interrogation circuit is shown in Figure 12.14.

In the figure, integrated circuit U_1 is a CMOS Schmitt trigger, U_2 is a bilateral analog switch (or a transistor), and Q_1 is an N-channel enhancement-mode power MOSFET. When the output of U_1 is low, U_2 is open, and C_1 discharges only through R_1. When the output of U_1 is high, U_2 is closed, and C_1 charges more quickly through the parallel combination of R_1 and R_2. This pulse-generating circuit was invented by the author. A somewhat similar circuit is commonly used with a diode instead of U_2, but is temperature sensitive due to the diode change in forward voltage with temperature.

If R_1 is 100 k ohms, R_2 is 49.9 ohms, and C_1 is 0.02 μF, then pulses of about 1 μs duration will be produced at a rate of about 500 Hz. The exact amounts of time depend on the technology of Schmitt trigger used (such as CD4000 series, 74C, 74HC, 74HCT, 74LV, and so on).

R_3 charges up C_2 in between interrogation pulses. C_2 may be a parallel connection of two or more capacitors, in order to achieve lower equivalent series resistance (ESR) and/or a larger value of capacitance. The pulse generator controls Q_1. When Q_1 is switched on, C_2 is discharged into the waveguide. The energy discharged into the waveguide depends on the waveguide impedance, the energy available in the capacitor, and the time during which Q_1 is turned on. The energy available in the capacitor depends on the voltage across it and the amount of its capacitance (for example,

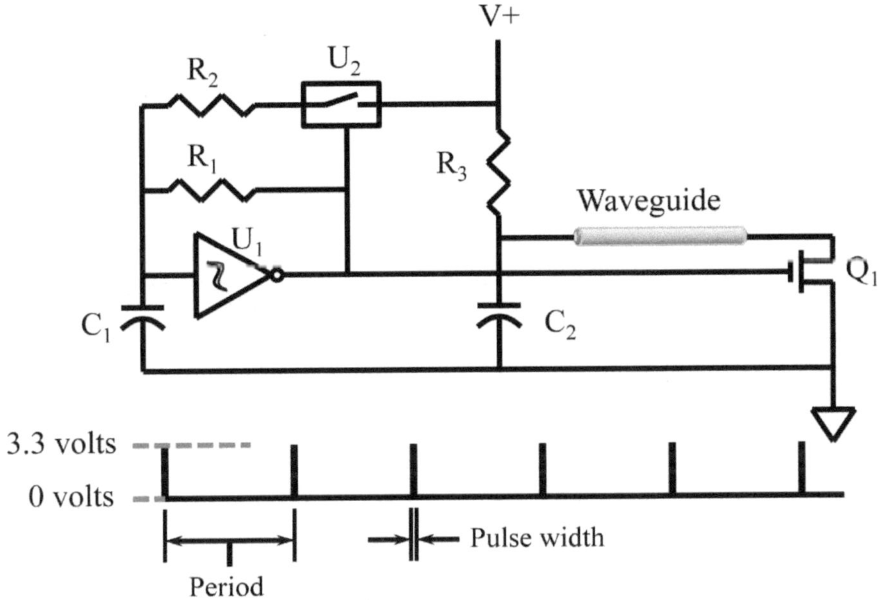

FIGURE 12.14 Oscillator and interrogation circuit.

doubling the capacitance doubles the energy available, whereas doubling the voltage increases the available energy by four times). The current developed depends on the capacitor voltage and the impedance of the waveguide and waveguide-driving circuit.

After interrogating the waveguide, the sonic pulse is received by the pickup. It is a low level voltage of tens of mV_{peak}, so it will be amplified to a few hundred mV. Then it is filtered to remove unwanted noise. A low-pass filter is needed to remove electrical noise, but a high-pass filter could remove the effects of mechanical noise. So a bandpass filter can be utilized to remove both. The signal pulse (a decaying sine wave pulse) has a frequency of somewhere between 100 kHz and 400 kHz, depending on many factors of the waveguide, pickup, and circuit design. But once the design is settled, then the frequency of the signal pulse remains constant. If, for example, the signal pulse is 300 kHz, then a band pass filter could reject below 200 kHz and also above 350 kHz.

After filtering, the signal pulse is detected, or changed so that it switches between logic zero and one (or low and high, respectively) instead of the analog voltage of the signal pulse. This can be done with a Schmitt trigger, or with a comparator circuit. The comparator circuit may have an adjustable threshold voltage and may have some hysteresis built in. The output of the detector drives a latch. The latch is *set* when the interrogation pulse is applied, and the set is applied for a duration long enough so that any ringing of the interrogation pulse at the pickup has subsided (for example, the set may be applied for about 25 μs). When the set is released, the latch is ready to receive a *reset* signal. The reset is applied when the sonic wave is detected. So, a timed period (latch output) is generated whereby the logic one time represents the time

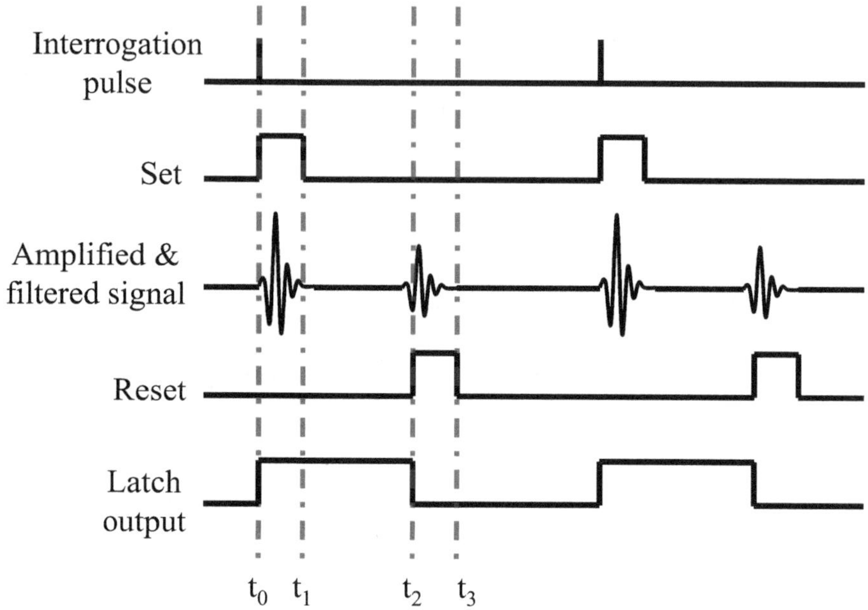

FIGURE 12.15 Timing of the interrogation pulse, set, and reset.

period between application of the interrogation pulse and detection of the sonic wave. The relative timing of the various pulses and signals can be seen in Figure 12.15.

When designing a sensor that could be as short as a few mm or as long as several meters, the waveguide current must be controlled to give sufficient current for a good signal amplitude and low hysteresis, but not so large that ringing in the pickup limits the usefulness near the head end. Sometimes a constant current source circuit is used to drive the waveguide for the interrogation pulse period, and sometimes a microprocessor adjustable voltage or current source is used. Since the interrogation pulse usually recurs at a rate somewhere between 10 Hz and 5 kHz, but the pulse duration may be only 1 to 2 microseconds, energy for the interrogation pulse is normally stored in a holding capacitor as mentioned earlier, reducing the peak current demand on the sensor power supply leads. Normally, a capacitor is charged up in one of various ways and then a field effect transistor (FET) discharges the capacitor into the waveguide. The FET is used because it can have a low on-resistance at high current (2 to 5 amperes), but only requires micro amps at its control pin (gate).

Figure 12.13(b) shows the functional blocks that may be implemented in a magnetostrictive sensor that utilizes a μC. All of the same functions must be accomplished as with the non-μC design, but much of the work can be done within the μC. The μC provides the interrogation rate, as well as the interrogation pulse width. A D/A converter can be used along with a power amplifier to supply an adjustable voltage for charging the capacitor that provides the interrogation energy. Each of these three parameters (pulse rate, pulse width, pulse voltage) can be adjustable as needed for a particular type of sensor. For example, when needing a low-power supply current, the rate of interrogation can be slower. When interrogating a longer sensor, the capacitor voltage can be higher (because sufficient current is still needed to avoid hysteresis, but a longer waveguide has greater impedance, so a higher voltage is needed for a longer waveguide in order to produce the same current as with a shorter waveguide). The μC can wait for a programmed amount of time (blanking timer) after interrogation before looking for a signal pulse. Once a signal pulse is detected, the μC can control its latch to ignore any further signal pulses for an amount of time equal to the width of a typical signal pulse (digital one-shot). The μC can also implement filtering and scaling, and then drive a D/A converter to provide an analog output, or a line-driver to implement a digital protocol.

The μC should also have two means of digital communication. One is for programming the μC chip with the operating firmware during assembly at the factory. It is often best to retain this firmware programming capability (by populating the PCB with a programming connector) in case any further firmware improvements are developed after the sensor is in production. Then it is easy to update the firmware by plugging in a programmer and revising the code. The other digital communication means is for calibration and accessing other factory or user settings. This is usually implemented in an RS-485 or USB port. But on occasion has been implemented by the author with a proprietary circuit to allow communication over one wire, to reduce the number of wires in a cable. Optical or other wireless communication is also possible.

12.11 ANGULAR/ROTARY MAGNETOSTRICTIVE SENSORS

Most magnetostrictive position sensors are for measuring linear position. A linear sensor may also be curved for some applications. But an angular/rotary sensor configuration is also possible, and was invented by the author, US patent 6,501,263. A representative configuration is shown in Figure 12.16, but other configurations are also possible.

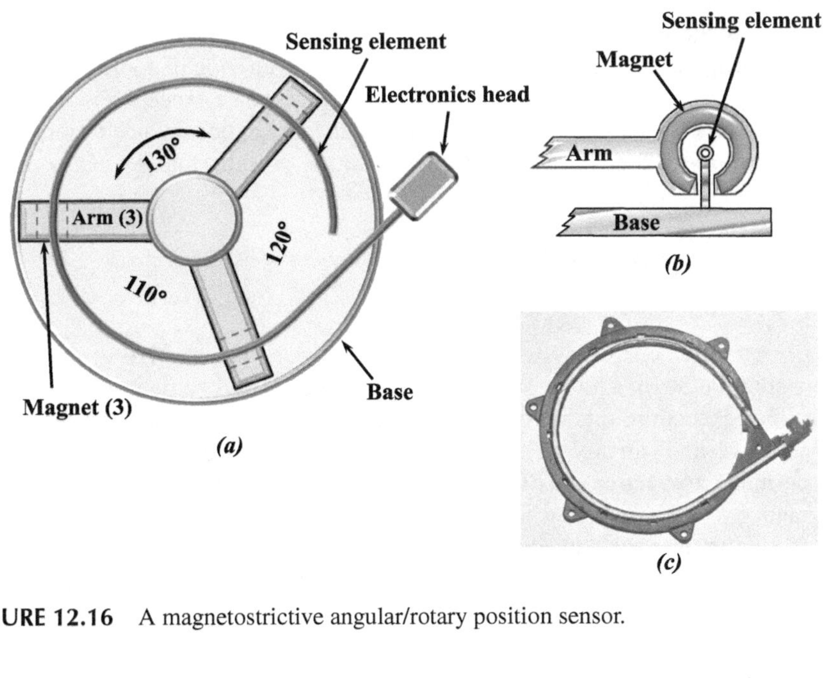

FIGURE 12.16 A magnetostrictive angular/rotary position sensor.

Figure 12.16(a) is a top view of a sensing element configuration comprising three arms and a central hub. A C-shaped magnet is attached near the end of each arm. The arms have respective angles to one another of 110, 120 and 130°. The sensing element contains a waveguide fitted inside a mechanical isolation sleeve, and is curved around an angular measuring range. An electronics head contains the pickup and may contain enough electronics to interrogate the waveguide and provide signal pulses. Figure 12.16(b) is a side view of the end of one arm, showing a C-shaped magnet as it travels along the sensing element, which is supported by the base. Because there are three position magnets, there are always at least two magnets aligned along a usable portion of the sensing element and able to be read. The different angles among the magnets allow the μC to interpret which magnets are being read. So, implementing the proper algorithm, it is possible to measure the angle of rotation throughout the full 360°. Figure 12.16(c) is a top view photo of a production version of the curved waveguide and pickup assembly being supported by a molded thermoset plastic base, not showing the arms or magnets. (Note: a *thermoset* plastic starts as liquid, then hardens when heated in the mold. A *thermoplastic* is heated to a liquid, then hardens as it cools in the mold. Only thermoplastics can be re-melted and used again.)

12.12 ADVANTAGES

Position sensors based on magnetostrictive technology are non-contact, absolute reading, and have extremely high resolution. Sensor lengths in production include a full-scale range (FSR) of less than 10 mm, up to a FSR of over 20 meters.

Analog and digital outputs are available to indicate position, displacement, velocity, and acceleration. Competing technologies include LVDTs, inductive sensors, distributed impedance, encoders, ultrasonic sensors, and potentiometers.

Due to the non-contact nature of magnetostrictive position sensors, there are no mechanical parts to wear out. This is an advantage over contact-type sensors, the most popular being the potentiometer. For example, even though some potentiometers list an operating life of over 10 million cycles, this can be exceeded in only a few months when the system experiences a continuous application-induced vibration or a control system-induced dithering. With a vibration or dithering frequency of only 10 Hz, one finds:

$$\begin{aligned}&\textit{10 cycles/sec} \times \textit{3600 sec/hr} \times \textit{24 hr/day} \times \textit{30 days/mo} = \\ &\textit{25 million cycles/month}\end{aligned} \tag{12.4}$$

Typically, a motion control system will have a few spots in the active stroke that are used most often. So, in a few months, these often-used spots can be traversed a number of times exceeding the rated lifetime. When a spot is worn on a potentiometer element, the result is an area of unstable or inaccurate readings.

Since magnetostrictive position sensors read the location of a position magnet, performance is not affected when non-magnetic materials are placed between the position magnet and the waveguide. This leads to novel applications requiring the position magnet to be on the outside of a housing or other functional member, and the waveguide on the inside (such as in a hydraulic cylinder). For example, the non-magnetic materials can be aluminum, plastic, some stainless steels, and others.

The absolute measurement provided by magnetostriction-based position sensors offers an advantage over the incremental (or displacement) measurements of magnetic or optical linear scales, encoders, and so on. This is because an absolute position sensor measures the distance between a fixed datum and the point of interest, whereas a displacement sensor measures the distance between a previous measurement point and the current point of interest. An incremental sensor (such as an incremental encoder) counts the number of increments accumulated since the last reset of the count. The size of the increment determines the measurement resolution. If the previous position or count is forgotten (due to a power failure, or a mechanical or electrical glitch, for example), then the displacement or incremental sensor must be reset to zero or to a known position.

Since the actual measurement of a magnetostrictive position sensor is presented as a variable time period, digital circuits can be used for all the electronic functions after detection of the return pulse. The sonic velocity of the torsional strain wave in the waveguide does not change appreciably with temperature or with time, so the position output is extremely stable.

The time period representing the measured position is analog, with infinite resolution, although the timing of digital pulses is used to indicate it. The resolution of the sensor signal output is dependent virtually only on the resolution of the timing circuitry and its ability to recognize the signal over any noise that may be present. Some industrial sensors have a resolution as fine as 1 micron. Table 12.1 compares some essential features of magnetostrictive linear position sensors with other technologies. In the table, high resolution is desirable and means finer increments of resolution. Low nonlinearity is desirable and means a smaller amount of nonlinearity error.

TABLE 12.1
Magnetostrictive Linear Position Sensors versus Other Linear Position-Sensing Technologies.

Technology	Resolution	Nonlinearity	FSR Available	Ruggedness	Cost
Resistive	Medium	Medium	10 mm–0.5 m	Medium	Low
CET	Low	High	50 mm–45 m	Medium	Low
Capacitive	Low-High	Med-High	1 mm–50 mm	Medium	Medium
Inductive	Medium	Medium	2 mm–0.5 m	High	Medium
LVDT	High	Medium	2 mm–0.2 m	High	High
Dist. Imp.	High	Low	10 mm–10 m	High	Medium
Hall effect	Medium	Med-High	1 mm–50 mm	Med-High	Low
MR	Medium	Med-High	1 mm–50 mm	Medium	Low
Magnetostrictive	High	Low	10 mm–20 m	High	High
Encoder	Medium	Low	10 mm–2 m	Low	High
Optical triang.	Low	Low	1 mm–50 mm	Medium	Medium

12.13 TYPICAL PERFORMANCE SPECIFICATIONS

Listed next are some of the specifications of a higher grade magnetostrictive linear position sensor with a digital output, such as CANbus:

Measured variables:	Position, displacement, velocity, set points
Range:	50 to 500 mm
Power supply:	8 to 32 volts DC, at ≤ 80 mA
Resolution:	0.1 mm
Repeatability:	0.005% FRO (full-range output)
Nonlinearity:	0.01% FRO (relative to a least-squares straight line)
Hysteresis:	0.1 mm
Update time:	≤ 1 ms
Operating temp:	−40 to 105°C
Output style:	CANbus

Lower performance magnetostrictive linear position sensors are also available and have lower cost. When testing the performance of a magnetostrictive linear position sensor, or calibrating it, a suitable test fixture is necessary. The test fixture should provide for fixing the device under test (DUT) firmly in place, while moving the position magnet along the measuring range. A test fixture designed and built by the author is shown in Figure 12.17.

In the figure, a rack is mounted on top of an aluminum channel base. A stage rides along a dove-tail track built into the edges of the rack, and a pinion gear on the stage engages the rack gear. A hand crank turns the pinion gear, and moves the stage along the track. A position magnet is mounted into the top of a part of the stage. As the position magnet is moved by turning the hand crank, its position is measured against a reference scale and shown on a digital readout. The reference scale and digital readout used in the figure are part of a standard set of digital calipers. This scale measures up to 1,000 mm, with a resolution of 0.01 mm. A similar fixture could be built with a motor to move the stage, to implement an automated calibration system. The caliper readout has a digital output that could be used with an automated system, or a linear encoder could be used instead of the calipers.

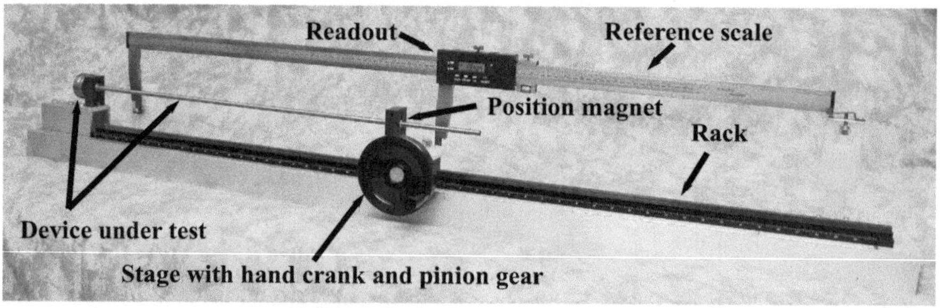

FIGURE 12.17 Test fixture for accurately locating the position magnet of a magnetostrictive linear position sensor.

12.14 APPLICATION

Magnetostrictive position sensors are made in many configurations for use in a wide variety of applications. Often, the position magnet is separate from the sensing element housing, and is attached to the member to be measured. This configuration is used in general machinery applications, primary and secondary woodworking equipment, hydraulic cylinders, shock absorbers, and many others. Figure 12.18 shows an example of a typical configuration for mounting a magnetostrictive linear position sensor and position magnet within a hydraulic cylinder.

As can be inferred from the figure, a hole is bored into the piston and its rod so that a pressure tube may be installed. The position magnet is shown mounted to the piston. Any materials in contact with and in close proximity to the magnet must be non-magnetic, so they won't interfere with the performance of the position magnet function. Such materials may include non-magnetic stainless steels, aluminum, and various plastics. The probe (sensing element assembly) part of the sensor is inserted into the pressure tube, and the electronics housing is threaded into a flange on the pressure tube. The pressure tube flange is sealed to the cylinder end to prevent leakage of the hydraulic oil. This configuration allows the hydraulic cylinder to be built, and then the position sensor can be added without disturbing the cylinder oil or relieving the oil pressure.

An additional advantage of the magnetostrictive position sensor, mentioned earlier, is that it can be used with multiple position magnets. On some standard models, up to 32 position magnets may be used simultaneously. An even greater number of magnets can be used in special applications. When an interrogation pulse is applied to a magnetostrictive waveguide having multiple position magnets, a torsional sonic pulse is produced at the location of each magnet. The pickup senses each return pulse in sequence. Each set of timing data, relating to the respective return pulse, is stored in an appropriate memory location. A μC can then retrieve the individual position data, representing each magnet, as needed.

In the late 1990s, large volume production of magnetostrictive sensors began for use in automotive applications. Automatic body control of passenger vehicles is one such use. To accomplish this, a magnetostrictive position sensor is mounted inside

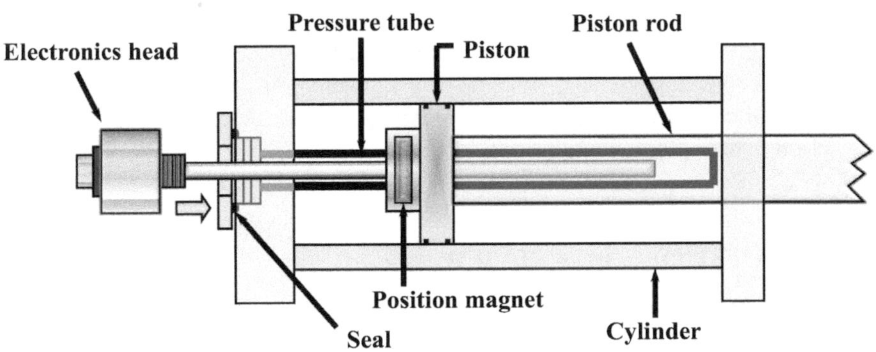

FIGURE 12.18 A magnetostrictive linear position sensor installed into a hydraulic cylinder.

TABLE 12.2
Applications of Magnetostrictive Position Sensors

Industry	Application
Automotive	Production machinery, onboard suspension, steering, transmission
Semiconductor	Chip and wafer handling
Electric actuators	One sensor can measure both linear and rotary position (patented)
Cylinders	Piston position in hydraulic and pneumatic cylinders
Food	Measuring liquid level using a float with magnet installed in the float
Level	Process control, leak detect, inventory control with magnet in float
Medical	Positioning hospital bed and diagnostic equipment, syringe measurement
Metalworking	Forges, presses, bending and cutoff machines
Mobile equip.	Waste disposal trucks, agriculture, grading and paving
Paper	Slitters and presses
Plastics	Injection molding, blow-molding
Primary metal	Walking beams and ladle control
Primary wood	Sawmills, lathes, cutoff saws, positioning knees, presses
Secondary wood	Saw positioning and tenoners
Textiles	Carpet tufters

one sliding member of a shock tower or strut. The position magnet is mounted to the other sliding member. This position data is used with a controller and hydraulic system to improve ride and handling.

Because of the ruggedness, temperature stability, reliability, high performance, multiple magnet capability, and wide range of lengths, magnetostrictive position sensors are used in many different applications. Some of those are listed in Table 12.2.

12.15 MANUFACTURERS

Some manufacturers of magnetostrictive position sensors include the following:

Ametek	www.ametek.com
Balluff	www.balluff.com
Gefran	www.gefran.com
MTS Systems Corporation	www.mtssensors.com
Novotechnik	www.novotechnik.com
Penny and Giles (Curtis Wright)	www.pennyandgiles.com
Revolution Sensor Company (design)	www.rev.bz
Santest	www.santest.co.jp/en

12.16 QUESTIONS FOR REVIEW

1. **In a magnetostrictive waveguide, a torsion wave is formed at the location of the:**

    a. Pickup
    b. Damp

c. Weld
d. Position magnet
e. Interrogation pulse

2. This may reduce position hysteresis in a magnetostrictive waveguide:

a. Weaker magnet
b. More gain
c. Better filter
d. Less gain
e. More interrogation current

3. The factor that relates the amount of strain produced in a magnetostrictive material to a given amount of magnetic flux density is the coefficient of:

a. Elasticity
b. Damping
c. Strain induction
d. Expansion
e. Magnetostriction

4. The velocity of the torsion wave as it travels in the waveguide is often called the:

a. Wave rate
b. Torsal speed
c. Torque moment
d. Sonic velocity
e. Pulse time

5. Gas inside the waveguide housing should be dry, in order to prevent formation of:

a. Magnetic domains
b. Ice
c. Villari effect
d. Beaded wire
e. Torsion

6. Regarding a capacitor, ESR is an acronym for:

a. Extra slow resonance
b. Easy set recovery
c. Equivalent series resistance
d. Full-scale result
e. Equal static reactance

7. Sonic velocity variation along the length of the waveguide is a source of:

a. Nonlinearity
b. Attenuation
c. Damping
d. Temperature error
e. Cold work

8. The interrogation pulse width is usually:

a. 0.5 to 5.0 μs
b. A sine wave
c. 500 Hz
d. 0.2 to 0.5 ms
e. Resistive

REFERENCES

[1] R. M. Bozorth, *Ferromagnetism.* New York: D. Van Nostrand, 1951.
[2] E. D. Tremolet de Lacheisserie, *Magnetostriction: Theory and Application of Magnetoelasticity.* Boca Raton: CRC Press, 1993.
[3] L. H. Van Vlack, *Elements of Materials Science.* Reading: Addison-Wesley, 1964.
[4] D. S. Nyce, Magnetostriction-Based Linear Position Sensors. *Sensors,* 11(4), 1994.
[5] D. Craik, *Magnetism Principles and Applications.* New York: Wiley, 1995.

13 Encoders

13.1 LINEAR AND ROTARY

Linear and rotary position or displacement can be measured and communicated by a device called an encoder without using any form of analog-to-digital (A/D) conversion, because the basic output signal is already in a digital format. Although the term *encoder* has also been applied to devices based on Moiré patterns from diffraction gratings, as well as laser interferometers, the terms *encoder*, *linear encoder*, and *rotary encoder* will be used here in reference to standard industrial sensors based on geometric patterns applied along a linear, angular, or circular scale and detected by any one of several methods. Encoders are available as incremental or absolute reading, and encompass various detection techniques, including brush-type, optical, magnetic, and capacitive types. The most common detection methods for industrial encoders are magnetic and optical. Besides selecting whether the output will be absolute or incremental, encoder designers make trade-offs among important product features, including ruggedness, resolution, physical size, and cost.

13.2 HISTORY OF ENCODERS

The earliest type of encoder was the brush-type, a linear version of which is depicted in Figure 13.1. In the figure, flexible mechanical contact fingers rub along a metal pattern printed onto an insulating base. The path of the brush moves over conducting and insulating segments. The conductor pattern is formed onto the base in the same way as printed circuit boards are made for connecting electronic circuits.

The figure shows a straight binary code. An alternative, the Gray code, is explained later in this chapter. In the figure, the brush holder and four brushes (flexible metallic contact brushes) move along a left-to-right measuring axis. Dark segments represent the electrically conductive areas that would normally be connected to a positive voltage, such as +3.3 or +5 volts DC, to represent a logic one. Light areas are insulated or not electrically connected, and represent a logic zero. So, when a brush is on a light area, the signal from that brush is at logic zero (zero volts), and when a brush is on a dark area, the signal from that brush is at logic one (+3.3 or +5 volts). The brush holder shown is an insulating material, so wires can be connected with each brush to bring their voltages to a measurement or readout circuit. With the brushes in their rightmost position as shown, all four are at logic one, and indicate binary 1111, or decimal 15 (or hexadecimal F). At the leftmost position, binary 0000, or decimal 0 would be indicated. As the brushes move to the right from zero, the indicated position increments by 1 for each new position. These positions can be called digital positions zero through fifteen, or numeral positions one through sixteen. This simple pictorial, having only 4 bits, is shown for simplicity. Normal encoders will usually have at least eight, and up to 18 bits. Eight bits can indicate positions 0 through 255, for a resolution of about 0.4% (that is 100% × 1/256), while 18 bits can indicate

DOI: 10.1201/9781003368991-13

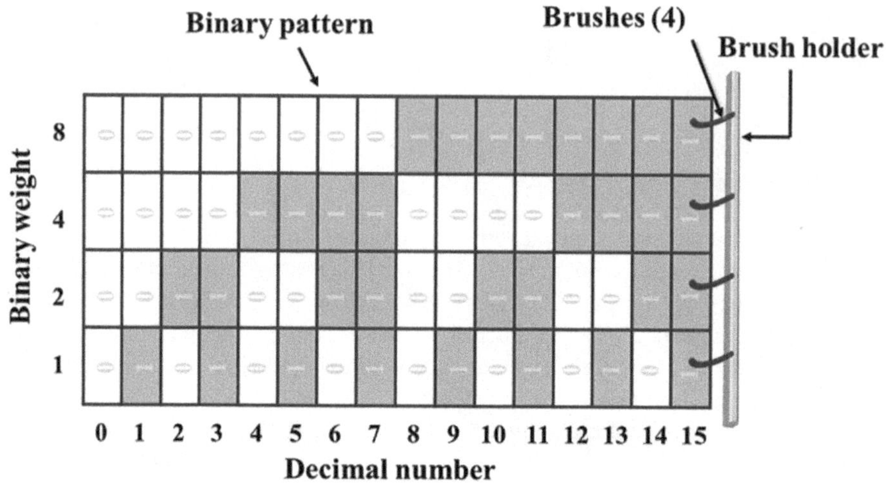

FIGURE 13.1 Binary pattern of a brush-type of absolute linear encoder.

0 through 262,143 (note: 2^{18} = 262,144), for a resolution of about 0.0004% (that is, 100% × 1/262,144).

The major drawback of the brush-type encoder technique was the wear of the metal pattern and fingers, sometimes resulting in errors, and eventually resulting in failure. Non-contact encoders based on optical and magnetic techniques took over in the 1960s and 70s because of their increased life and reliability as compared with the brush-type. In the 1990s, the brush-type largely became no longer acceptable for industrial measurement systems due to their wear-out, and the resulting shorter life span. For this reason, additional details of brush-type encoders are not included here, except where expedient in order to illustrate switching and coding theory.

13.3 CONSTRUCTION

Linear and rotary encoders can take many physical forms. Some linear encoders have a rod and housing similar to the potentiometric linear position sensor that was depicted in Figure 4.3. This type would typically comprise a housing, rod, bushing, wiper, scale, detector, and electronics. The rod is usually made from hardened and polished steel, and moves linearly into and out of the housing from one end. The bushing supports the rod, reducing wear and resisting any force due to side loading. One or more wipes clean particles away from the rod before it goes through the bushing. This reduces wear of the rod and bushing. Also, if it is an optical encoder, particles must be avoided because they may interfere with the optical sensing element. The scale of an optical encoder is an opaque member with slots (or reflective areas), to indicate the zeroes and ones. The scale of a magnetic encoder is a non-magnetic base with a magnetizable coating that may be magnetized in spots for coding the zeroes and ones, or a coded magnetic strip. Some encoders use a capacitive transducer, with the scale comprising an insulating base with conductive coding. As the

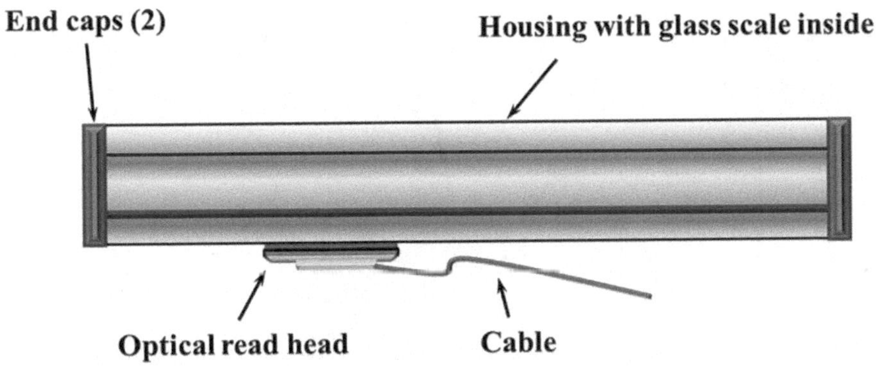

FIGURE 13.2 An optical type of linear encoder.

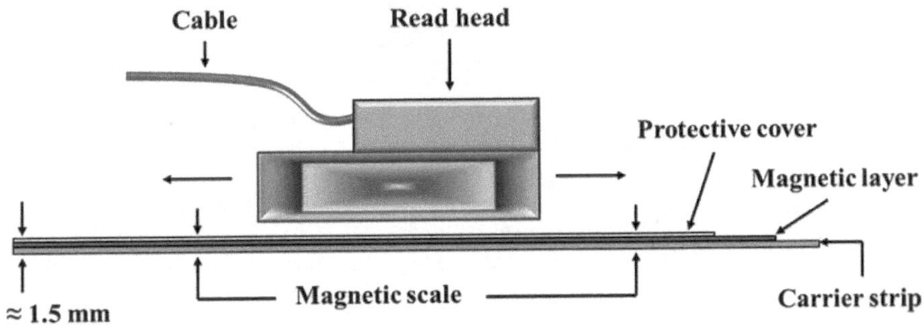

FIGURE 13.3 Incremental magnetic linear encoder with separate magnetic scale.

rod moves in or out, displacement or absolute position information is read from the scale via the optical, magnetic, or capacitive detector.

One configuration of an optical type of linear encoder is shown in Figure 13.2. The housing provides the base for combining the component parts, but also includes means for mounting of the encoder to the application.

An alternative linear encoder construction is shown in Figure 13.3. In this magnetic encoder, the magnetic tape is supplied to the user as a separate component. The tape is attached to a stationary member, along the axis to be measured. The moving member is fitted with a magnetic pickup device. The combination of the coded tape and the pickup device comprises the linear encoder.

In an incremental encoder, there is not a binary equivalent of an absolute position as was shown in Figure 13.1. Instead, there is a series of ones and zeroes along a straight or circular path for a linear or rotary encoder, respectively. The ones could be magnetized segments for a magnetic encoder, or slits or reflective segments for an optical encoder. The read head counts the number of zero/one transitions that pass by, and keeps a record of the count. The pulse-count is zeroed at a starting position, often at one end of the measuring range. Then, as the read head moves along the measuring axis, variations on the scale are counted and used to indicate the position as a distance from the starting position. As in all incremental encoders, this is actually a

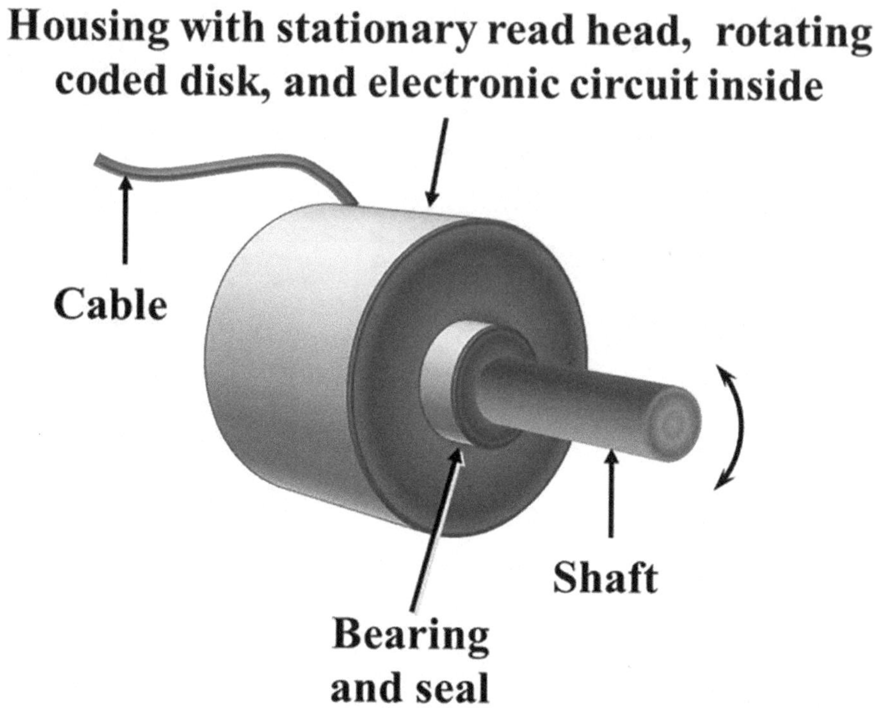

FIGURE 13.4 Magnetic rotary encoder.

measurement of displacement, rather than one of position. This is further explained later in this chapter.

A magnetic type of rotary encoder is shown in Figure 13.4. It functions in the same way as a linear encoder, except that the digital patterns are laid out in a circular instead of linear fashion. Magnetized and non-magnetized areas represent the logical ones and zeros. Or, ones and zeros could be magnetized in two different ways (such as different polarization or other readable difference).

A simple coded disk for a rotary encoder is shown in Figure 13.5. This one, like the linear pattern shown in Figure 13.1, has an absolute binary code.

In the figure, each dark area represents a logical one, a light area represents a logical zero. A read head would remain in a fixed position as it reads the rotatable disk. A logical one could be a slit in an opaque disk, or could be a reflective area, depending on the configuration of the sensing elements in the read head. The figure shows the innermost track as the least significant bit and the outermost track as the most significant bit, but this order can be reversed if desired.

13.4 ABSOLUTE VERSUS INCREMENTAL ENCODERS

The binary patterns shown in Figures 13.1 and 13.5 were for absolute encoders, in which the position is always known with regard to a zero reference position (the zero reference position is normally 00 . . . 0). Conversely, an incremental encoder

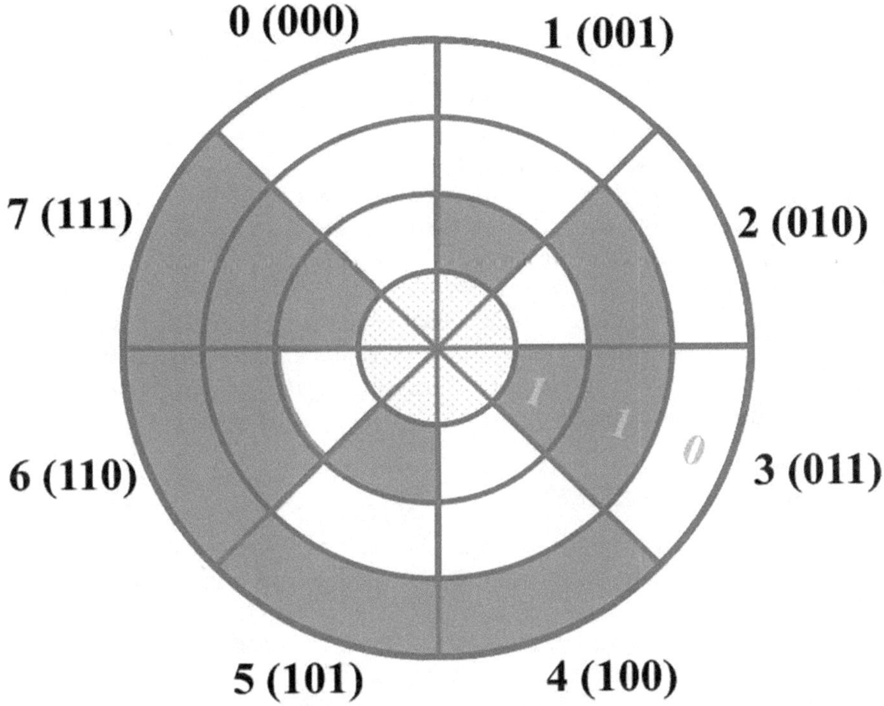

FIGURE 13.5 Binary pattern for an absolute rotary encoder (simplified).

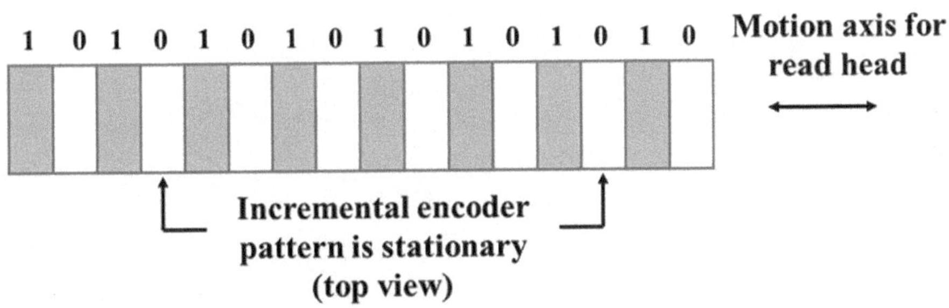

FIGURE 13.6 Pattern of an incremental linear encoder.

measures a change in position with respect to the last known position. If the record of the last known position is lost or corrupted, such as through a power loss or electromagnetic interference, then the position information is lost. In this case, an incremental position sensor must be sent back to a known position so the count can be reset. An example of a pattern for an incremental linear encoder is shown in Figure 13.6.

It may be obvious that the information encoded onto the track of an incremental encoder is simpler than that of an absolute encoder, and can have a finer pitch than that of an absolute encoder. This is because only 1 or 2 bits must be toggled while the position is changing with an incremental encoder. With an absolute encoder, sufficient information must be read at each position to totally represent the position at that point in the sensor stroke, with respect to the reference datum. So, if the position count is 2046 counts, an absolute encoder must encode the number 2046 (requiring 11 binary bits) at that exact location. Whereas, with an incremental encoder, there is only one bit of information located at that point (or two, counting the quadrature bit at 90° separation from the first, as presented later). In addition to the track needed to supply the A quad B increments, an incremental encoder may also have an index track. The index track has a mark at one point that indicates the index, reference, or starting point. Then the count can be re-zeroed automatically each time the index point happens to be passed.

So, an advantage of an incremental encoder is that of higher resolution for a given spacing of the detected feature (optical slots, for example), due to its simpler pattern. An absolute encoder, however, has the advantage of instant start-up after a corruption of power or data, without needing to re-zero or reset the count.

13.5 OPTICAL ENCODERS

An optical encoder uses a transmitter/receiver pair of optoelectronic devices. The transmitter part of the pair is a light source, usually one or more light-emitting diodes (LEDs). The receiver part usually comprises two or more phototransistors. (Two for quadrature output with an incremental encoder, or a number of phototransistors equal to the number of bits of resolution, for an absolute encoder.) A track having opaque and transparent sections arranged in series is passed through the optoelectronic pair as the position changes; see Figure 13.7. (Alternatively, reflective and non-reflective sections can be used instead of transparent and opaque sections.)

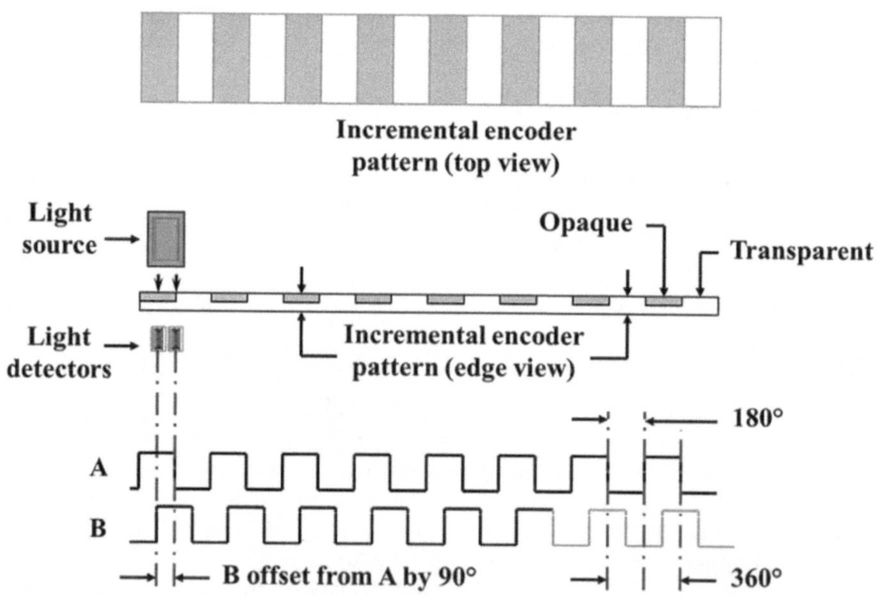

FIGURE 13.7 An incremental optical encoder with LED and two phototransistors arranged to provide quadrature output.

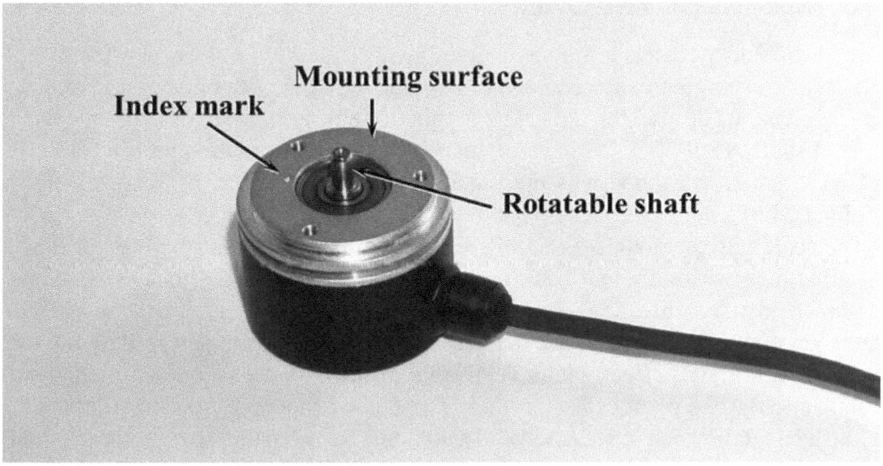

FIGURE 13.8 An incremental optical rotary encoder.

The light source includes a means for focusing the light onto the detector(s). This may include a focusing lens, collimator, and/or a slit or pinhole. If the encoder is incremental, only two detectors are needed. The second detector is spaced from the first by a set distance in order to yield a 90° shift between the two outputs, resulting in a quadrature output (see Section 13.7). In this case, the alternate dark/light areas are counted to obtain the present position. A closer spacing between adjacent dark or light areas will yield a higher resolution. A third detector may be used to detect an index mark. An incremental optical rotary encoder is shown in Figure 13.8.

The encoder of Figure 13.8 has a housing diameter of 58 mm, and a resolution of 5,000 increments per revolution. An index mark can be seen that is machined into the mounting surface, indicating the angular position at which the index signal will be sent. This encoder has sine wave-shaped A and B outputs (plus an index output), rather than a square wave as previously shown. The sine wave signal shape allows interpolation between the 5,000 increments. On the encoder in the figure, 5X interpolation is allowed, according to the manufacturer specifications. With interpolation, positions can be determined that are in between the increments, based on the relative amplitudes of the A and B output signals. Thus, 5X interpolation enables a higher resolution to be determined, at the equivalent of 25,000 increments per revolution in this particular encoder.

In an optical absolute encoder, a sufficient number of light detectors are needed to represent the maximum number of bits to be indicated. The reciprocal of this number of bits is the resolution. For example, a 12-bit encoder requires 12 detectors and represents 4096 counts, having a theoretical resolution of 0.024% or 1/4096. In this example, there will be 4096 sets of position data along the full-range stroke of the sensor. Twelve parallel data bits will be read by twelve phototransistors. It is possible to use one long light bar to drive all of the phototransistors, or multiple LEDs can be used.

13.6 MAGNETIC ENCODERS

A magnetic encoder uses a magnetic tape or other medium onto which the position information is magnetically recorded (the scale), together with one or more magnetic sensing elements to read the data. The magnetic sensors in modern magnetic encoders are usually Hall effect or magnetoresistive types. Information on the Hall effect and on magnetoresistance was presented in Chapters 10 and 11, respectively. An incremental magnetic encoder is shown pictorially in Figure 13.9.

The scale may be a rigid assembly, for mounting by screws or rivets, or may be a flexible tape with adhesive backing for mounting by sticking to a clean surface. As shown in the figure, the scale incorporates reversals of magnetic polarization uniformly along its length. These magnetic field variations are detected by the magnetoresistive (or Hall effect) pickup elements. If the encoder is an incremental type, as shown, only two pickups are needed. The second pickup is spaced from the first pickup by a set distance in order to yield a 90° shift between the two outputs (A and B), resulting in a quadrature output (see Section 13.8). In this case, the pulses from the pickups are counted to obtain the present position, and the 90° phase shift is used to indicate the direction of travel. The highest resolution that is possible is limited by the smallest distance that can be achieved between two adjacent pole reversals, and the ability of the receiving detector and circuit to read the relative signal levels. A Hall effect sensing element can directly provide indication of the north and south magnetic poles as shown in the waveform in the figure. Obtaining the response shown with a magnetoresistive sensing element requires that the MR sensing elements be

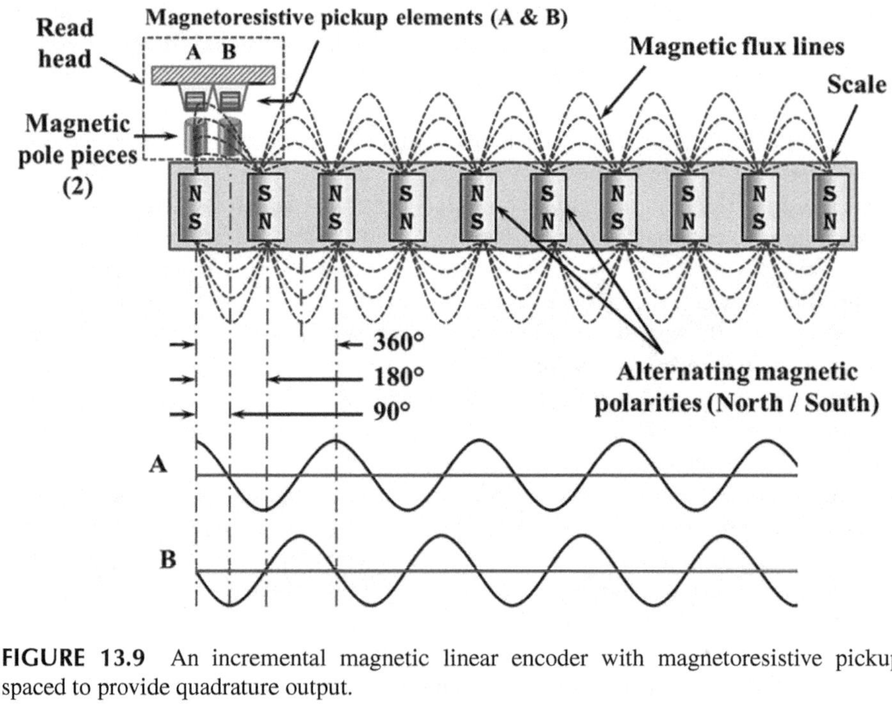

FIGURE 13.9 An incremental magnetic linear encoder with magnetoresistive pickups spaced to provide quadrature output.

manufactured with a magnetic bias so that the north and south poles can be distinguished. An MR sensor normally provides the same resistance with either a north or south magnetic pole at a given flux density. But if the MR element has a north bias (with zero externally applied magnetic field), for example, it can be able to sense both north and south applied poles. With a north pole bias, applying a north magnetic field will further increase the north flux density and provide a further resistance change from the north-biased zero state. Applying a south pole will reduce the flux density below the north-biased zero state.

Incremental magnetic encoders are commonly used, but absolute magnetic encoders are also available, following the same coding theory as presented for the optical type. An absolute magnetic linear encoder may have a similar configuration and form factor as an incremental version (as was shown in Figure 13.3), but the scale of an absolute encoder may be wider so that a greater number of parallel bits may be encoded. Also, with the greater number of parallel bits being encoded, an absolute version will include effective means to accurately track the individual lines of encoded bits. This tracking may be provided by mechanical members such as with the read head being enclosed in a car that is guided to travel along the scale while not being able to move laterally.

13.7 CAPACITIVE ENCODERS

Although magnetic and optical versions have historically been the most common types of encoders, there is also an encoder technology that utilizes capacitive sensing. They can be rugged, as magnetic encoders, but also may have resolution capability similar to optical encoders. There are several configurations of capacitive encoders in use, both rotary and linear. An absolute rotary encoder will be described and illustrated here.

A preferred arrangement for the capacitive members of an absolute rotary encoder incorporates means for a fine measurement coupled with a coarse measurement, such as taught in Netzer US patent 6492911.

The sensing element comprises a rotor and a stator (although there are also rotor and two-stator versions). A movable rotor having coarse and fine rotor plates is mounted parallel to a fixed stator, such that the rotor rotates about a central shaft. The coarse rotor plate can be an eccentric circle, so that its capacitance changes in a sinusoidal way with rotation. The fine rotor plate has a sinusoidal pattern of two or more lobes (with a four lobe version being shown in the figure). Coarse and fine transmitting plates on the stator are energized with oscillating voltages. A receiver plate, also on the stator, enables measuring capacitance that varies with rotation angle of the rotor. The fine measurement provides the desired high resolution by implementing a number of lobes or fingers about the 360° of a single rotation. A coarse measurement provides the selection of which half, quadrant, or other order of division the fine position is to be coupled. With four lobes, for example, the coarse position selects the quadrant of the measurement, and the fine position provides the position within that quadrant. See Figure 13.10.

Four oscillating voltages, V_1 through V_4, energize the fine transmitting plates, with each fourth plate connected together, so that all of the 16 plates shown are energized

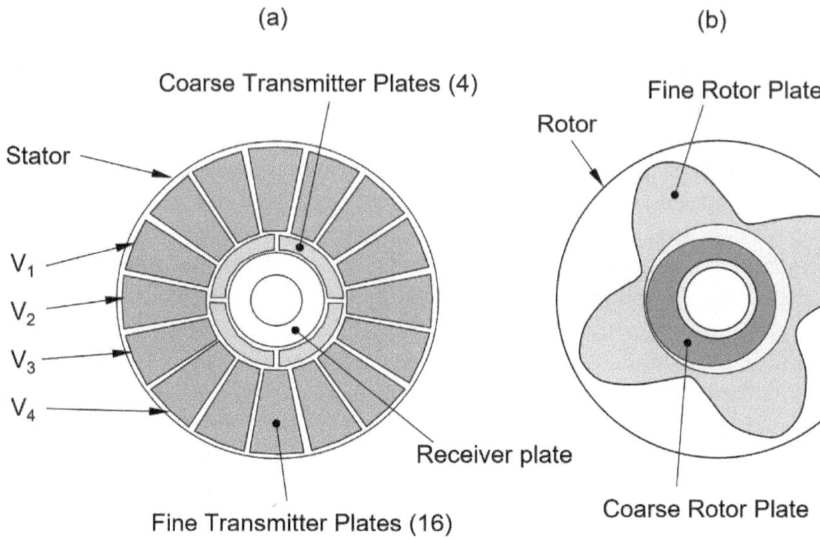

FIGURE 13.10 Capacitive rotary encoder stator (a), and rotor (b).

at appropriate times. The oscillating voltages impressed on each of V_1 through V_4 are separated in phase one from the next by 90°. Likewise, the four coarse transmitting plates receive voltages separated by 90°. The fine transmitting plates are energized at a higher frequency, such as 40 kHz. The coarse transmitter plates are energized at a lower frequency, such as 10 kHz. The large difference in the two frequencies aids in demodulating and filtering the signals individually. The fine transmitting plates are energized, then the coarse transmitting plates are energized, alternating so that each can be detected separately. See Figure 13.11.

Each voltage is received into a voltage converter with virtual ground, so that the plates are held at virtual ground. Signals are detected with respect to the associated oscillator phase, and then filtered (such as by using a low-pass filter). Resulting signals include coarse sine and cosine, and fine sine and cosine. The coarse signals define an absolute position quadrant (if using a four-lobe rotor pattern), while the fine signals provide the more exact position within that quadrant.

In normal operation, the sensing element operates in coarse mode on power-up, or at any time that re-zeroing may be required. Then the fine mode is used to keep track of the present position for ongoing operation.

13.8 QUADRATURE

Incremental encoders have two outputs, called A and B. These are arranged in quadrature, which means they are separated in phase by 90°, as shown in Figure 13.12.

This arrangement and type of signal are also called *A quad B*. It is called *quadrature* because "quad" means four, and a transition of one or the other data line (A or B) occurs four times per count cycle. A square wave shape is shown in the figure, but a sine wave shape can also be implemented. Whether the encoder is linear or rotary, a

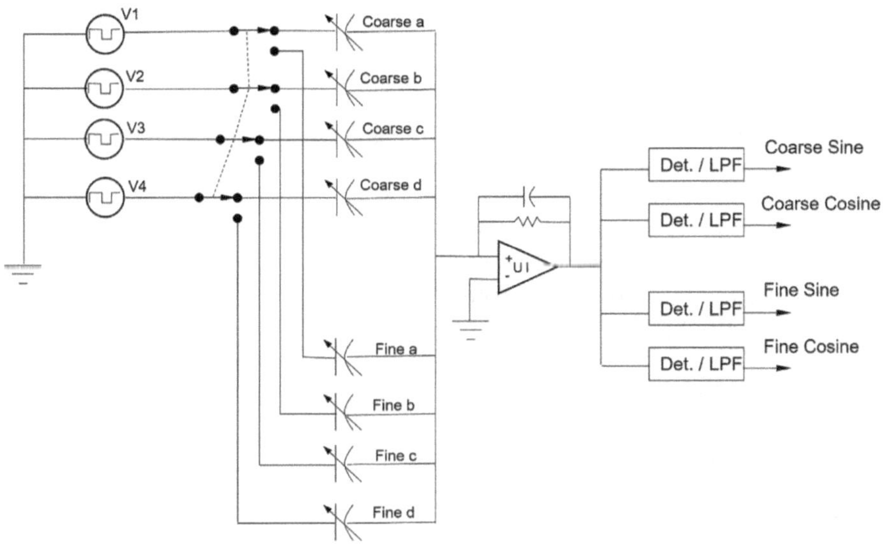

FIGURE 13.11 Circuit illustration for a Capacitive rotary encoder.

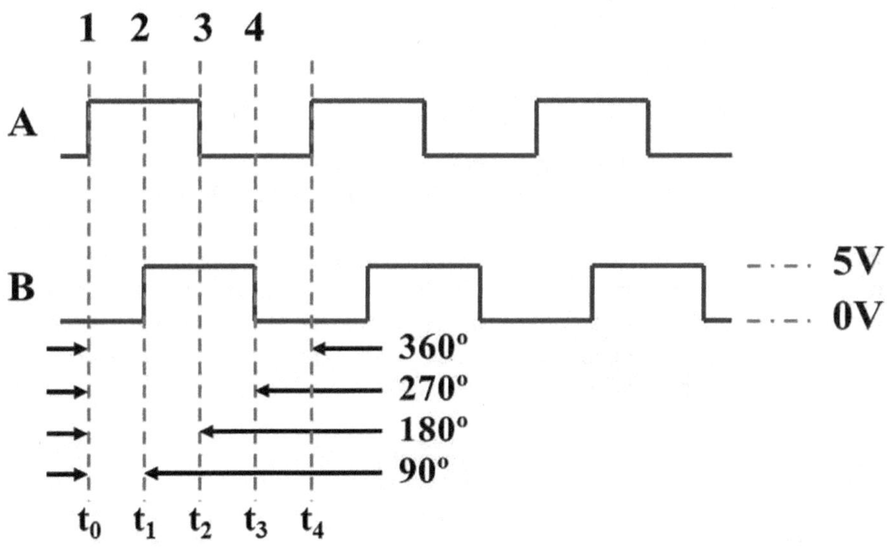

FIGURE 13.12 A quad B outputs are separated by 90°. Transitions 1 through 4 occur during each count cycle.

complete count cycle in most situations is generally considered to comprise 360° (the 360° is a count cycle, not the angle of rotation of a rotary sensor). So, with two phases having four transitions each per count cycle, one phase is delayed when compared to the other by 90°, or one-fourth of 360°. The four transitions (describing the status just after the transition takes place) as shown in the figure are: (1) A high, B low; (2)

A high, B high; (3) A low, B high; (4) A low, B low. At t_4, the same cycle starts over again, with A going high as it did at t_0. So, there are four time periods per cycle, each of 90° duration, the time periods being $t_1 - t_0$, $t_2 - t_1$, $t_3 - t_2$, and $t_4 - t_3$. So that $t_4 - t_0$ is equivalent to 360°.

If the pulses from either A or B are counted, the change in position is indicated by the number of counts multiplied by the distance per count (for example, 1,000 counts with a resolution of 10 microns would be 1000 counts × 10 microns = 10 mm). The reason for having the other output (B or A) is to indicate the direction of motion, either incrementing or decrementing the count, so that the current count may represent the actual position. In Figure 13.10, for example, reading the chart normally from left to right, these transitions would be *increments*:

t_0: B is low when A goes high
t_1: A is high when B goes high
t_2: B is high when A goes low
t_3: A is low when B goes low

Reading the chart backwards, from right to left, these transitions would be *decrements*:

t_3: A is low when B goes high
t_2: B is high when A goes high
t_1: A is high when B goes low
t_0: B is low when A goes low

So, any transitions of either A or B signal will increment or decrement the count. Whether it is an increment or a decrement is determined by the relationship between the logic states of A and B. Instead of only counting the transitions of A or B as the magnitude of the count, it is also possible to count all transitions (going high or going low) of both A and B, and to still use the previously mentioned relations to determine the direction of movement (increment or decrement of the count). Counting of all transitions of A and B serves to improve the measurement resolution. As mentioned earlier, some sensing elements provide sine wave signals instead of high-low logic states. In that case, it is also possible to determine a more fine resolution by interpolating based on the relative signal amplitudes.

Some modern circuits that read the A quad B signals from a position sensor do not actually wait for a transition to occur, and then count that transition. Instead, the states of A and B may be continuously monitored at a much higher sampling rate than the transitions are expected to occur. With this (state, rather than transition) information and, usually, a microcontroller, smoother operation can be obtained with less likelihood of error.

13.8.1 Burst Mode

Normally, the quadrature signals are used continuously to update a count that represents the measured displacement. But in some incremental sensors, another method may be available, called burst mode. With burst mode, an external device (such as a controller) may ask for displacement data, and the sensor responds by sending a burst of pulses equal to the number of counts that represent the present displacement. This way, the controller doesn't need to keep a running count (that is done within the sensor), and can instead ask for a total count at any time that the measurement information is required.

13.9 BINARY VERSUS GRAY CODE

In an absolute encoder, the output is typically either in straight binary or in Gray code, but Binary Coded Decimal (BCD) is also sometimes supported. The patterns and codes presented earlier in this chapter have all been binary. We will call that natural binary, to distinguish it from BCD. BCD is similar to natural binary, except that the BCD data bits are arranged into sets of 4 bits each. Four bits of BCD data equal one character, and can have a value of zero through nine (instead of zero through fifteen as in natural binary). For reference, hexadecimal is also shown in Table 13.1 because the reader may be familiar with this binary number representation. An 8-bit *byte* or *word* can be viewed as two 4-bit *nibbles*. In natural binary, the 4-bit nibble can have 16 possible values (0 through 15). These are represented as the decimal numbers zero through nine, followed by letters A through F (for a total of 16 characters).

Natural binary or BCD are easy to interface directly to standard digital circuits, but have the disadvantage that a change of only one increment involves the simultaneous change of more than one bit. This means that a large error can be indicated if all of the bits do not switch at the exact same time. For example, when the count changes from 7 to 8, it is a change from 0111 to 1000 in binary or BCD. But, if the *most significant bit* (MSB) changes a few milliseconds before the rest of the bits change, there will be a reading of 15 (1111 in binary, F in hexadecimal, or an error in BCD) for those few milliseconds. This can represent a large error and cause stability problems in a servo control system. A system called the Gray code was developed to solve this problem.

The Gray code is arranged so that an increment of one bit always changes only one of the bits in the whole set of output bits [1, pp 6–126]. The differences among BCD, hexadecimal, binary, and a Gray code are shown in Table 13.1.

Figure 13.13 is a corresponding Gray code pattern with the decimal equivalent number along the abscissa and the bit weight along the ordinate.

One can see that a change between any two adjacent numbers requires only the change of one bit in the Gray code. When the sensor operates using the Gray code, the controller or other device to which it is connected will usually incorporate a Gray to binary conversion. A Gray to binary converter can be built in hardware with gates as shown in Section 13.10, or implemented in software by storing the conversion constants in a lookup table in memory. Then the controller can use the binary numbers for operation as usual after they are converted from the Gray code.

13.10 ELECTRONICS

The light source of an optical encoder is usually one or more LEDs driven by a simple regulated voltage with current limiting resistor, or by a constant current source. Temperature compensation capability may be added in order to maintain a constant

TABLE 13.1
Decimal Equivalents of Hexadecimal, Binary Coded Decimal (BCD), Natural Binary, and a Gray Code

Decimal	Hexadecimal	BCD	Natural Binary	Gray
0	0	0000 0000	0000	0000
1	1	0000 0001	0001	0001
2	2	0000 0010	0010	0011
3	3	0000 0011	0011	0010
4	4	0000 0100	0100	0110
5	5	0000 0101	0101	0111
6	6	0000 0110	0110	0101
7	7	0000 0111	0111	0100
8	8	0000 1000	1000	1100
9	9	0000 1001	1001	1101
10	A	0001 0000	1010	1111
11	B	0001 0001	1011	1110
12	C	0001 0010	1100	1010
13	D	0001 0011	1101	1011
14	E	0001 0100	1110	1001
15	F	0001 0101	1111	1000

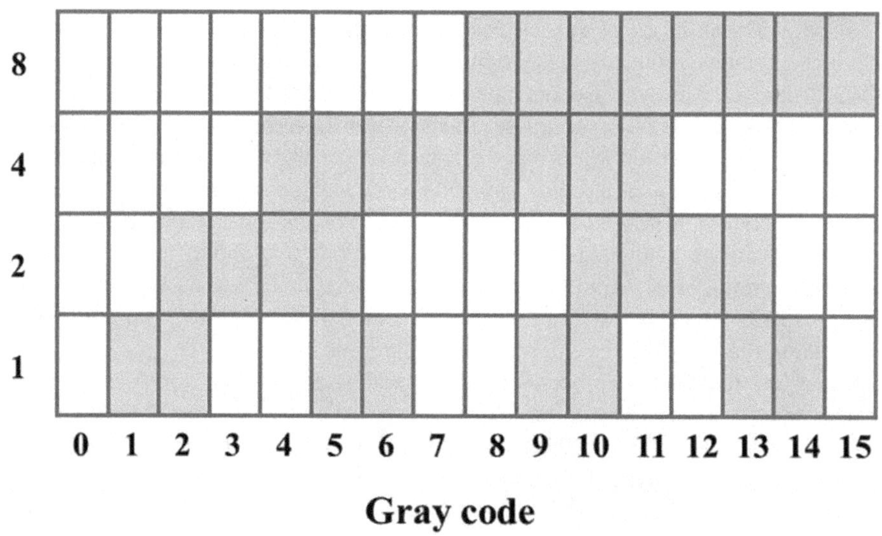

FIGURE 13.13 A Gray code pattern corresponding to Table 13.1.

light output with temperature variations. The light from the LEDs is normally detected by phototransistors.

The LED and phototransistor are typically connected as shown in Figure 13.14.

The resistor in series with the LED can be a combination of a thermistor and one or two resistors in order to compensate the LED output intensity over a temperature

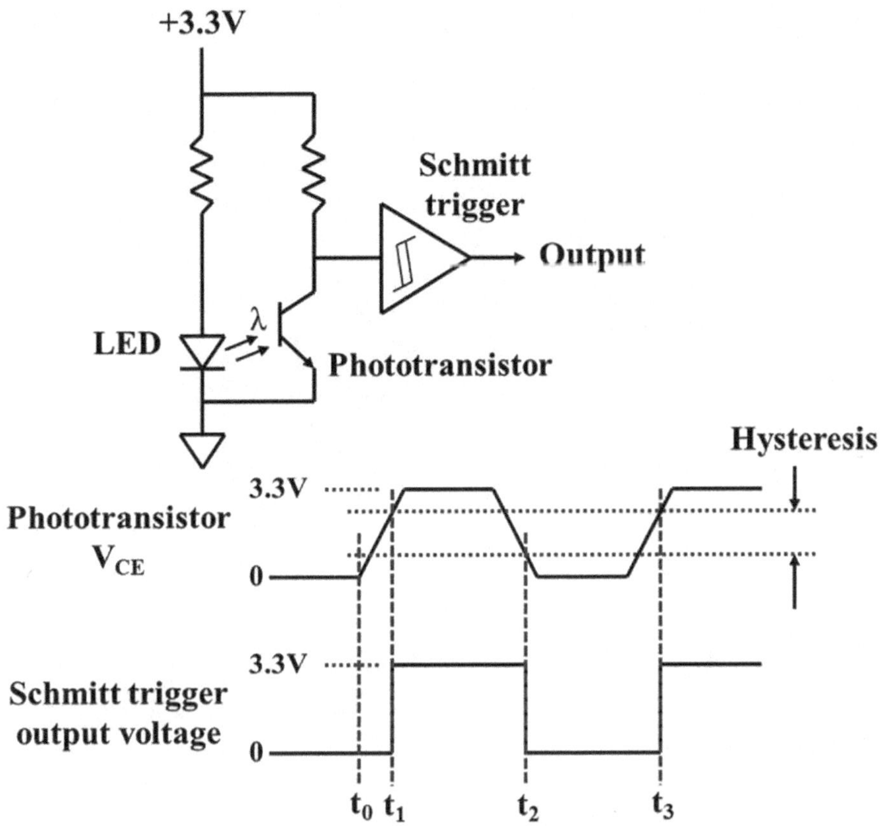

FIGURE 13.14 LED and phototransistor connection circuit, with Schmitt trigger.

range (application notes on thermistor linearization are available on the webpages of thermistor manufacturers). The signal directly from the phototransistor will not be a waveform with sharp transitions. The sharp transitions needed for a square wave A quad B output can be generated by a Schmitt trigger circuit. A Schmitt trigger transitions from one state to the other when the input crosses a voltage threshold. It has hysteresis built in, so that the opposite transition will not take place until the input moves substantially past that same threshold. So the Schmitt trigger has a positive-going threshold and a negative-going threshold that differ by a fixed percentage of the supply voltage. In Figure 13.14, the positive-going threshold at t_1 is approximately 2.5 volts (with a 3.3 volt power supply voltage). The negative-going threshold at t_2 is approximately 0.8 volt. This is a hysteresis of 1.7 volts, or about 51% (of the 3.3 volts). Various integrated circuit technologies have their specified Schmitt trigger hysteresis percentages or levels. The Schmitt trigger shown is non-inverting. Other Schmitt triggers may be an inverting type, and that's OK, but the output voltage will be reversed regarding high and low.

If an absolute encoder uses a Gray code pattern, the resulting output must be converted to natural binary before the data can be easily handled by a microcontroller.

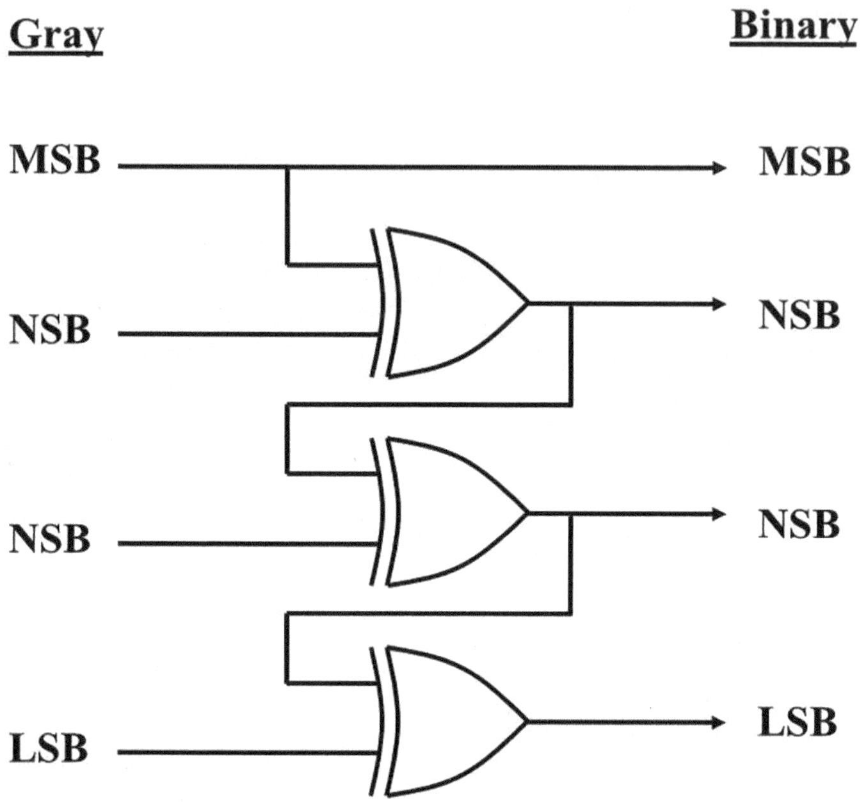

FIGURE 13.15 Circuit schematic for converting Gray code to natural binary.

This conversion can be done by using a mathematical formula in the microcontroller, or by utilizing a lookup table for the microcontroller, or it can be implemented in hardware. A schematic for implementing a Gray code to natural binary conversion is shown in Figure 13.15.

The schematic accommodates 4 bits, but it is easy to visualize how it can be expanded by just duplicating more of the same basic element for a desired number of bits. At left in the figure, *MSB* means most significant bit, *NSB* is next significant bit, and *LSB* is least significant bit. The three logic elements in the figure are Exclusive OR gates (also called XOR). With an Exclusive OR gate, if either of the two inputs is high then the output is high, except that when both inputs are high, the output is low. If both inputs are low, the output is low. Understanding this hardware logic also enables one to achieve the same result through implementation of a software algorithm.

13.11 ADVANTAGES

With some glass scale optical linear encoder models up to several meters long having resolution as fine as 0.01 μm (μm can be pronounced as "microns", or "micro meters", each meaning millionths of a meter), linear encoders are often the preferred method of position sensing in the precision environment of machine tools. The highest resolution sensors tend to be the incremental ones. A disadvantage to the incremental mode is that data can be corrupted due to electromagnetic noise or power fluctuations.

Magnetic linear encoders, with a magnetic tape and a read head, can have a long measuring range of up to 40 meters. The user performs the installation by attaching the magnetic tape to a fixed surface, with the read head traveling along the length of the magnetic tape. Such magnetic linear encoders retain the advantage of non-contact measuring, and so, have a virtually unlimited life.

Shorter linear encoders are often magnetically or optically coupled, and may be self-contained with the magnetic tape or optical scale contained within a housing. Magnetic, optical or capacitive types may have a read head that is part of a car that slides within a track on the housing. Typical industrial encoder models can have a resolution of better than 0.1 micron. The bushings and wipers, used with models having an actuator rod, have a finite lifetime. End of life, however, does not mean that the measurement accuracy is affected. It means that some mechanical drag is encountered as the rod rubs on worn bushings or bearings, sometimes accompanied by audible noise.

13.12 TYPICAL PERFORMANCE SPECIFICATIONS AND APPLICATIONS

Figures 13.2 and 13.7 depicted optical linear encoders. A typical specification of an incremental optical linear encoder is listed in Table 13.2.

It would be desirable that all position encoder manufacturers (and all other position sensor manufacturers, as well) listed the same specifications and used the same format, but they do not. There should not be a single specified item called accuracy, but this should be separated into the several components of nonlinearity, repeatability, hysteresis, temperature sensitivity, and so on. But specifications are listed here as actually found on data sheets from manufacturers, even though, in the author's opinion, their formats could be improved. Next, a specification of a typical incremental optical rotary encoder is listed in Table 13.3.

TABLE 13.2
Typical Specification of an Incremental Optical Linear Encoder

Measuring range:	1,000 mm
Accuracy:	± 10 μm
Increment signals:	TTL (transistor-transistor logic)
Power supply:	5 V DC, ±10%, at up to 100 mA
Max speed:	≤ 60 m/minute
Moving force:	≤ 5 N
Operating temperature:	0 to 55°C

Note: Nonlinearity, repeatability, hysteresis, temperature sensitivity not listed in specification.

TABLE 13.3
Typical Specification of an Incremental Optical Rotary Encoder.

Measuring range:	360° rotation
Line counts:	3,600/revolution
Reference mark:	one
Maximum frequency:	100 kHz
Power supply:	5 V DC, ±10%, at up to 155 mA
Starting torque:	≤ 0.001 Nm
Operating temperature:	−30 to 70° C

TABLE 13.4
Typical Specifications of an Absolute Optical Rotary Encoder

Measuring range:	360° rotation
Line counts:	4,096/revolution
Output type:	SSI
Maximum SSI clock:	1 MHz
Power supply:	4.6 to 26.4 V DC, at up to 80 mA
Starting torque:	≤ 2 ounce-inch at 25°C
Operating temperature:	−20 to 85°C
Output type:	natural binary or Gray code

There are many incremental or absolute, linear or rotary encoders that are available on the market. Many utilize optical technology, and many are magnetic. Some are capacitive, and other technologies. Although the specifications listed here are from actual data sheets, they are not necessarily indicative that one type may be better than another type for a selected application. Other similar encoder models may also be available that have better specifications for a particular application. So, the specifications listed in this book are intended to provide general knowledge for the reader to become familiar with various aspects to be considered regarding encoder specification. Table 13.4 lists some specifications of an absolute optical rotary encoder:

Table 13.5 lists some specifications of an absolute magnetic rotary encoder having an SSI output and that can be ordered with either natural binary or Gray code type of output configuration.

Table 13.6 shows specifications of an incremental magnetic linear encoder. The read head rides above a scale, as depicted in Figure 13.3. The scale is attached to a rigid base with an adhesive backing. The distance between the scale and the read head is called the ride height.

Coding patterns for optical encoders can be photographically reproduced onto the measuring medium; or for magnetic encoders, magnetically written as tracks onto the measuring medium. Any nonuniformity of the pattern on the measuring medium is a source of error. In a linear encoder, such nonuniformities may include the width and spacing of the optical, magnetic, or conductor tracks that represent the individual bits.

TABLE 13.5
Typical Specifications of an Absolute Magnetic Linear Encoder

Measuring range:	360° rotation
Accuracy:	0.35° rotation
Line counts:	16,384/revolution
Output type:	SSI
Maximum SSI clock:	2 MHz
Power supply:	10 to 30 V DC, at up to 10 mA
Starting torque:	≤ 2 N cm
Operating temperature:	−30 to 85°C
Output type:	natural binary or Gray code

TABLE 13.6
Typical Specifications of an Incremental Magnetic Linear Encoder

Measuring range:	10 m
Precision:	± 40 μm/m
Resolution:	5 μm
Hysteresis:	≤ 4 μm (with ride height up to 0.5 mm)
Maximum ride height:	1.0 mm
Output type:	AB quadrature
Count frequency:	1 MHz
Power supply:	4.7 to 30 V DC, at up to 35 mA
Operating temperature:	−10 to 80° C
Output type:	AB quadrature
Maximum read head speed:	2 m/s

Incremental encoders generally have the possibility of finer resolution, in a given technology, compared with absolute versions. Magnetic scales are somewhat more rugged than optical scales, because high-resolution optical scales are made from glass.

Typical encoder applications include process machinery feedback and control, robotics, and measuring equipment. Equipment utilizing an incremental encoder will often implement a zeroing function when the machine is first turned on, and additional re-zeroing cycles at opportune times during operation. This is to avoid, as much as possible, prolonged error of the position count if it becomes corrupted by erratic motions or externally induced noise.

13.13 MANUFACTURERS

Some manufacturers of linear and angular/rotary position encoders, having incremental or absolute configuration, and optical or magnetic technology include the following:

BEI Sensors	www.beisensors.com
Baumer	www.baumer.com
Curtis Wright (Penny & Giles)	www.cw-industrialgroup.com
Dynapar	www.dynapar.com
Encoder Products Company	www.encoder.com
Gurley	www.gurley.com
Heidenhain	www.heidenhain.com
Megatron	www.megatron.de
Netzer	www.netzerprecision.com
Revolution Sensor Company (design)	www.rev.bz
Renishaw	www.renishaw.com
Sick (Stegmann)	www.sick.com
Turck	www.turck.us
US Digital	www.usdigital.com

13.14 QUESTIONS FOR REVIEW

1. A micron is:

a. 1×10^{-6} newtons
b. 1×10^{-6} meters
c. 1×10^{-4} inches
d. an electron
e. a square wave

2. Hexadecimal characters are labeled as:

a. double precision
b. red or blue
c. significant
d. 0–9 plus A–F
e. Moiré patterns

3. On a rod-style linear encoder, wipes are used to:

a. clear memory
b. rub the LED
c. focus the LED
d. provide support
e. remove debris

4. After a power interruption, this type of encoder may require re-zeroing:

a. incremental
b. Gray code

c. absolute
d. BCD
e. SSI

5. With an A quad B signal, the B signal is offset from the A signal by:

a. 90 degrees
b. 45 degrees
c. 180 degrees
d. 360 degrees
e. twisting it

6. Regularly spaced magnetic, optical, or other variations are on the encoder's:

a. slide
b. car
c. scale
d. phototransistor
e. top

7. When an encoder sends a number of pulses representing the displacement, it is in:

a. reset mode
b. burst mode
c. index mode
d. pulse mode
e. quadrature mode

8. This circuit's output switches states when the input crosses a voltage threshold:

a. filter
b. demodulator
c. Exclusive OR
d. MSB
e. Schmitt trigger

REFERENCE

[1] J. G. Webster, *The Measurement, Instrumentation and Sensors Handbook.* Boca Raton, FL: CRC Press, 1999.

14 Optical Triangulation

14.1 LINEAR SENSING

Optical triangulation is a technology sometimes utilized for linear position sensing, and is not typically used for rotary or angular sensing. It can be very effective in linear position measurement applications where physical contact with the object to be measured would be undesirable. This includes measuring fragile parts that can be damaged by contact, surfaces that may be wet with lubricant or adhesive, and objects that are moving, such as measuring the thickness of a moving web. In this context, a web is a continuous roll of thin, flat material, such as paper, textile, foil, or plastic film, that is rolled up for convenient handling and transport. The converting industry processes a web by feeding it through a machine that may coat, slit, laminate, or apply another process. Such a machine is a web-processing machine, and includes slitters, coaters (to make adhesive tape, for example), and printing presses, among others. It is often important to monitor the web thickness. In order to measure a web thickness, two sensors may be used: one above and one below the web, focused onto the same spot from top and bottom.

Optical triangulation sensors may also be used in many other applications that would otherwise wear out a measuring device that operates in contact with a moving surface. In addition to having an advantage in being totally non-contacting, optical triangulation sensors also have a very fast response time. So that, for example, the thickness of a vibrating target can be measured by two sensors that are synchronized so that the two measurements are made at precisely the same time during a cycle of vibration.

They can also provide the average distance to a somewhat rough surface, when the spot size of the illuminating beam covers a sufficient area to provide an average over the surface peaks and valleys. Since the sensing element is mounted in a fixed location, measuring the distance to a movable target provides an absolute position measurement of the target with respect to the sensing element. Such sensors on the market are often called distance sensors. So, in this chapter, the terms distance and position may be used interchangeably, and are regarding an absolute measurement of the position of a target or the distance to the target from the sensing element.

14.2 HISTORY

The triangulation method for determining the distance to a target has long been used, and is based in geometry and trigonometry. Optical triangulation has been utilized for centuries, as in use of the sextant for navigation. Distance or position sensors based on optical triangulation have been marketed since at least the 1970s, and many US letters patent reading on this technology have issued since the 1960s.

DOI: 10.1201/9781003368991-14

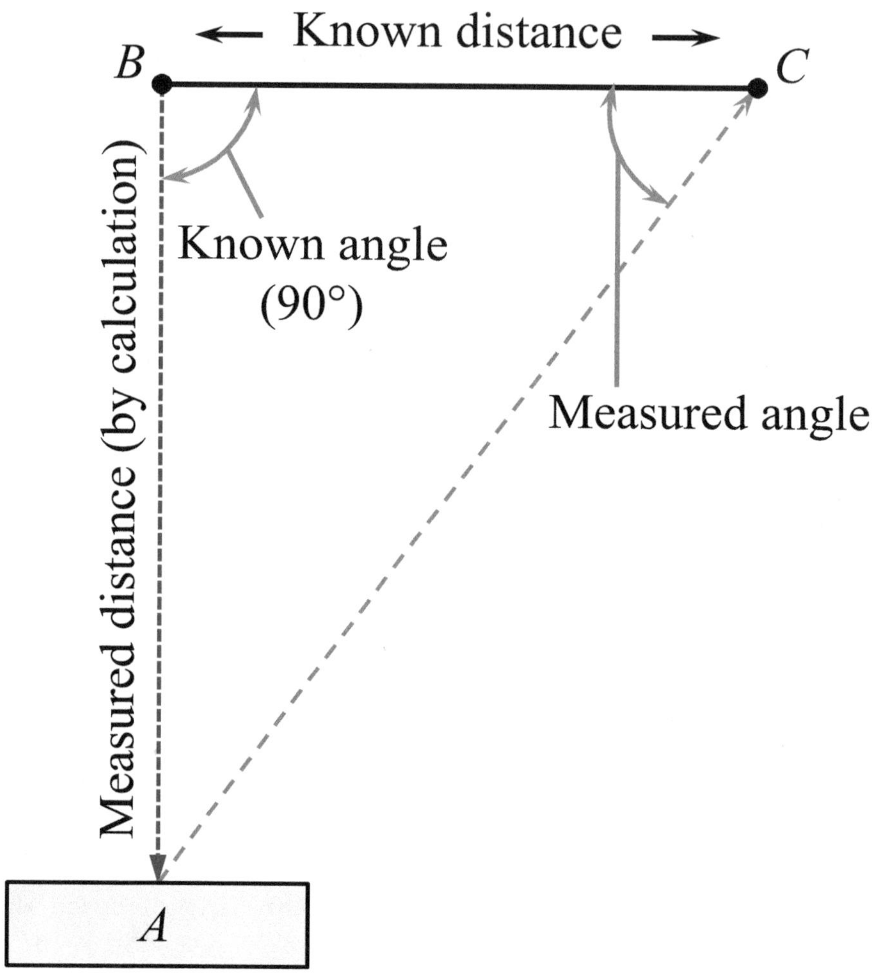

FIGURE 14.1 Triangulation method to determine the distance to a target by using a known angle, a known distance, and a measured angle.

A distance *AB*, from a starting point, *B*, to a target *A*, may be triangulated by measuring the angles to the target from two points, *B* and *C*, that are separated by a known distance *BC*. See Figure 14.1.

Generally in triangulation, there would be a base line of known length (the known distance) and known angles for each of the two lines connecting at its ends. Then the third angle and other line lengths could be determined. But in the figure, the diagram is simplified by setting a known angle at *B* to 90°, to emulate an optical triangulation distance/position sensor emitting its light beam at a right angle to the line *BC*. Then, once the measured angle at *C* is known, the measured distance can be found as:

$$AB = BC^{*} Tan\, C \tag{14.1}$$

TABLE 14.1
With a Known Base Length, BC, the Measured Distance Varies as the Tangent of the Measured Angle

Known Length BC (mm)	Measured Angle at C (°)	Tan C	Measured Distance AB (mm)
20	**10**	0.176	**3.52**
20	**20**	0.364	**7.28**
20	**30**	0.577	**11.54**
20	**40**	0.839	**16.78**
20	**50**	1.192	**23.84**
20	**60**	1.732	**34.64**
20	**70**	2.747	**54.94**
20	**80**	5.671	**113.42**

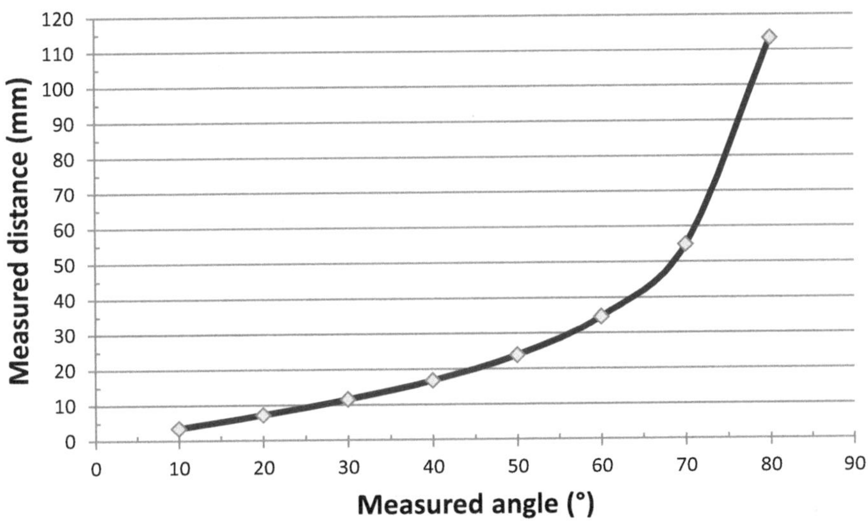

FIGURE 14.2 Measured distance versus measured angle.

where *AB* is the distance to be measured, *BC* is a known, fixed distance, and *Tan C* is the tangent of the angle in degrees at point *C*. For example, with the known distance, *BC*, being a length of 20 mm, the measured distance varies as shown in Table 14.1.

Figure 14.2 is a plot of the measured distance versus measured angle data from Table 14.1.

It can be seen that the distance versus angle curve has a nonlinear characteristic, and will be addressed later in this chapter. To further simplify, if the known distance *BC* is defined as a distance of 1 unit, then the measured distance would be just the *tangent of the measured angle* number of units. So, in the table, the *Tan C* column would be multiplied by one instead of by 20 mm. Then, the units can be re-scaled later in a μC as needed to provide a desired output.

Some simple light-sensing elements (photodiode and PSD) were used in the earlier optical triangulation sensors. CCD and CMOS light sensors were invented in the 1960s, and were then also used. These light sensor types are explained later in this chapter.

14.3 CONSTRUCTION

An optical triangulation distance sensor will almost always employ a laser as the light source. That's because a laser can be focused to illuminate a small spot on the target, such that the spot remains small over a relatively large range of distance, and because the illuminated spot can have a high intensity. Laser light is also monochromatic and coherent (an explanation of monochromaticity and coherence is presented at www.worldoflasers.com/laserproperties.htm). Other types of lasers besides a laser diode, such as helium neon (HeNe) could be used, but most optical triangulation distance sensors utilize a solid state laser diode because they are simple and robust. These are similar to the laser diodes used in a common laser pointer or laser sight.

Figure 14.1 provided a geometrical pictorial of measuring a distance by using a known fixed angle, a known fixed length, and a measured angle. This geometry is utilized in the optical triangulation position sensor of Figure 14.3, comprising the main active components of a laser diode, lens, light sensor, and electronics module.

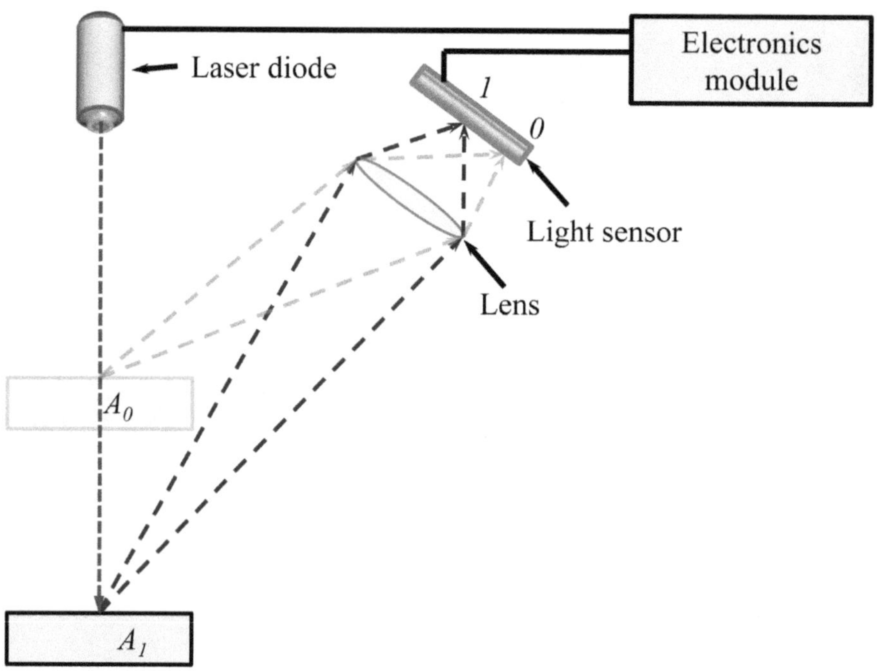

FIGURE 14.3 A laser diode, lens, light sensor, and electronics module comprise the main active components of an optical triangulation position sensor.

A small spot is illuminated on the surface of the target by a laser diode. A lens or set of lenses (not shown in the figures) may be inserted in front of the laser diode to collimate the laser light into a beam so that it exhibits a uniform diameter at various distances. A lens or set of several lenses is used to focus an image of the illuminated spot onto an elongated surface of a light-sensing element. Since the laser light is monochromatic, a narrow band optical filter lens may be used to ensure that only the laser light is admitted to the light sensor. The light-sensing element (or light sensor) provides a variable output signal, depending upon where the light spot image is falling onto its light-sensing surface. An electronics module drives the laser diode current, and receives the signals from the light sensor.

When the target is in the position as A_1 in the figure, the spot illuminated on the target by the laser diode is then focused onto the sensing element by a lens, and an image of the spot is formed approximately at the center of the light sensor. The light sensor signal to the electronics module indicates that the light spot is at the center of the light-sensing surface. When the target is moved closer to the laser diode, to position A_0, an image of the spot is now formed approximately near one edge of the light sensor. The light sensor signal to the electronics module indicates that the light spot is near one edge of the light-sensing surface. If the target would be moved farther from the laser diode than position A_1 (not shown), an image of the spot would be formed nearer to the other edge of the light sensor, and the light sensor signal to the electronics module would indicate that the light spot is nearer to the other edge of the light-sensing surface. So, the signal from the light sensor to the electronics module regarding the location of the light spot is indicative of the distance between the target and the laser diode.

As in photography, there is a parameter regarding the lens imaging system that is called depth-of-field. When a lens system is adjusted so that an image is formed on the desired focal plane (such as a film or photo sensing array), there is a limited range of distance over which the image will remain in focus to a desired degree. For example, a subject may be 2.0 m from the lens, and remain in focus from 1.5 to 2.5 m, so that the depth-of-field is then 1.0 m. In a camera, reducing the aperture size (that is, increasing the *f*-stop number) will increase the depth-of-field. Likewise, increasing the aperture diameter will decrease the depth-of-field. With a large aperture, such as *f* 1.4, the depth-of-field may be only a few cm, depending on the lens type and size. In an optical triangulation sensor, reducing an aperture size might reduce the amount of light available by too great an amount. But since the image is only moving across one dimension, an adjustment can be made to the lens and focal plane angles to somewhat increase the range of measured distance over which the spot image will remain sufficiently in focus to operate the sensor, as shown in Figure 14.4.

In the figure, *N* is an imaginary line passing through the spot on the target and the centers of the lens and the light sensor. The light sensor is shown as being tilted at an angle N_1, from being normal to line *N*. The optimum angle depends on the characteristics of the lens and light sensor, as well as the desired measuring range. In addition, it may also be possible to slightly improve the measuring range even further by a slight tilting of the lens, in addition to tilting the light sensor, the amount and direction depending on many variables of design, so it is more expedient to determine by test than trying to calculate an optimum tilt angle.

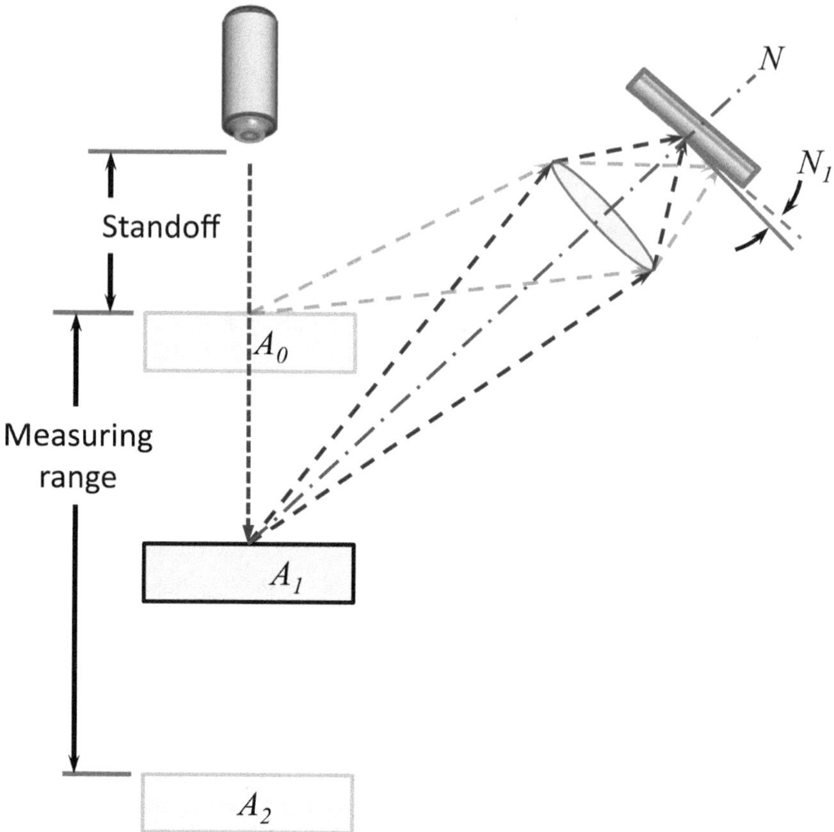

FIGURE 14.4 Alignment of the lens and focal plane may be adjusted to improve focus of the spot image over the measuring range.

A measuring range and a standoff are depicted in the figure. The standoff is the minimum distance between the target and the optical triangulation sensor. The measuring range is the total distance over which the target may be positioned and still obtain accurate measurement of its position.

14.4 LIGHT SENSOR

The light sensor may be one of three types: position-sensing detector (PSD), charge-coupled device (CCD), or complementary metal-oxide semiconductor (CMOS). The PSD is essentially an analog device, and the other two types are digital.

14.4.1 PSD

The PSD is also sometimes called a lateral-effect photodiode (LEP). A light sensor for use in an optical triangulation position sensor would normally be a one-dimensional or single-axis type (also called a linear array), such as PSD model S3270 from

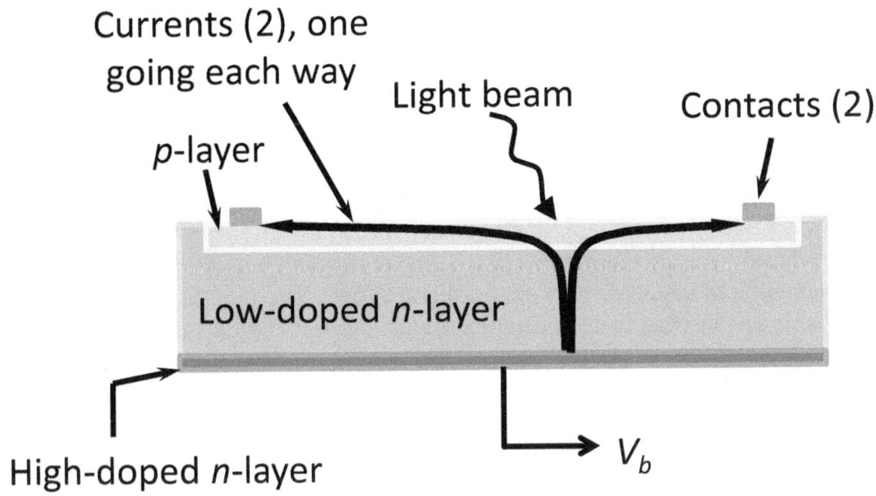

FIGURE 14.5 Construction of a position-sensing detector.

Hamamatsu (www.hamamatsu.com) having a light-sensing area of 1 mm × 37 mm. Such a PSD comprises a low-doped *n*-type semiconductor base material that has a high-doped *n*-type semiconductor layer diffused into the bottom, and a *p*-type semiconductor layer diffused into the top, as shown in Figure 14.5.

The figure is a side view of a PSD sensing element, with left to right being the measuring direction, and could be, for example, 30 mm long. Its width (going into the paper) would be much smaller, for example 1 mm. Two metallic contacts are attached to the *p*-type semiconductor near the ends, to provide two signal outputs. The high-doped *n*-type area is electrically conductive and may be implemented in two strips along the long edges of the sensing element, providing electrical contact with the low-doped *n*-layer. One or more metallic contacts attached to the high-doped *n*-layer(s) at the bottom provide connection means for application of a bias voltage V_b.

When a light beam (a spot illuminated on the target by the laser, reflected from the target, and focused on the PSD by the lens) falls on a particular location on the PSD, a current flows up from V_b, through that spot and to the two contacts in the *p*-layer. The resistance of the *p*-layer is very uniform, and divides the current going to the two contacts based on the length of *p*-layer through each current flows. If the light beam falls in the middle of the PSD, then equal output currents are produced. If the light beam is closer to one contact, the current from that contact is proportionally greater than the current from the other contact. Thus, the location of the light beam can be determined by comparing the relative amounts of the two currents.

Since the light beam has a non-zero diameter and a shape that may not be perfectly circular, the PSD outputs are a function of the center of light density over the entire spot area. This center of light density is sometimes referred to as the light "center-of-gravity". So, if a spot is strongly malformed (nonuniform shape, or nonuniform intensity), it may affect the spot location being reported by the PSD.

The simplicity of a PSD offers some advantage of reliability and stability. The electronics can be relatively simple, and implemented at low cost. They also have a fast response time of much less than 1 ms.

14.4.2 CCD

A one-dimensional CCD light sensor, such as a model IL-P3–0512 from Teledyne-Dalsa (www.teledynedalsa.com), has a linear series of measuring cells. The amount of light sensed by each cell forms a pixel. The construction of a cell is similar to the construction of an insulated gate field-effect transistor (FET). The active parts comprise a silicon substrate and a polysilicon gate, separated by a thin layer of silicon dioxide (an insulator). An individual cell is pictorially represented in Figure 14.6.

The silicon substrate in the figure has been lightly doped as a *p*-type. A thin layer of silicon dioxide is applied to the top of the substrate. A polysilicon gate electrode is deposited onto the silicon dioxide in the desired pixel shape and size. Polycrystalline silicon is a very pure grade of silicon (but not a single crystal as would be the usual monocrystalline silicon substrate material), and is usually called polysilicon. It becomes relatively transparent to light in wavelengths longer than about 450 nm. In the cell, a photon of light passes through the polysilicon gate, the energy associated with the photon wavelength is absorbed by the silicon substrate, forming electron-hole pairs. Electrons migrate to the area below the polycrystalline gate due to its bias at a positive voltage. The cell acts as a capacitor, with electrons building up a charge, and forming a potential well below the polysilicon gate. The CCD cell measures the light intensity distribution over the area of each pixel. The amount of

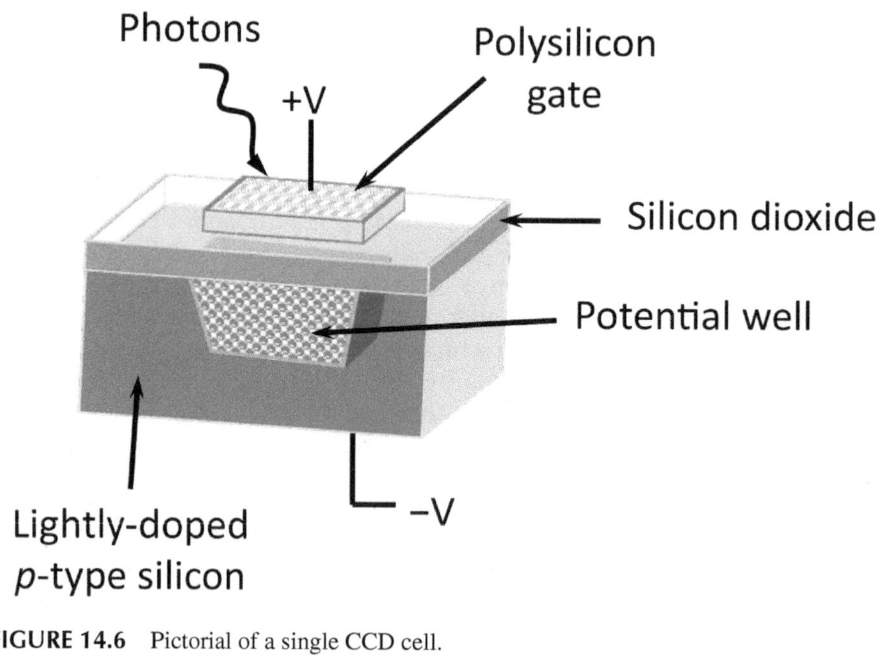

FIGURE 14.6 Pictorial of a single CCD cell.

the stored charge is approximately proportional to the light intensity (for a given light wavelength) up to the capacity of the well. So when the well is full, no further charge can be accumulated. This defines the upper limit of the cell's dynamic range. The lower limit is defined by the photon energy versus any noise present in the system.

At a controlled time interval, the many cells in a CCD light sensor are operated as a shift register, each cell shifting its charge to the next cell, by applying voltages at the gates. At the last gate, the charges are fed into a charge-to-voltage converter circuit, then to an analog-to-digital converter, then to a memory register that holds the latest data from all of the cells.

14.4.3 CMOS

A CMOS one-dimensional light sensor, such as a model LFL1402 from IC Haus (www.ichaus.de), is somewhat similar to a CCD light sensor in that it has a linear series of measuring cells. The amount of light sensed by each cell forms a pixel. Each CMOS cell comprises a photodiode and switching circuitry, as shown in Figure 14.7.

Rather than the capacitive cell of a CCD, the CMOS cell uses a photodiode. A *p*-type silicon substrate is doped in one area to make that area *n*-type. This forms a *p*-*n* junction, that acts as a photodiode. The photodiode usually takes up a little over half of the surface area of the cell. The remaining area is used to implement the amplifying and switching circuitry.

14.5 ELECTRONICS

Typical packaging of the three types of single-axis light sensors are depicted in Figure 14.8.

In the figure, (a) is a PSD, (b), a CCD, and (c), a CMOS single-axis light sensor array package. The PSD only needs three pins, as shown in Figure 14.7(a). One is the

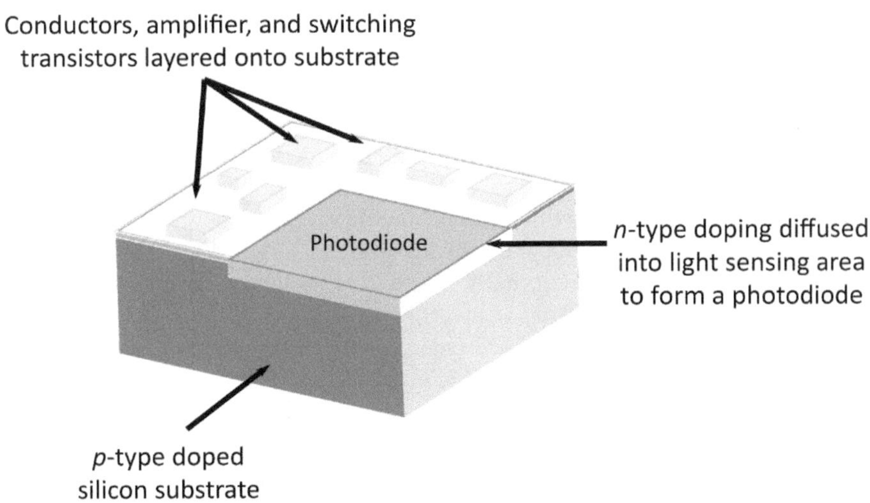

FIGURE 14.7 Construction of a CMOS light-sensing cell (pixel).

(a)

(b)

(c)

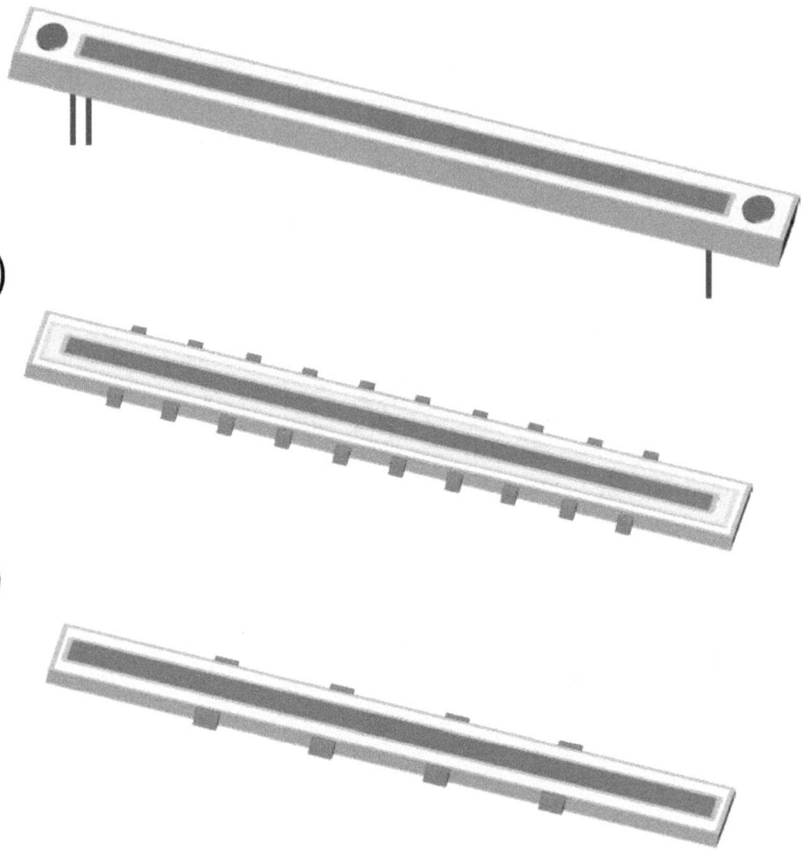

FIGURE 14.8 Packaging of three types of light sensor arrays: (a) PSD, (b) CCD, and (c) CMOS.

common pin (substrate), and the two other pins provide variable currents depending on which end of the array is closer to the light spot.

A PSD may be connected with a simple circuit, like that shown in Figure 14.9.

A fixed positive voltage is applied the substrate at $+V_{bias}$. A light spot falling on the PSD produces a first current that flows toward the inverting input of U_1, and a second current that flows toward the inverting input of U_2, the ratio of the two currents depending upon the position of the light spot with respect to the two ends of the PSD. U_1 and U_2 are inverting current-to-voltage converters, providing output voltages proportional to the respective input currents (inverting in this case means providing a negative voltage output in response to a positive current input). The amount of output voltage from U_1, for example, is equal to $-R_1$ times the U_1 input current. U_3 is configured as a differential voltage amplifier, having an output proportional to the difference between the outputs of U_1 and U_2. The value of R_3 is equal to that of R_4,

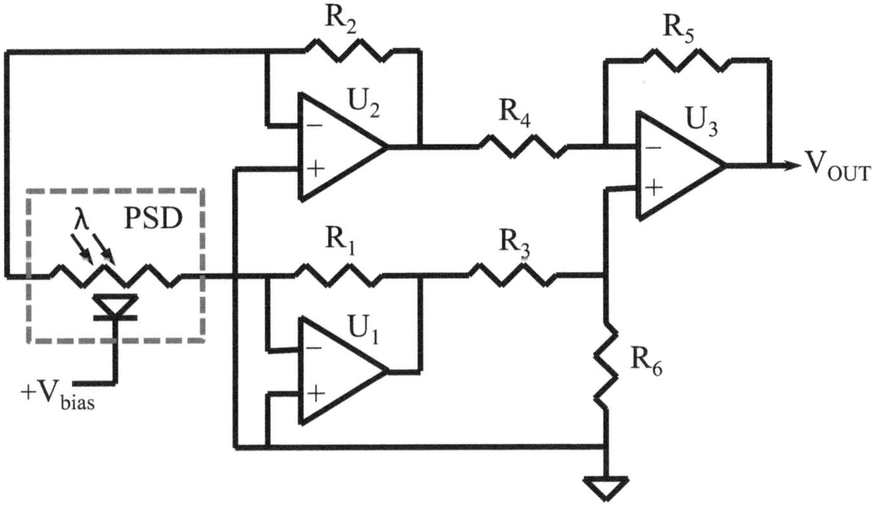

FIGURE 14.9 Circuit to connect with a PSD.

and the value of R_5 is equal to that of R_6. Voltage gain of the U_3 circuit is equal to the value of R_5/R_4. The output voltage (V_{OUT}) is indicative of the position of the light spot falling on the PSD. For example, with the light spot centered on the PSD, the currents into U_1 and U_2 are equal, so the output voltages from U_1 and U_2 are equal. That means the output from differential amplifier U_3 would be zero volts. If the light spot moves away from the center so that the U_1 output voltage is higher (more positive) than the U_2 output voltage, then V_{OUT} would be more positive. If the spot moves so that the U_1 output voltage is lower than the U_2 output voltage, then the U_3 output would be more negative. This simplified circuit assumes that the op amps have both positive and negative supply voltages. Of course, it could be modified to operate with a single supply voltage.

Following the circuit of Figure 14.9, V_{OUT} would then be digitized by an analog-to-digital converter (A/D), and the data supplied to a microcontroller (μC). The μC would filter and scale the signal, and then provide the desired output type. For an analog output, the μC would utilize a D/A converter. For a digital output, the μC would connect with appropriate line-driving circuits, and apply any data requirements dictated by the selected communication protocol.

A single-axis CCD array requires a greater number of pins than the three pins of a PSD. Enough pins are supplied to decode and allow access to each of the cells, depending on the configuration of the selection method. Figure 14.10 shows a circuit example of how an array of CCD pixels may be connected when configured with a two-phase pixel selection cycle.

Alternatively, a pseudo-single-phase, a three, or four phase cycle may be used. In the figure, a μC provides a varying voltage to the pixels arranged into two sets, or phases (*φ1* and *φ2*). The doping of the substrate under the polysilicon gates is non-uniform, so that providing a voltage pulse on one gate pushes the electrons from its well into the well under the next gate. By alternately pulsing the voltages of the two

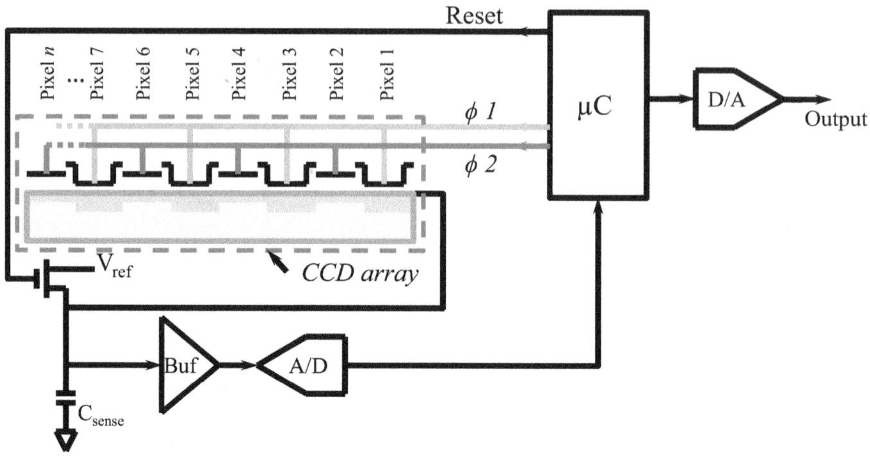

FIGURE 14.10 Circuit to connect with a CCD array.

phases, the charges are shifted down the array as in a shift register. The charge in the well under the last gate in the line (pixel 1) is pushed out into the sense capacitor, C_{sense}. In between one cycle to the next, the μC sends a reset pulse to bring the voltage at C_{sense} back to be equal to a reference voltage. When a charge is pumped into the fixed capacitance of C_{sense}, it appears as a voltage across the capacitance. A high impedance buffer circuit (Buf) passes that voltage to an A/D converter. The A/D sends digital data to the μC that indicate the position of the light spot on the CCD array. The μC filters and scales the digital data, and then provides the desired output type. For an analog output, the μC would utilize a D/A converter as shown in the figure. For a digital output, the μC would connect with appropriate line-driving circuits, and apply any data requirements as dictated by the selected communication protocol. Sometimes the CCD array requires all timing and control logic functions to be supplied by the external μC. But some CCD models include varying amounts of this circuitry inside the CCD chip, and may also include the signal buffer and reference circuits on-chip.

A single-axis CMOS light-sensing array normally requires a fewer number of pins than a single-axis CCD array (unless being compared with a CCD model that includes the pixel selection logic, buffer, and reference on-chip). A CMOS array includes a desired number of photodiodes, along with the conditioning and control logic on-chip, as shown in Figure 14.11.

In the figure, D_1 and D_2 are the first and second photodiodes in an array of hundreds or thousands of photodiodes. Each photodiode generates an amount of current proportional to the amount of light falling on it. Amplifier and switching circuitry as shown is repeated for each photodiode in the array. The photocurrent generated by D_1 is converted to a voltage by U_1 and R_1. The pixel multiplexer receives timing signals from an external μC (not shown), and controls the various switches. When switch U_2 is closed, capacitor C_1 charges to a voltage indicative of the amount of light falling on D_1 at that instant. Buffer U_5 has a high input impedance and low

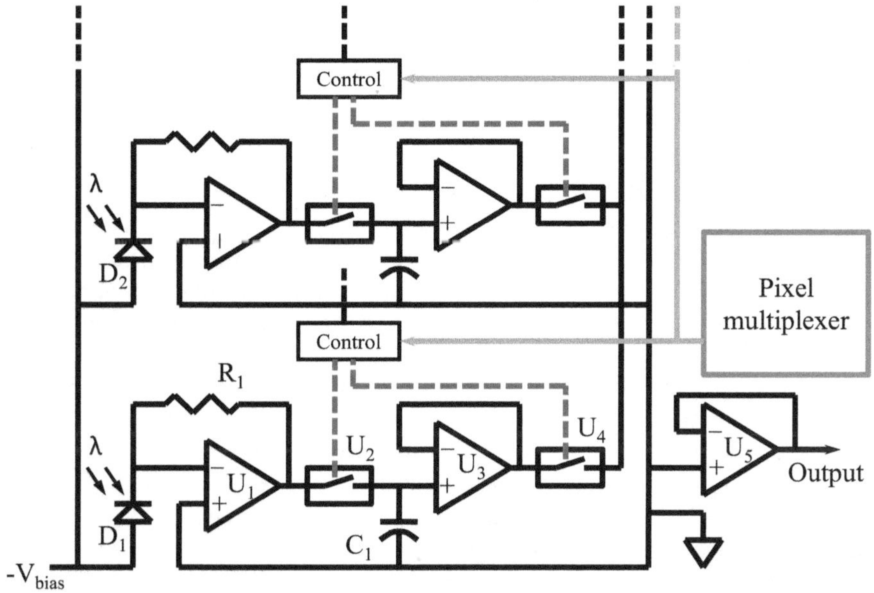

FIGURE 14.11 Single-axis CMOS array with electronics integrated on-chip.

input current to ensure that C_1 holds its voltage while U_3 presents the voltage to switch U_4. When switch U_4 closes, a voltage representative of the D_1 pixel appears at the output. The μC controls the pixel multiplexer until all pixels have been read. The CMOS array output signal goes to an external A/D (not shown, but sometimes, the A/D converter may be included on the CMOS chip). The A/D provides digital data to the μC. The μC filters and scales the digital data, and then provides the desired output type. For an analog output, the μC would utilize a D/A converter as shown in the figure. For a digital output, the μC would connect with appropriate line-driving circuits, and apply any data requirements as dictated by the selected communication protocol.

As illustrated in Table 14.1 and Figure 14.2, the position signal derived from the optical triangulation angle will be nonlinear. Additional nonlinearity may occur, due to some changing in beam focus throughout the measuring range, or to other factors. An algorithm should be implemented within the μC to correct the nonlinearity, as well as make some additional adjustments. These adjustments will normally include correction of any temperature-induced error, calibration for unit-to-unit differences, and scaling for the desired output. In addition, the electronic circuitry must also power the laser, and provide means to ensure the proper amount of light is being presented to the light sensor. In an old-fashioned film camera, for example, an aperture size and a shutter time control the amount of light falling on the film. But in the case of an optical triangulation sensor, intensity and timing control of the laser is available, as well as control of the switches and/or multiplexer in the light sensor.

14.6 LASER

The term laser is an acronym for light amplification by stimulated emission of radiation. As mentioned earlier in this chapter, an optical triangulation position sensor will almost always utilize a solid state laser diode. Some lower performance sensors may use an LED light source, but the laser diode type will be presented here. A laser diode has a threshold current, below which the light emitted is similar to that of an LED (light emitting diode). At the threshold current, and higher, the laser diode generates laser light that is coherent and monochromatic. Above the threshold current, the output power quickly increases with diode current until reaching the maximum operating current. The required threshold current increases as the diode temperature increases. So, it is important to make sure the laser diode has enough current to operate over the expected operating temperature range. This must be accomplished while also ensuring that the laser diode is not damaged by exceeding its maximum power or temperature ratings.

The wavelength of light produced by the laser diode varies with its temperature. While a GaAlAs laser typically emits light at a specific wavelength somewhere between 750 to 880 nm, the wavelength may vary with temperature at a rate of about 0.25 nm/°C on average, with some steps, rather than being a continuous change. So for best results, the laser diode temperature should be monitored by the μC, and the temperature information used in the control loop powering the laser diode.

The intensity of the laser light can be controlled somewhat by the amount of current supplied to it, and the laser can be pulsed on and off, together allowing the desired amount of light to fall onto the light sensor. Also, the laser diode may be pulsed to synchronize with reading of the light sensor. The laser diode is normally powered by a constant current or constant power circuit, such as the two constant current circuits shown in Figure 14.12.

In Figure 14.12(a), an external μC provides digital data to a D/A converter. For example, the analog voltage from the D/A could be 0 to 3.3 V. With R_1 = 20 k and R_2 = 5 k, the voltage supplied to the non-inverting input of U_1 would be 0 to 0.66 volts. 1 μF filter capacitor C_1 removes any higher-frequency noise. The output of U_1 drives FET Q_1 through current limiting resistor R_4. Phase-compensating 0.01 μF capacitor C_2 provides AC stability. 20 k resistor R_3 provides feedback from 10 ohms current-sense resistor R_3. Q_1 passes a current I through laser diode D_1, causing a sense voltage to appear across R_5. For example, with a D/A analog voltage of 2.5 volts, the voltage across R_2 is 0.5 V, and the voltage across R_5 is also 0.5 V, for a constant current of 50 mA. Voltage divider R_1/R_2 is included so that the D/A output voltage can have a full swing of 0 to 3.3 V, while still keeping the voltage range across the current-sense resistor low. (If the voltage across the current-sense resistor was too high, then it would dissipate a lot of heat, and would also limit the voltage available to drive the laser diode.)

The circuit of Figure 14.12(b) operates in very much the same fashion, but uses a digital potentiometer (pot) instead of a D/A. A digital pot has a fixed number of possible wiper settings, such as 256 positions for an 8-bit pot, or 1024 positions for a 10-bit pot. Digital pot U_1 is controlled by an external μC. When the μC talks with U_1, the chip select input (CS) is active. While the up/down (U/D) direction is held in

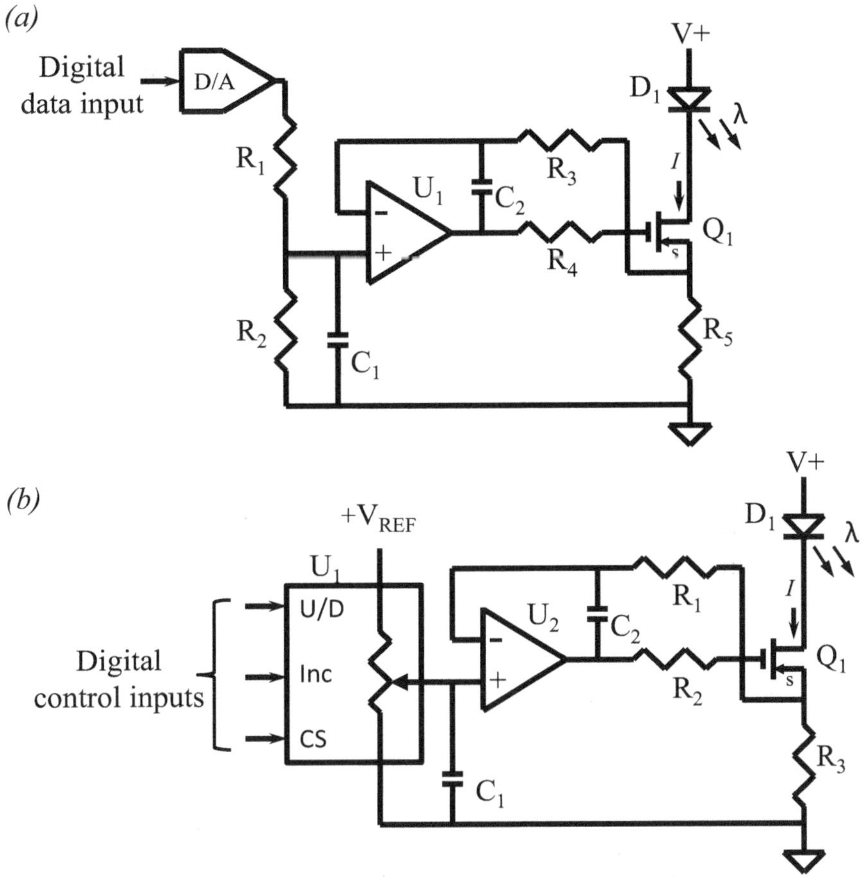

FIGURE 14.12 Constant current circuits to power a laser diode.

the desired direction, the increment (Inc) is pulsed the number of times required to move the wiper of the pot to the desired position. Some pots allow the present wiper setting to be read back by the μC, but others require that the μC itself keep track of the count after reset.

To operate with constant power, both the current through and the voltage across the laser diode are measured and controlled, while keeping the current level within the specified amounts. The amount of laser diode current or power that is set by the μC should be adjusted as the ambient temperature changes. The amount of temperature compensation should be determined so that the same laser light output is produced throughout the sensor's specified operating temperature range.

Only constant current circuit configurations were shown in the figure. A constant power implementation would also require measurement of the voltage across the laser, so that the μC can calculate the laser power from the current and voltage. The power supply of the constant current or constant power circuit for the laser diode should be well filtered so that the laser light intensity and wavelength are very stable.

14.7 ADVANTAGES

A main advantage of an optical triangulation sensor over other position-sensing technologies is that there is no contact with the target being measured. The laser beam is the only thing that touches the object being measured. See Figure 14.13 for a pictorial representation of an optical triangulation position sensor configuration similar to many available on the market.

The laser light being the only thing that touches the object being measured avoids wear of any contact probe that may be used in another technology, and also allows measurement of fragile objects that could be damaged by contact, such as a web of thin paper or film. Non-contact sensing also allows measurement of a surface that may be wet with sealant or adhesive.

Some optical triangulation sensors can measure at high speed, having a response time of 1 ms or less. The optical measurement also does not add any mass to a moving object being measured. But resolution, repeatability, and nonlinearity are generally not as good in an optical triangulation sensor as compared with some other position-sensing technologies. Optical triangulation sensors have a minimum distance to the target, called standoff, that is greater than in the other technologies.

14.8 TYPICAL PERFORMANCE SPECIFICATIONS AND APPLICATIONS

A typical set of specifications for an optical triangulation linear position sensor is listed in Table 14.2.

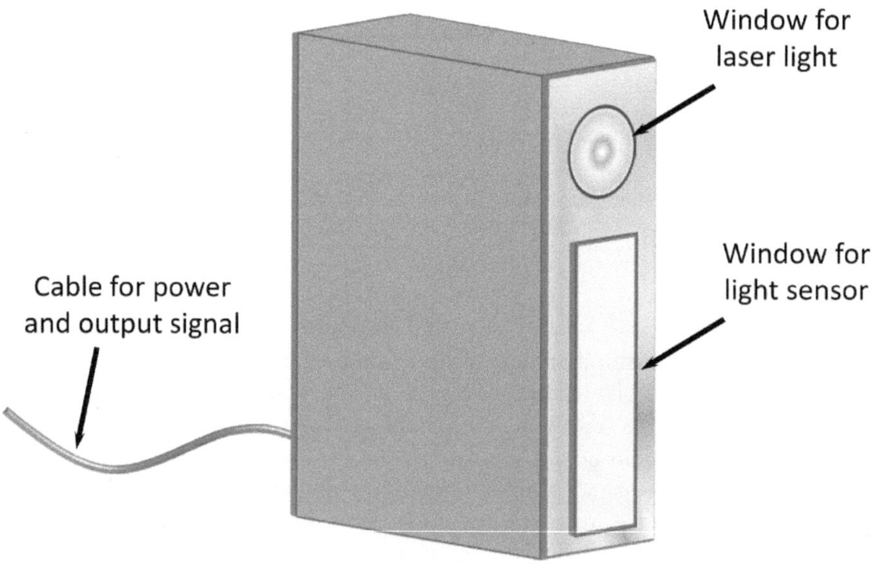

FIGURE 14.13 An optical triangulation position sensor configuration that is representative of some products commonly available on the market.

TABLE 14.2
Typical Specification of an Incremental Optical Linear Encoder

Measuring range:	1,000 mm
Minimum standoff:	245 mm
Nonlinearity:	≤ 0.1% FRO (full-range output)
Resolution:	≤ 0.02% FRO
Power supply:	9 to 36 V DC, at 70 mA
Output:	4 to 20 mA
Max frequency:	9 kHz
Operating temp:	−10 to 60°C
Temp sensitivity:	≤ 0.02%/°C

Due to the fast response of optical triangulation sensors, real-time adjustments can be made to a moving web processing machine. Also, thanks to the non-contact sensing, web sagging or wrinkles in the web fabric or film can be detected and measured. Because of the speed and non-contact sensing method, a rotating cutting disk can be measured when slicing the thin veneer used to make laminated wood. The ability to measure a varying thickness with no contact allows an optical triangulation sensor to measure the quality of raw wood (a log) before sawing. Many other applications that benefit from fast response and non-contact commonly use optical triangulation sensors.

14.9 MANUFACTURERS

Some manufacturers of optical triangulation linear position sensors include the following:

Acuity	www.acuitylaser.com
Banner Engineering	www.bannerengineering.com
IFM	www.ifm.com
Keyence	www.keyence.com
Micro Epsilon	www.micro-epsilon-laser.com
MTI Instruments	www.mtiinstruments.com
Optex-Ramco	www.optex-ramco.com
Revolution Sensor Company (design)	www.rev.bz
Sick	www.sick.com

14.10 QUESTIONS FOR REVIEW

1. **In a context of the industrial converting industry, a web is a continuous roll of:**

   a. pastry
   b. wire

c. waste material
d. catalyst
e. thin material

2. In an optical triangulation sensor, the measured distance versus beam angle is:

a. nonlinear
b. inverted
c. coherent
d. asymptotic
e. parabolic

3. The range of distance over which an image will remain in focus is called:

a. field of view
b. depth-of-field
c. focus ring
d. focal length
e. perspective

4. A position-sensing detector (PSD) typically has this many electrical leads:

a. three
b. four
c. two
d. sixteen
e. twelve

5. At a controlled time interval, the cells of a CCD light sensor are operated as a:

a. flash memory
b. filter
c. current source
d. shift register
e. switch

6. The light-sensing element of each cell in a CMOS array operates as a:

a. variable resistance
b. lens
c. photo diode
d. LED
e. capacitor

7. Voltage and current are both measured when operating a laser diode in this mode:

a. squared
b. scalar
c. constant power
d. thermal control
e. current rate

8. In Figure 14.12(b), U_1 provides this to U_2:

a. current steering
b. adjustable voltage
c. filtering
d. time delay
e. more cowbell

15 Ultrasonic Sensing

15.1 ULTRASONIC POSITION SENSING

An ultrasonic system for measuring a distance to a target operates by generating an ultrasound pulse, and then listening for an echo as the pulse bounces back from the target. The elapsed time between transmission of the ultrasonic pulse to receiving the echo is called time-of-flight (ToF). The time-of-flight measurement is converted to indicate distance from the ultrasound transducer to the target by dividing the total travel time (T_{tot}) by two, and multiplying by the speed of sound (V_s). The measured distance from the transducer to the target is $D_{m,}$ as shown in Equation 15.1.

$$D_m = T_{tot} / 2xV_s \qquad (15.1)$$

Ultrasonic measurement is often implemented in a linear position sensor. Angular or rotation sensing can also be accomplished by measuring one or more distance(s), with a mechanical arrangement so that measured linear distance varies with rotation of the member of which the angle is to be measured. Two pictorial examples of such an arrangement are shown in Figure 15.1.

In Figure 15.1(a), the transducer indicates distance 1 when in the neutral position. When rotated to the right in Figure 15.1(b), a shorter distance 2 indicates the amount of rotation. In the alternative configuration of Figure 15.1(c), distances 1 and 2 are equal. The two distances can be measured by one transducer, as shown, or two separate transducers can be used. When rotated to the right in Figure 15.1(d), unequal distances 1 and 2 indicate the angle of rotation.

As with optical systems, ultrasonic ones can have an advantage where physical contact with the object to be measured (the target) would be undesirable. Such applications include measuring a wet, sticky, or very fragile surface, or one that may be difficult to access with a sensor that contacts the target.

Of course, the non-contact nature of ultrasonic measurement provides for long life of the position sensor, since there are few mechanical parts to wear out. It is possible for the ultrasonic transducer to wear out if not robustly designed, but these parts are typically designed well so that they will not wear out or fail in service.

Ultrasonic measurements are not repeated as quickly as optical measurements, typically in the range of about 10 to 100 measurements per second. (Optical measurements can be completed at a rate that is more than ten times faster than this.) Ultrasonic measurements can also provide the average distance to a somewhat rough surface, like the surface of a powder.

DOI: 10.1201/9781003368991-15

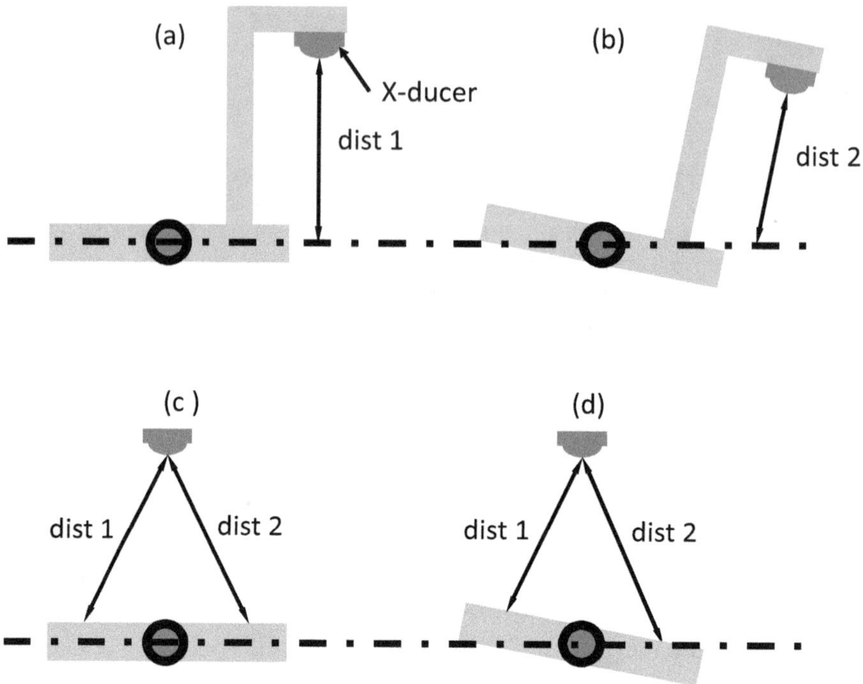

FIGURE 15.1 Two methods of measuring rotation angle using ultrasonic sensing.

15.2 HISTORY

The word SONAR stands for SOund NAvigation and Ranging. Most are familiar with sonar systems that have been used in submarines since the early twentieth century. An ultrasonic position sensor works in a similar way as a submarine sonar system, but is implemented somewhat differently. Also, a submarine system reports the compass angle to the target, in addition to the distance, by rotating the transducer aiming direction.

With sonar, a transmit transducer directs a sonic pulse toward a target, the distance to which is to be measured. Some of the sonic pulse bounces from the target back toward a receiving transducer (the echo). The sending (transmitting) and receiving transducers may be separate units, or one transducer can be used for both functions. A timer is started when the pulse is transmitted, and the timer stopped upon detection of the echo. Knowing the speed of sound in the medium (water, for a submarine; air, for most position sensors), the distance to target can be determined by multiplying the speed of sound by the measured time period. Then this result must be divided in half to get distance to target, because the measurement was of a round trip flight of the ultrasonic pulse. For example, with a measured transmit-to-echo time of 5 ms, and a speed of sound in air of 340 m/s, the distance is $(0.005 \text{ s} \times 340 \text{ m/s})/2 = 0.85$ m. The division by 2 is due to the time measured being that of a round trip from the transducer, to the target, back to the transducer.

For a submarine sonar system, either an audible or an ultrasonic frequency can be used. Ultrasonic frequencies are those above 20 kHz. For the purpose of an industrial type position sensor, and ultrasonic frequency is used, most often between 25 and 500 kHz, but can be higher in some applications. The higher frequency (shorter wavelength) enables a more narrow beam angle, so that a relatively small target will reflect a significant amount of the transmitted beam energy.

Although sonar had already been known for some time, the possibility of developing a small, inexpensive system for industrial position sensing first became viable with the development of the autofocus camera by Polaroid™. Their small, inexpensive ultrasonic transducer and integrated circuits were used in an autofocus Polaroid Land camera that was introduced in 1978. An ultrasonic distance measurement was used in order to drive a motor to focus the camera lens. Two years later, their transducer and integrated circuits were available for engineers to design into other products. In 1980, the author purchased the Polaroid Ultrasonic Ranging System Designer's Kit, shown in Figure 15.2.

With some modifications, the author used the Polaroid transducer and integrated circuits of the kit to design an ultrasonic position sensor, with a rod and piston configuration, having a calibrated range of zero to one meter. Later versions of transducers and integrated circuits have been developed by many manufacturers, and several of these components are readily available on the market. There are even some complete transducer and electronics modules for sale, although some are of a lower-resolution type that may not handle some industrial applications. See Figure 15.3.

The widely available model shown in Figure 15.3 is designed to interface simply. A measurement sequence is initiated by supplying a 10 μs trigger pulse to the Trig input of the module. The output from the Echo pin of the module is a pulse width of a period that is proportional to the measured distance. With this simple interface, it can be easily integrated with Arduino and Raspberry Pi platforms, as well as with any microcontroller. The PCB is marked with T and R, to identify the transmit and receive transducers, respectively. The transducers are each about 16 mm in diameter.

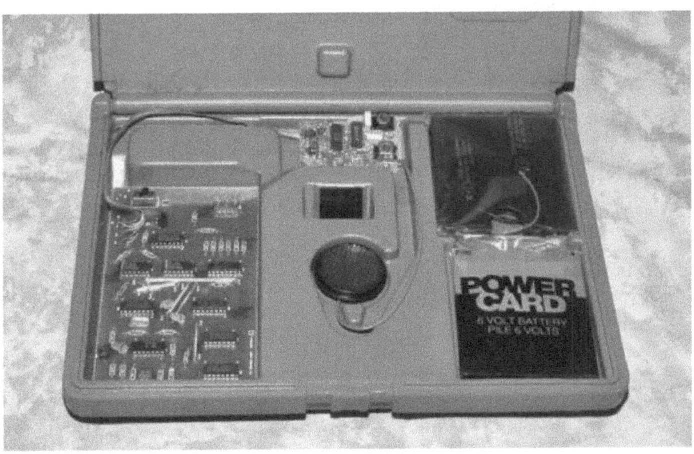

FIGURE 15.2 The Polaroid Ultrasonic Ranging System Designer's Kit.

FIGURE 15.3 Complete module with two transducers and electronics.

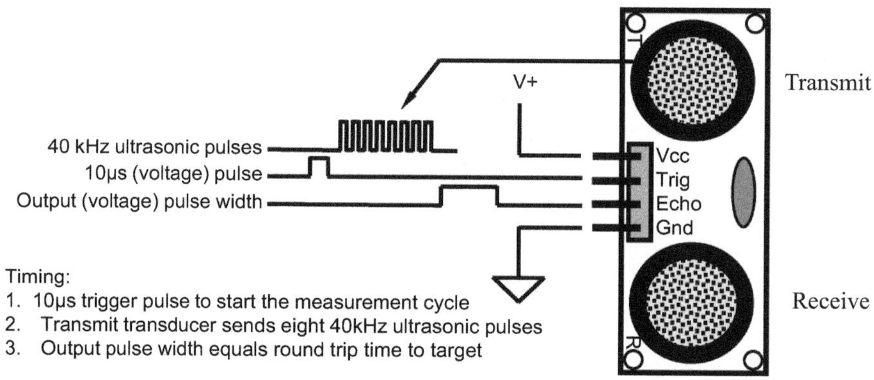

FIGURE 15.4 Function diagram of dual transducer ultrasonic module.

The diagram shows a stream of 8 ultrasonic pulses, in which the uptime represents the times during which the 40 kHz wave is being produced.

The ultrasonic module of Figures 15.3 and 15.4 is specified to measure distance to a target in a range from 2 to 400 cm. Rather than listing the errors independently, the specification shows only an "accuracy" of 3 mm. So, for a full-scale range (FSR) of 300 cm, an error of 3 mm would equate to 0.1% FSR. Other specifications of this module include:

- Power supply: 5 VDC, at 2 mA;
- Input trigger signal: 10 μs TTL pulse (high for 10 μs);
- Output signal: pulse width, TTL, high level ≈ 5 VDC, low level ≈ 0 VDC;
- Ultrasonic beam angle: not more than 15 degrees;
- Detection distance: 2 cm to 400 cm;
- Distance to target: pulse width in seconds × sound velocity/2.

15.3 CONSTRUCTION

An ultrasonic position sensor for industrial use may utilize a single transducer, or may use one for transmit and another transducer for receive. An example of an early single transmit/receive transducer is shown in Figure 15.5. This one is about 43 mm in diameter.

FIGURE 15.5 An early single ultrasonic transducer: Polaroid model 616341.

The single transducer type has an advantage that the complete assembly can be housed in a relatively small diameter cylinder. The separate transducer type has the advantages that the transmitter and receiver transducers can be optimized for their specific function, and that the receiver does not get overpowered by the transmit signal being directly applied to it. When the transmit signal is applied to a single transducer, the receiving circuit must be clamped to avoid overdriving during the transmit time. Then, after the transmit is completed, the receive circuit waits an amount of time for signals to quiet down before opening up to listen for the echo. An ultrasonic position sensor having one transducer and electronics all mounted into a cylindrical housing is depicted in Figure 15.6.

The configuration of Figure 15.6 is easy to install, and is therefore a popular one. For installation, a mounting hole is required in a panel or bracket. The sensor housing is threaded on the outside, so that installation can be completed by mounting through the hole with one retaining nut on each side of the hole. The nuts fix the sensor in place, and align it perpendicularly with respect to a target. If distance to the target must be adjusted, that can be accomplished by adjusting the positions of the retaining nuts.

15.4 TRANSDUCER

Since a single transducer configuration is popular for position sensing, this type of transducer will be described here. A single transducer configuration is one that uses the same transducer for both the transmit and the receive functions. The original

FIGURE 15.6 A single transducer type of ultrasonic position sensor mounted into a cylindrical housing.

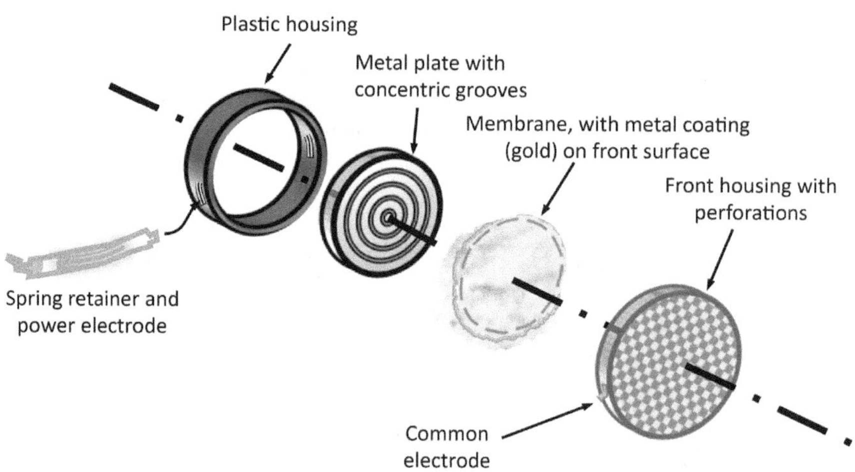

FIGURE 15.7 Construction of an ultrasonic dual function, transmit/receive transducer.

Polaroid camera autofocus transducer of Figure 15.5 was this type. Such a transducer is constructed approximately according to Figure 15.7.

In the figure, all of the components are mounted within a (rear) housing. The housing has slots into which a spring retainer is clipped. The spring retainer also contacts a grooved metal plate, and thus provides one of the two electrical contacts required to drive the transducer. The metal plate has concentric grooves to provide effective backing to the piezoelectric membrane, allowing the membrane to vibrate freely at an ultrasonic frequency. The membrane is made of a piezoelectric material, such as polyvinylidene fluoride (also known as Kynar™). The membrane must have a conductive coating on its front surface in order to enable electrical contact, thus facilitating connection of the second of the two required contacts. So, a 0.1 to 10 microns thick film of gold is applied to one surface of a 0.1 to 0.2 mm thick PVDF membrane. The front housing holds the edges of the membrane against the metal plate, and has perforations so that the ultrasound can pass through it. With the housing pressing against the conductive surface of the membrane, an electrical contact is attached to provide connection means for the second contact. In transmit mode, the common electrode connects with circuit common, and the power electrode is driven with a pulse at an ultrasonic frequency. In receive mode, the "power" electrode is connected with the receiving electronic circuit.

The transducer construction of Figure 15.7 is one type, but there are others. Another construction is shown in Figure 15.8.

The ultrasonic transducer version shown in Figure 15.8 implements a planar element of a piezoelectric crystal, such as one made of lead zirconite titanate (PZT). The P is for the atomic symbol of lead (Pb), the Z for zirconium (Zr), and the T for titanium (Ti). PZT is the most commonly used piezoelectric ceramic material for an ultrasonic transducer. When this ceramic is fired, it develops a perovskite crystal structure, with each crystal therein having a dipole moment. This structure facilitates the piezoelectric property. The crystal plane is thin, around 0.2 mm thick, and is sandwiched between two circular conductors, forming the two required electrodes.

Alternatively, a thin membrane of silicon, silicon nitride, or silicon dioxide may be used, with a thin film of PZT applied to the surface.

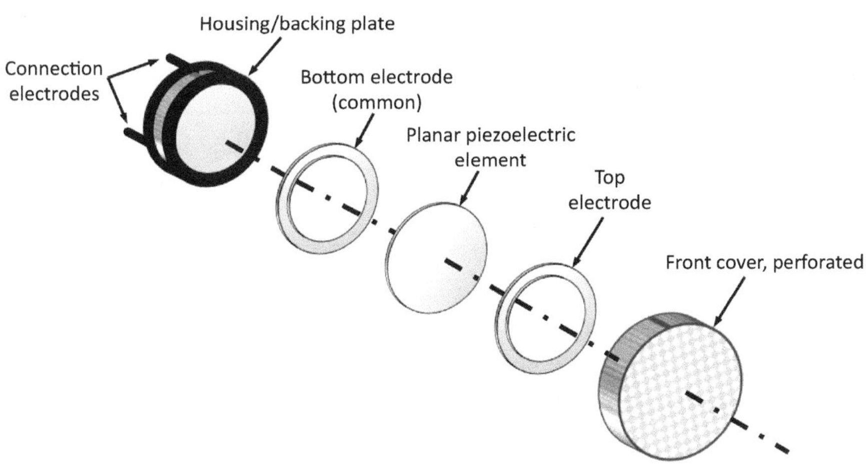

FIGURE 15.8 Ultrasonic transducer with planar piezoelectric crystal.

15.5 DESIGN CONSIDERATIONS

Ultrasonic position sensors for industrial use typically operate with the ultrasonic pulse transmitted in an air medium. The speed of sound (or ultrasound) in dry air at 20°C is approximately 343 m/s (meters per second) or 1,125 ft/s (feet per second). This speed increases as the temperature of the air increases, by about 0.55 m/s per °C. The speed of sound also increases as humidity (RH) increases. At 20°C, sound travels about 0.37% faster in 100% RH than it does in dry air (dry air = 0% RH). There are calculators online in which the air temperature and RH may be entered, and the speed of sound answer is returned. Although mostly unrelated to this description of ultrasonic position sensors, sound travels much faster in liquids than in gasses, and even faster in solids.

As mentioned, most ultrasonic position sensors transmit a pulse having a frequency of between 30 kHz and 500 kHz. However, only a small portion of the transmitted wave energy makes it back to the receiving transducer, as an echo. A smaller target or an irregularly shaped target can cause an even smaller amount of wave energy to be received. But for a given transducer and target, other qualities can also cause a reduction in the signal amplitude. As the pulse of ultrasound travels through the air, the energy of the ultrasound pulse is reduced due to friction losses as it interacts with the air through which it is propagating. This reduction continues as long as the pulse is traveling through the air, and is called attenuation. So, a longer measurement distance results in greater attenuation. The rate of attenuation increases as the ultrasonic frequency increases, and as the temperature increases. If the operating frequency is below about 100 kHz, there is also a value of relative humidity at which a particular frequency exhibits maximum attenuation. For example, a sonic wave of 50 kHz traveling in air at 20°C will experience maximum attenuation at around 65% RH. A sonic wave above 100 kHz, traveling in air at 20°C, will experience maximum attenuation at 100% RH.

Since attenuation is greater at higher frequencies, a lower frequency (≤ 100 kHz) should be used for longer ranges. But for shorter ranges, where attenuation is not a major problem, a higher frequency can be more effective due to the shorter wavelength having the possibility of a more narrow beam (lower beam angle). A shorter wavelength (higher frequency) also supports having a shorter minimum-detection range, by reducing the amount of time that must be allowed for decaying ringing of the receive transducer.

15.5.1 ECHO AMPLITUDE

If the target is a large (larger than the beam width), flat surface, at an angle perpendicular to the transmitted beam, the entire beam can be reflected back toward the transducer, producing a strong echo. However, the continual spreading out of the beam as it travels will still mean that only a small portion of the transmitted energy will fall onto the receiving transducer. See Figure 15.9 (transducer is often written as x-ducer).

The area of the receiving transducer (two-dimensionally depicted as receiving x-ducer area in the figure) is much smaller than that of the complete returning echo

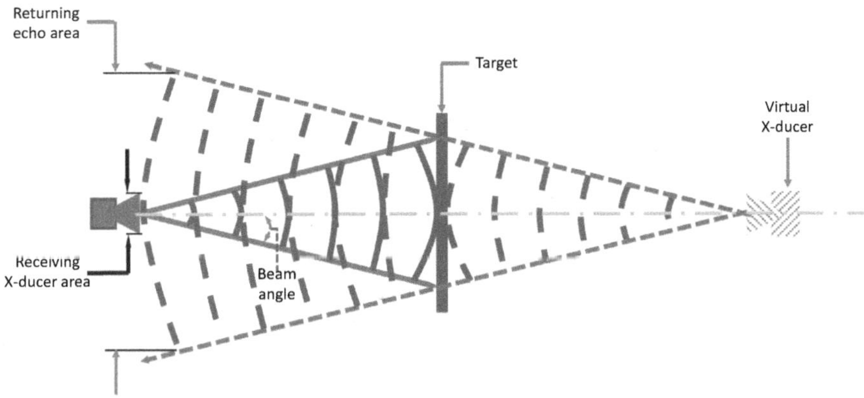

FIGURE 15.9 Spreading out of the ultrasonic beam.

(returning echo area, shown in the figure). This is called the beam-spreading loss, not counting the attenuation previously described. To be considered large, the target area must be larger than the sound beam area where it falls onto the target. If there is little loss at reflection from the target, the sound beam can be considered the same as one generated by a virtual transmitting transducer at an equal distance behind the target. So, in the figure, the target and illustrations to the left of the target are real. To the right of the target are shown virtual ultrasonics and a virtual x-ducer for the purpose of visualizing the shape of the real echo.

15.6 ELECTRONICS

An electronic circuit to implement an ultrasonic position sensor comprises various features to accommodate the transmit, receive, and signal processing functions. When the ultrasonic pulse is transmitted from the transducer, the receiving circuit must be clamped down to avoid being overpowered by the transmit signal, since the transmit signal is very much greater. When the ultrasonic pulse has been sent, then the receiving circuit can again become active after a blanking time period. The blanking time allows the transducer and circuit to settle down, as it needs to be quiet in order to receive the echo pulse from the target. The blanking time equates to a distance from the transducer over which measurements are not possible, and can be called a minimum standoff distance. As the receiving circuit is waiting for the echo pulse, the expected pulse amplitude will continue to become smaller, the longer the time period waiting for the echo (as the target becomes farther away). Accordingly, the amplifier gain of the receiving circuit must be steadily increased as the time becomes longer.

When the transmit pulse is produced, a latch is set. When an echo pulse is received, that latch is reset. Therefore, the ON time of the latch (the time during which the latch was set, or logic high) is representative of the distance to the target.

There are also many other details that must be implemented into the circuit, including communication with a microcontroller or other interface. So, it can be most

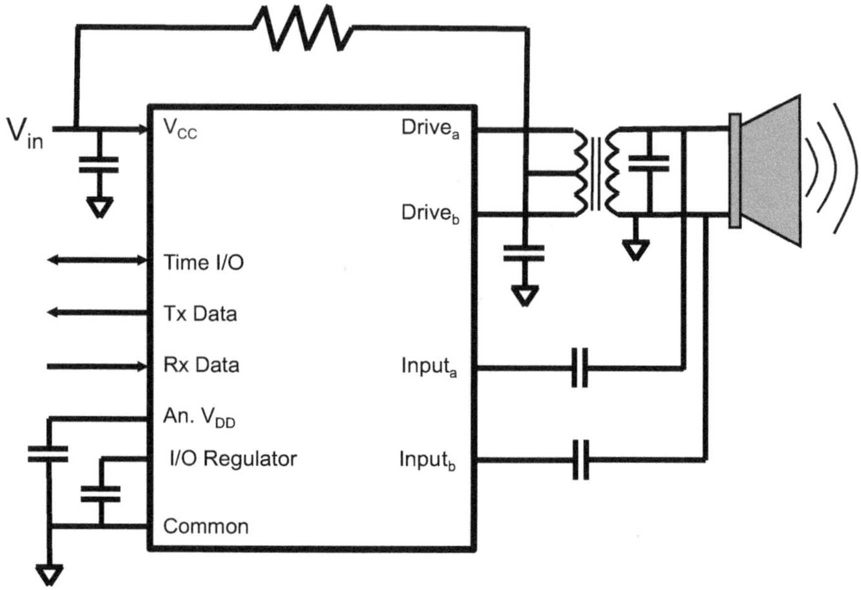

PGA460 Ultrasonic Signal Processor and Transducer Driver from Texas Instruments

FIGURE 15.10 Basic schematic for TI ultrasonic integrated circuit.

effective to utilize an integrated circuit that is designed specifically for an ultrasonic transducer. One such IC is the PGA460, manufactured by Texas Instruments (TI), a basic schematic for which is shown in Figure 15.10.

The PGA460 will operate from a 3.3 or 5 VDC power supply. The Tx data, Rx data, along with Sclk (serial clock, not shown) provide communication with a microcontroller (μC). Timing info about the ultrasonic pulse is communicated on a separate connection, Time I/O. The two drive pins power the transducer through a transformer. The resistor provides a bias voltage for the circuit driving the transformer. The echo is received through the two DC blocking capacitors, into the two receive pins. Three capacitors reduce AC that may appear on the V_{CC}, Analog V_{DD}, or I/O regulator pins. Similar functioning integrated circuits are also available from other manufacturers.

15.7 ADVANTAGES

A main advantage of an ultrasonic position sensor over other position-sensing technologies is that there is no contact with the target being measured. The ultrasonic beam is the only thing that touches the object being measured.

The ultrasonic wave being the only thing that touches the object being measured avoids wear of any contact probe that may be used in another technology (besides optical ones), and also allows measurement of fragile objects that could be damaged by contact, such as a web of thin paper or film. Non-contact sensing also allows measurement of a surface that may be wet with sealant or adhesive. And ultrasonic position sensors are easy to mount through a hole, with two nuts.

Most ultrasonic position sensors, however, are relatively slow when compared with other technologies, at around 30 ms per reading. Their best resolution of about 0.025 mm is adequate for many industrial applications. Their nonlinearity of about 0.3 to 1.0% is on the less precise side as compared with other technologies. But still, they are easy to use and can be inexpensive.

15.8 TYPICAL PERFORMANCE SPECIFICATIONS

A typical set of specifications for one of the higher resolution ultrasonic position sensors is listed in Table 15.1.

Ultrasonic position sensors are often used to measure uneven surfaces. They are also used to measure the height of liquids and powders.

15.9 MANUFACTURERS

Some manufacturers of optical triangulation linear position sensors include the following:

Banner	www.bannerengineering.com
Leuze	www.leuze.com
Massa	www.massa.com
Microsonic	www.microsonic.de
Peperl-Fuchs	peperl-fuchs.com
Revolution Sensor Company (design)	www.rev.bz
Sick	www.sick.com
WayCon	www.waycon.de
Wenglor	www.wenglor.com

TABLE 15.1
Typical Specification of an Ultrasonic Linear Position Sensor

Measuring range:	350 mm
Minimum standoff:	30 mm
Mechanical configuration:	threaded cylinder, 30 mm diameter
Transducer frequency:	320 kHz
Resolution:	0.025 mm
Repeatability	0.015%
Nonlinearity:	1.0%
Update time:	32 ms
Output:	4 to 20 mA and 0 to 10 VDC
Operating temperature:	–25 to 70°C
Power:	9 to 30 VDC at 80 mA

15.10 QUESTIONS FOR REVIEW

1. An ultrasonic frequency is generally considered to be one that is above:

a. water
b. 20 kHz
c. beam dispersion
d. 2 kHz
e. a thickness of 2 mm

2. The elapsed time for a signal to travel to a target and back is called:

a. linear window
b. reflexed
c. target recognition
d. pulse time
e. time-of-flight

3. A type of plastic film that has the property of piezoelectricity is:

a. polyvinylidene fluoride
b. polytetrafluoroethylene
c. polyvinyl chloride
d. polybutylene
e. polypropylene

4. As the distance to target increases, the receiver circuit is adjusted for:

a. DC blocking
b. phase stability
c. micronics
d. higher gain
e. lower current

5. A Kynar membrane may be coated with a thin layer of gold, to provide:

a. thermal conductivity
b. filtering
c. electrical conductivity
d. reflection
e. insulation

6. Reduced ultrasonic pulse amplitude as it travels through the air is called:

a. elasticity
b. reciprocity

c. conjugation
d. attenuation
e. amalgamation

7. The minimum distance at which a target can be accurately measured is called:

a. standoff
b. near-scale
c. span
d. calibration
e. beam angle

8. The best technical author whose book I have enjoyed reading is:

a. Albert Einstein
b. David S. Nyce
c. Stephen Hawking
d. Charles Darwin
e. Otto Lilienthal

References and Bibliography

D. Askeland, *The Science and Engineering of Materials*. Boston: PWS-Kent, 1989.

L. K. Baxter, *Capacitive Sensors Design and Applications*. Piscataway: IEEE Press, 1997.

R. Boll, *Soft Magnetic Materials*. London: Heyden & Son, 1977.

R. M. Bozorth, *Ferromagnetism*. New York: D. Van Nostrand, 1951.

H. Burke, *Handbook of Magnetic Phenomena*. New York: Van Nostrand Reinhold, 1986.

J. J. Carr, *Sensors and Circuits*. Upper Saddle River: Prentice Hall, 1993.

J. R. Carstens, *Electrical Sensors and Transducers*. Upper Saddle River: Regents/Prentice Hall, 1992.

D. Craik, *Magnetism Principles and Applications*. New York: Wiley, 1995.

B. D. Cullity, *Introduction to Magnetic Materials*. Reading: Addison-Wesley, 1972.

Elcon Instruments, *Introduction to Intrinsic Safety*. Annapolis: Elcon, 1989.

O. Esbach, *Handbook of Engineering Funda*mentals. New York: Wiley, 1975.

J. Fraden, *Handbook of Modern Sensors*. New York: Springer-Verlag, 2010.

E. Herceg, *Handbook of Measurement and Cont*rol. Pennsauken: Schaevitz Engineering, 1976.

D. Jiles, *Introduction to Magnetism and Magnetic Materials*. London: Chapman & Hall, 1991.

D. E. Johnson and J. L. Hilburn, *Rapid Practical Designs of Active Filters*. New York: Wiley, 1975.

Z. Kequian, *Electromagnetic Theory for Microwaves and Optoelectronics*. New York: Springer Science, 2007.

R. Lerner and G. Trigg, *Encyclopedia of Physics*. New York: VCH Publishers, 1990.

P. Lorrain and D. Corson, *Electromagnetic Fields and Waves*. San Francisco: W. H. Freeman, 1962.

E. C. Magison, *Intrinsic Safety*. Research Triangle Park: Instrument Society of America, 1984.

F. Mazda, *Electronics Engineer's Reference Book* (6th ed). London: Butterworth, 1989.

W. Nawrocki, M*easurement Systems and Sensors*. London: Artech House, 2005.

P. Neelakanta, *Handbook of Electromagnetic Materials*. Boca Raton: CRC Press, 1995.

J. C. Newson, *Operational Amplifier Circuits*. Wobwin: Butterworth Heinemann, 1995.

H. Norton, *Handbook of Transducers*. Upper Saddle River: Prentice Hall, 1989.

NVSB series datasheet: Nonvolatile Electronics, Eden Prairie, MN March 1996.

D. S. Nyce, *Distributed Impedance Sensor*, U.S. patent 7,340,951, 2008.

D. S. Nyce, *Electromagnetic Apparatus for Measuring Angular Position*, U.S. patent 7,583,090, 2009.

D. S. Nyce, *Electromagnetic Method and Apparatus for the Measurement of Linear Position*, U.S. patent 7,216,054, 2007.

D. S. Nyce, *Folded Magnetostrictive Waveguide*, U.S. patent 6,919,779, 2005.

D. S. Nyce, *Linear and Rotary Magnetic Sensor*, U.S. patent 6,600,310, 2004.

D. S. Nyce, *Linear Position Sensors, Theory and Application*, *Sensors*. Hoboken: John Wiley & Sons, 2004.

D. S. Nyce, *Low Duty Cycle Oscillator*, U.S. patent 6,791,427, 2004.

D. S. Nyce, *Low Power Magnetostrictive Sensor*, U. S. patent 5,070,485, 1991.

D. S. Nyce, Magnetostriction-Based Linear Position Sensors. *Sensors*, 11(4), 1994.

D. S. Nyce, *Magnetostrictive Linear Displacement Transducer for a Vehicle Strut Suspension*, U.S. patent 6,607,290, 2004.

D. S. Nyce, *Position Sensor with Improved Signal to Noise Ratio*, U.S. patent 9,389,061, 2016.

D. S. Nyce, *Position Sensors*. Hoboken: John Wiley & Sons, 2016.

D. S. Nyce, *Simplified Inductive Position Sensor and Circuit Configuration*, U.S. patent 7,135, 855, 2006.

D. S. Nyce, *Slow Wave Structured Sensor with Zero-Based Frequency Output*, U.S. patent 6,647,780, 2004.

D. S. Nyce, *Vehicle Suspension Strut Having a Continuous Position Sensor*, U.S. Patent 6,401,883, 2002.

H. Olson, *Dynamical Analogies*. New York: D. Van Nostrand, 1943.

R. Pallas-Areny and J. G. Webster, *Sensors and Signal Conditioning* (2nd ed). New York: Wiley, 2001.

R. Philippe, *Electrical and Magnetic Properties of Materials*. Norwood: Artech House, 1988.

E. Ramsden, *Hall Effect Sensors*. Cleveland: Advanstar Communications, 2001.

R. Rose, L. Shepard and J. Wulff, *The Structure and Properties of Materials*. New York: Wiley, 1966.

J. Shackelford, *Introduction to Materials Science for Engineers*. New York: Macmillan, 1985.

E. D. Tremolet de Lacheisserie, *Magnetostriction: Theory and Application of Magnetoelasticity*. Boca Raton: CRC Press, 1993.

W. J. Tompkins, *Interfacing Sensors to the IBM PC*. Upper Saddle River: Prentice Hall, 1988.

L. H. Van Vlack, *Elements of Materials Science*. Reading: Addison-Wesley, 1964.

J. G. Webster, *The Measurement, Instrumentation and Sensors Handbook*. Boca Raton: CRC Press, 1999.

J. Williams, *Analog Circuit Design*. Wobwin: Butterworth-Heinemann, 1991.

J. S. Wilson, *Sensor Technology*. Boston: Newnes, 2005.

Glossary of Sensor Terminology

Note: Definitions are not in general terms, but are provided with respect to their use in sensor technology.

Absolute reading—Measurements are provided with respect to a fixed datum.

Accuracy—The degree to which a measurement agrees with the standard or desired amount, sometimes expressed as an amount of error as a percent of full-range output. It is a general term, and should not be used in a specification. A specification should list errors instead.

Active transducer—A transducer that does not require external application of energy (other than the energy that is being transduced) in order to produce a desired output.

A/D or ADC—Analog to digital converter. Usually to convert an analog voltage or current to a digital representation of same, and to a specified number of binary bits. For example, a voltage input 12 bit A/D converter accepts an input voltage that can be variable over a determined range and changes it to a number of output binary bits (equivalent to 4096 base ten, 0 through 4095). In this case, there would be 12 binary bits in which each can be a 1 or a 0. If all 1s, the output is 4095, base ten.

Algorithm—A set series of mathematical steps applied to an input variable to obtained a desired output.

Asynchronous—(1) Hardware—A property of an event that occurs at an arbitrary time, without synchronization to a reference clock. (2) Software—A property of a function that begins an operation and returns prior to the completion or termination of the operation.

Beat frequency—The difference between two frequencies, such as excitation frequencies.

Best Straight Line (BSL)—A straight line drawn through a plot of measurand versus output signal, having a slope and *Y*-intercept such that the maximum error between the data and the line are minimized.

Calibration—A test during which a series of known measurand values are applied to a sensor so that adjustments can be made to the sensor in order to minimize errors.

CAN (Controller Area Network)—A serial bus developed by Bosch for automotive use, but now is common for industrial automation. Employs two conductors (twisted-pair wires) for power and two for digital signals, utilizing priority-level resolution of bus contention.

Capsule—See diaphragm.

Cermet—A composite material used to fabricate a resistance element, usually applied onto a ceramic substrate. The composite comprises ceramic and metallic particles.

Compensation—The application of an algorithm to measurement data in order to counteract undesired influences or tendencies. For example, a position sensor that has undesired thermal sensitivity may be compensated to make it less temperature sensitive; a position sensor that has a nonlinear position response can be compensated to be less nonlinear.

Current loop transmitter—A sensor that communicates its output signal through means of a variable current, usually 4 to 20 milliamperes. A measurand of zero produces a loop current of 4 mA, and a full-scale measurand produces 20 Ma, with linear current levels between zero and full scale. A transmitter is most often implanted as a two-wire system in which the 4 mA is sufficient to power the sensor and additional current is drawn to indicate the signal level of up to 20 mA, so only two wires (a single twisted pair) are needed to provide power and to transmit the signal.

D/A or DAC—Digital to analog converter. Converts a binary number (usually from a microcontroller) to an analog voltage or current. For example, a voltage output 12 bit D/A converter accepts a binary input (up to 4095 base ten) and changes it to a voltage output over a determined range.

Damping—A characteristic of decreasing amplitude of an oscillating system as a result of energy being drained from the system. With a measurand step change having been applied to a sensor, the sensor is *under-damped* if the output level overshoots the final output before settling. *Critical damping* provides the shortest amount of time until the sensor output reaches its final value without overshoot. The output of an *over-damped* sensor also comes to the final value without overshoot, but has more damping than is necessary to avoid overshoot.

Damping Ratio—Damping ratio is a unitless mathematical expression of the ratio of the level of damping in a system relative to critical damping.

Diaphragm—A thin barrier between two spaces, moving up or down in relation to a pressure difference. Often made of a circular metal sheet having surface convolutions to extend the range of linear motion. Two diaphragms edge-welded together may form a pressure capsule.

Displacement sensor—A sensor that measures the distance between the present position of a target and a previously recorded position of the target.

Drift—An undesired change in a sensor output over time, not related to a change in the measurand. Short-term drift is usually in minutes or hours, and may be sometimes positive and sometimes negative. Long-term drift is usually per month or per year, and is often cumulative in one direction.

EMI (Electromagnetic Interference)—When electromagnetic radiation is emitted from one device so that it affects the performance of another device, it is called Electromagnetic Interference (EMI).

Encoder—A device that converts linear or angular position into analog or digital pulses or square wave-in-quadrature signals. They may be absolute or incremental reading.

Error—The difference between the actual measurement and an ideal measurement, such as the difference between a sensor output and the output of an ideal sensor under the same conditions. Error is usually expressed in percent of full-range output.

Excitation—Power source for a sensing element or transducer, usually having a closely controlled voltage, current, frequency, amplitude, wave shape, or other parameters.

Frequency response—The frequency range over which a signal amplitude is usable, usually the –3 dB points, such as voltage being reduced to 0.707 of the normal amount.

FSK (Frequency Shift Keying)—Bell 202 standard for digital communication with logical zero = 2,200 Hz and logical one = 1,200 Hz.

Full scale—The maximum or 100% of the measurand in the operating range of a transducer or sensor.

Full-scale range/full range—In a unipolar output sensor, it is the difference between full scale and zero. In a bipolar output sensor, it includes the sum of positive and negative portions of the range.

Gage head—An LVDT assembly that includes a spring-loaded or air return probe shaft that is guided within a sleeve, roller, or ball bearing, the probe shaft having a replaceable tip or stylus, and the overall housing usually incorporating a mounting thread.

Hall effect—When a semiconductor, through which a current is flowing, is placed in a magnetic field, a difference in potential (voltage) is generated between the two opposed edges of the conductor in the direction mutually perpendicular to both the field and the conductor. Typically used in the sensing of magnetic field magnitude and polarity.

HART (Highway Addressable Remote Transducer)—Analog and digital communication protocol originally developed by Rosemout. Implements FSK digital signals on top of a 4 to 20 mA transmitter loop. Utilizes a leader command and follower response over two wires (twisted pair).

Hysteresis—The measure of a sensor's ability to represent changes in the input parameter, regardless of whether the input is increasing or decreasing.

Incremental-reading—A sensor that indicates only the changes in the measurand as they occur. An electronic circuit is used to keep track of the sum of these changes (the count) since the last time that a reading was recorded and the count was zeroed. If the count is lost due to a power interruption, or the sensing element is moved while power has been interrupted, the count when normal operating conditions are restored will not represent the present magnitude of the measurand.

Input transducer—A device that produces a usable output that is representative of the input measurand. Its output would then typically be conditioned (i.e. amplified, detected, filtered, scaled, etc.) before it is suitable for use by the receiving equipment (such as an indicator, controller, computer, or PLC, etc.). The terms "input transducer" and "transducer" can be used interchangeably.

Intelligent sensor—A sensor that includes such additional functionality as self-calibration, self-testing, self-identification, adaptive learning, taking a particular action when a pre-determined condition is present, etc.

Intercept—See *Y*-intercept.

Intrinsic safety (Intrinsically safe)— A method used to prevent combustion or explosion by removing the likelihood of the presence of ignition energy.

Least squares—A linear regression method to solve for a best straight line through a set of known points.

Linearity—See nonlinearity.

Measurand—The physical quantity that is to be measured by the sensor, and which causes a representative response in the sensor output.

Measuring range—A position sensor can have a measuring range specified from zero to full scale, or it can be specified as a ± full-scale range (FSR). It is common with an LVDT, for example, to specify bipolar ranges, such as ± 100 mm FSR. So, the measuring range encompasses the total possible values of the measurand that the sensor is capable of measuring.

Natural frequency—The frequency at which a sensing element or transducer tends to oscillate without any damping or driving force.

Nonlinearity—The maximum deviation between a best straight line drawn through the sensor output data and the actual data points.

Offset—A difference between a measured value and the expected value, that is constant at different levels of the measurand, that is, not a scale factor error.

Passive transducer—A passive transducer is one that requires an external supply of energy, and produces an output signal that is usually a variation in an electrical parameter, such as resistance, capacitance, inductance, and so on.

PWM (Pulse Width Modulation)—Generation of a pulse waveform with a fixed frequency and variable pulse width (the duty-cycle). PWM is used to control discrete devices such as DC motors and heaters by varying the pulse width (the ratio of on time to off time).

Quadrature (Square wave-in-quadrature)—An incremental sensor output having two signals, A and B, in which there is a 90 degree difference in phase between the two signals. The number of transitions of a signal indicates the displacement, and the phase relationship indicates the direction of displacement.

Quantizing error—An uncertainty in converting an analog value to a digital value, mostly due to resolution of the conversion process (such as being limited by a measurement clock frequency).

Range—The extent of values over which the measurand is intended to measure, bounded by zero and full scale.

Repeatability—The specified deviation that can be expected in consecutive measurements under the same conditions and using the same measuring instrument.

Resolution—The smallest change in the measurand that the sensor is able to detect.

RTD (Resistance Temperature Device)—A temperature sensitive resistor having a resistance that varies nearly linearly with temperature. Usually the resistive element is made of platinum, and the resistance is usually either 100 or 1,000 ohms at 25°C.

Safety barrier—For intrinsic safety systems, a safety barrier is placed in the nonhazardous area and connected in series with a wire going into the hazardous area in order to limit voltage and current to below such levels that could allow ignition of a flammable substance.

Sensitivity—A ratio of the amount of change in output signal re>sulting from an amount of change in the input measurand.

Sensor—Generally defined as an input device that provides a usable output signal or information in response to a specific physical quantity input. The physical quantity input that is to be measured is called the measurand, and affects the sensor in a way which causes a response that is represented in the output. For the purposes of this book, a sensor produces an electrical output which is representative of the input measurand. The output is conditioned and ready for use by the receiving electronics (such as an indicator, controller, computer, or PLC, etc.).

Signal conditioning—Electronic circuits as may be needed to utilize a transducer, such as excitation, amplification, filtering, analog-to-digital and/or digital-to-analog conversions, temperature compensation, linearization, and other functions to produce the desired output.

Smart sensor—A sensor that incorporates one or more microcontrollers in order to provide increased quality of information as well as additional information. This may include such functions as linearization, temperature compensation, digital communication, remote calibration, and sometimes the capability to remotely read the model number, serial number, range, etc.

Span—The difference between full scale and zero.

SSI (Serial Synchronous Interface)—A serial communication protocol commonly used for encoder applications, in which data are clocked out of a register within the sensor and into a register in the controller or other receiving device. Two wires are used for signal and two for power. The signal inputs are optically coupled.

Stability—The ability of a sensor or other device to maintain its performance (such as output voltage) over time.

Static error band—The net effect from nonlinearity, hysteresis, and repeatability at room temperature, ignoring other effects (such as temperature sensitivity). Usually summed as RSS (root sum of squares) of the three separate specifications.

Strain gage—A force sensing device that changes resistance when subjected to a change in force, accompanied by a very small amount of motion (strain) in the strain gage.

Synchronous (or Synchronized)—A relationship among two or more devices whereby the timing function of one device controls a timing function in at least one other device. With two synchronous LVDTs for example, the excitation of one LVDT controls the excitation of the second LVDT to avoid a beat frequency that would be caused if there were even a slight difference between their oscillation frequencies. A beat frequency is the difference between the two excitation frequencies.

Temperature compensation—When a sensor has some amount of sensitivity of its output signal due to temperature change, an opposite amount of correction can be applied, called temperature compensation.

Thermistor—A resistance device that indicates ambient temperature by the amount of its resistance.

Thermocouple—A bimetallic temperature transducer comprising a cold junction and a hot junction. A temperature difference between the two junctions produces a voltage difference at the ends of the thermocouple pair. Sometimes the cold junction is simulated by an electronic circuit so that there are only two wires forming the sensing thermocouple.

Transducer—A device that changes energy from one form into another, or more specifically, a device that converts input energy into output energy. Typically, the output energy may be in a different form from the input energy, but is related to the input.

Transmitter—See current loop transmitter.

***Y*-intercept**—When a best straight line is drawn through the data on a plot of output signal versus measurand, the *Y*-intercept is the point where the best straight line intersects the y-axis.

Zero—The minimum value of the measurand or of an output signal.

Index

H

I

L

M

Kolibri 360 GmbH
Ferdinand-Jühlke-Straße 7
99095 Erfurt, Deutschland
produktsicherheit@kolibri360.de